International marketing and trade of quality food products

International marketing and trade of quality food products

edited by:
Maurizio Canavari
Nicola Cantore
Alessandra Castellini
Erika Pignatti
Roberta Spadoni

Wageningen Academic
P u b l i s h e r s

ISBN 978-90-8686-089-0

First published, 2009

Wageningen Academic Publishers
The Netherlands, 2009

The individual contributions in this publication
and any liabilities arising from them remain the
responsibility of the authors.

The publisher is not responsible for possible
damages, which could be a result of content
derived from this publication.

Table of contents

Part 3 Food quality and consumers

Editorial

Food quality is a major topic in the scientific debate over agri-food products, but is generally given less prominence in discussions on international trade in food products. The economic analysis of trade, still focuses mainly on commodities, i.e. undifferentiated goods, where the problem of quality is usually resolved by standardization. However, food quality issues regarding differentiated and value-enhanced goods and related marketing problems are gaining the attention of food economists. Marketing and trade in quality food products is important for both developed and developing countries. For instance, in the less developed regions local food specialties may represent a starting point in enhancing the value of food production and gaining access to interesting markets, and thus promoting economic development. In developed areas increasingly sophisticated consumer preferences and the increased purchasing power of households is enabling the emergence of meaningful and actionable market segments, thus increasing the need for products that are differentiated on the basis of their unique sensory, cultural, functional, ethical, and other characteristics.

The concept of quality should be intuitive because to the layperson it is evocative of what is 'good' or 'better' or 'different'. However, from a scientific point of view this concept is very complex. It may incorporate several elements and represent various food characteristics. The variety of quality definitions that exist makes this concept even more fragmented and multifaceted than might at first sight appear.

The recent interest in functional food that incorporates new or augmented health-related characteristics, and the adoption of particular production methods aimed at reducing environmental impacts, such as organic production, are some examples of the different means used to substantiate the concept of quality into differentiated food products.

Also, these topics sometimes overlap. Interest among members of the public in organic food as a quality product has been increasing and organic food is attracting the attention of those who share an environmental friendly oriented way of life and of those keen to adopt a healthy lifestyle and have an awareness of food safety characteristics. However questions such as 'What is organic?', 'How can we promote organic?' and 'Is organic more healthy?' continue to be debated and are deserving of further discussion.

Discussion about the means used to promote quality is challenging because there is frequently confusion over what the term quality means. Indeed, it may mean something different to each of us since quality is generically defined as the ability of an entity (e.g., product or service) to satisfy implicit or explicit needs. The provision of quality involves meeting customer requirements or, in marketing parlance, meeting the needs, wants and expectations of customers.

In order to be able to assess and to assure quality, these requirements need to be specified in some way, either by a public authority or by some other market operator, according to the specific needs that are (or are deemed to be) relevant in specific situations.

Quality and information issues are very closely linked. Recent studies on the usefulness of labelling to communicate product characteristics to consumers, the ways that labelling should be applied and how the information should be provided are attracting the attention of scholars in the field of food product marketing. Labelling that does not directly indicate food characteristics but focuses perhaps on the product's geographical origin could implicitly transfer information about it based on the reputation that certain areas have acquired in relation to production processes and food quality.

Asymmetric information on product quality in terms of producer and consumer is often seen as a core issue in food marketing. If consumers are unable to feel confident about the properties of the goods that they are purchasing, the efforts of farmers and processors to incorporate desirable characteristics will be vain.

In the contributions contained in this book, several aspects of food quality, international marketing and international trade of quality food products are taken into consideration and analysed through a variety of approaches and methods. To outline the role of the different supply chain tiers the book contains three sections on production, trading and consumption.

Contributions in the 'Food Quality, supply network and competition' section focus mainly on the future evolution of farms and on the market characteristics crucial for the organisational aspects of the supply chain. The chapters in this section pay particular attention to the role of quality in international competition in agri-food production. New ways to propose quality labelling, management strategies and industrial relations are only some of the aspects investigated to highlight critical factors on the supply side.

Once quality goods incorporating quality characteristics are produced, it is important that the food can be traded in a context of minimal market frictions such as those arising from different rules, language, culture and other institutional factors. Many contributions in the 'Food Quality and trade section' focus specifically on gravity models that try to describe the volumes of imports and exports in world regions on the basis of standard economic factors such market size and people's willingness to pay for food quality and bear transaction costs. Recent work on gravity models extends the traditional models to include different kinds of market friction by taking account of product, country and sector characteristics.

When quality food is available in the consumer market, quality features must be value-enhanced and considered by the person purchasing the goods. The contributions in the 'Food Quality and Consumer' section focus mainly on the willingness to pay a price premium to obtain goods that incorporate quality characteristics. Research in this field is very important to understand whether the supply chain is properly oriented towards production processes, trading procedures and food features that match customers' preferences and tastes. Most importantly, sophisticated scientific techniques have been devised to calculate the amounts of money that consumers are willing to pay for 'quality'. This kind of research will help the food industry to conduct accurate cost-efficiency analysis to determine whether the implementation of quality characteristics is justified by market conditions.

In conclusion, this volume contains a sample of original and scientifically sound work aimed at addressing many facets related to quality in the agri-food industry. It represents an attempt to suggest new directions for research and to provide useful implications for policy makers who are trying to foster, improve, protect and value-enhance quality in food markets.

Part 1
Food quality and trade

Exports of Italian high quality wine: new empirical evidence from a gravity-type model[1]

A. Seccia, D. Carlucci and F.G. Santeramo

Abstract

Italian wine industry is facing two main challenges: a significant reduction in wine consumption and the growing competition on international markets. Consequently, market liberalisation and the aggressive international competition are threatening Italian wine producers. On the other hand, demand for high quality wines, which includes a large number of Italian wines, is increasing. The purpose of this study is to understand the determinants of trade flows' magnitude for Italian high quality wines to the main importing countries. In order to give a quantitative response, an econometric model derived from an extended version of the 'gravity model' was estimated. This model has been broadly applied to the analysis of international trade because it provides robust estimates. Nevertheless, applications referred to a single product are still limited in number. Estimates are useful to predict potential trends for trades of high quality Italian wines. Moreover, the model allows for the identification of growing markets that Italian firms could exploit, obviously using ad hoc promotional and communication strategies. Finally, relatively to Italian high quality wine, estimates evaluate the gains achieved by its exportation, which might be the consequence of a higher level of liberalisation of international trades.

Keywords: gravity model, high quality wine, export analysis, Italian wine

1. Introduction

Competition in the international wine market increased because of the progressive and substantial reduction in wine consumption jointly to the entry of new countries producing wine. Australia, Chile, the USA and South Africa (the so called New World wine producers) are threatening the international market shares held by the old wine producers such as Italy, France and Spain. The surprising increase in market shares of the New World producers has clear causes. Australia, the USA and South Africa, combining appropriate technologies, optimal climates and growing conditions, managed to increase their production rapidly. Their internal consumption, on the other hand, did not grow much. Contemporarily, South American wine production increased slowly in some countries while decreased in others. Their consumption decreased substantially (Zanni, 2004).

A significant fact is the decline of the aggregate world-wide wine consumption and the increase in demand of high quality wines, which interests a large number of Italian wines. Wine consumption, traditionally associated to dietary habits, has been changing with the modification of life-styles (urbanisation, decreasing caloric needs, increasing importance of leisure time and social activities, etc.). Thus, sensorial pleasure, symbolic value and psychological attitudes are becoming the most important determinants of wine consumption.

---

[1] The authors are jointly responsible for this paper; however, A. Seccia wrote paragraphs 1, 5 and 6, D. Carlucci wrote paragraphs 3 and 4, F.G. Santeramo wrote paragraph 2. The authors would like to acknowledge the many helpful comments of two anonymous referees.

Although Origin Appellation enables for quality differentiation, the appellations alone do not provide an effective competitive advantage at international level because they are not always known, especially if the product is not promoted through an effective marketing strategy. Thus, communicational and promotional planning plays a strategic role in exploiting the international trade, but yet it is often neglected or not applied because of financial feasibility constraints (Carbone, 2003).

This paper models the size of trade flows for Quality Wines Produced in Determined Regions (QWPDR) from Italy to its main importing countries using the 'gravity model' approach. Model and results are useful to predict potential trends in the export of high quality Italian wines. One additional feature is that the model allows for the identification of the main growing markets. Thus, Italian wine producers and suppliers (private wineries, joint-ventures, regional and national agencies, and producers' associations) can use this information to join their effort and succeed in the identified markets. Moreover, the model evaluates quantitatively the effects of international trade liberalisation on export performance of Italian high quality wine. Finally, this study represents an important contribution to the existing literature assessing the empirical validity of the 'gravity model' relatively to a specific product category and its international trade (Dascal *et al.*, 2002; Ševela, 2002).

The paper is structured as follows: Section 2 provides a general overview of Italian quality wine exports during recent years; Section 3 discusses the theoretical framework of the gravity model; Section 4 describes an extended version of the gravity model; Section 5 discusses the estimation and the results; and Section 6 presents considerations and conclusions.

2. General overview of high quality Italian wine exports

During the last decade, the value (at constant prices) of Italian wine exports has increased significantly, as illustrated in Figure 1. Nonetheless, in 2003 exports reduced considerably, and a moderate growth followed in 2004 and 2005. Italian QWPDR exports trend resembles the aggregated wine patterns except for the period after 2003. During that period, high quality wine exports dropped and never went back to the previous levels.

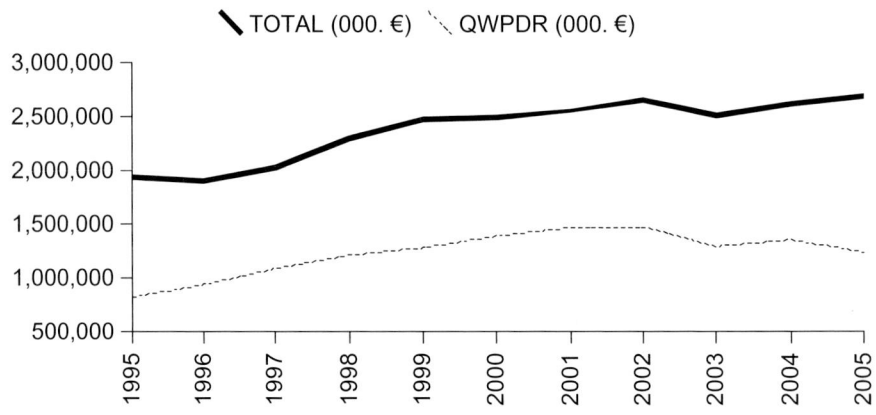

Figure 1. Italian wine export trends from 1995 to 2005 (at constant prices).
Source: ISTAT (www.coeweb.istat.it)

Furthermore, the last decade faced a modification of the composition of Italian wine exports: in 1995 high quality wine exports represented almost 40 percent of total wine exports and by 2001 they reached the 57 percent. Starting from 2002, the proportion of high quality wine in total exports has declined.

Relatively to the international trade, Italy exports quality wines to almost all the countries in the world (Table 1); however, 8 countries account for 80 percent of Italy's high quality wine exports: the USA, Germany, the United Kingdom, Switzerland, Canada, Japan, Denmark and Austria.

In the EU market, Germany imports approximately 23 percent of Italian QWPDR exports; the United Kingdom and Switzerland import 9 percent and 8 percent, respectively, of the Italian QWPDR exports. During the past few years, these European importing partners registered a reduction of their demand for high quality imported wine, but, contemporarily, new entered countries compensated for this reduction with their demand. Recently, wine demand of the

Table 1. Italian QWPDR exports towards main importing countries. Source: ISTAT (www.coeweb. istat.it).

Countries	Value[1]	Share	Countries	Value[1]	Share
USA	337,181	26.14%	China	1,104	0.09%
Germany	297,417	23.06%	New Zealand	914	0.07%
United Kingdom	114,135	8.85%	United Arab Emirates	909	0.07%
Switzerland	103,050	7.99%	Thailand	876	0.07%
Canada	76,516	5.93%	Israel	713	0.06%
Japan	46,267	3.59%	Latvia	625	0.05%
Denmark	33,054	2.56%	Venezuela	618	0.05%
Austria	27,356	2.12%	Estonia	492	0.04%
Belgium - Lux	23,046	1.59%	Costa Rica	489	0.04%
Netherlands	20,464	1.45%	Hungary	472	0.04%
France	18,703	1.45%	Cyprus	464	0.04%
Sweden	17,270	1.34%	Malaysia	413	0.03%
Norway	11,264	0.87%	Lithuania	376	0.03%
Russian Fed.	7,262	0.56%	Philippines	371	0.03%
Ireland	6,352	0.49%	India	355	0.03%
Brazil	5,289	0.41%	Dominican Republic	317	0.02%
Finland	4,980	0.39%	South Africa	251	0.02%
Spain	2,914	0.23%	Colombia	250	0.02%
Australia	2,777	0.22%	Ukraine	246	0.02%
Poland	2,709	0.21%	Portugal	246	0.02%
South Korea	2,111	0.16%	Romania	211	0.02%
Hong Kong	1,974	0.15%	Slovak Republic	194	0.02%
Czech Republic	1,782	0.14%	Kenya	167	0.01%
Singapore	1,668	0.13%	World	1,289,904	100.00%
Mexico	1,526	0.12%	UE(15)	566,845	43.94%
Malta	1,262	0.10%	UE(25)	575,354	44.60%
Greece	1,151	0.09%	North America	413,698	32.07%

[1] Values in thousands of Euros at constant prices (averaged values from 2003 to 2005).

Russian Federation and Ukraine also increased over time, becoming important trade partners to focus on.

During the past few years, the import rate for Italian wine also increased for North America. In particular, the USA leads the importing countries of Italian high quality wines, absorbing the 26 percent of the total Italian wine exported. Also Canada continues to increase its demand for Italian wines.

Central and South America show heterogeneous trends: Argentina, Brazil, the Dominican Republic, Ecuador, Guatemala and Peru reduced their imports, while Colombia, Mexico and Venezuela have increased their demand.

In recent years, the most dynamic Asian partners such as China and India registered a considerable growth of the demand for Italian high quality wines. On the other hand, Japan, which has historically been Italy's sixth largest importer, has shortened its consumption.

3. Theoretical framework of the gravity model

Many economists believe that the gravity model is a very powerful tool for analysing international trades. Tinbergen (1962) and Pöyhönen (1963) were first in proposing the idea, which only later was extended by several other researchers. After these decisive contributions, the gravity model was used in many empirical studies for bilateral trade analysis (Prentice et al., 1998) and for the estimation of the impact of a variety of policy issues relating to, for example, free trade blocs (Martinez-Zarzoso et al., 2003), multilateral commercial agreements (Rose, 2004), migration and tourism flows (Karemera et al., 2000), and foreign direct investment (Brenton *et al.*, 1999).

The basic concept of the gravity model for trade analysis follows the gravity equation from physics: the volume of trade between two countries is proportional to their economic 'mass' and inversely proportional to their respective distance.

The analytical relation of the basic gravity model is expressed as follows:

$$F_{ij} = G \frac{M_i^\alpha \, M_j^\beta}{D_{ij}^\gamma} \tag{1}$$

where, F_{ij} is the export flow from origin country i to destination country j, usually measured by its economic value; M_i and M_j are the economic size of the two countries, usually Gross Domestic Product (GDP) is considered; D_{ij} is the distance between the two countries, measured as physical distance between their first cities; G is a constant that depends on the units used to measure the other variables.

The multiplicative nature of the gravity equation means that it is possible to take natural logarithms and obtain a linear relationship between the log of trade flows and the log of economy sizes and distances as follows:

$$\ln F_{ij} = \alpha_0 + \alpha \ln M_i + \beta \ln M_j - \gamma \ln D_{ij} + \varepsilon_{ij} \tag{2}$$

This equation is estimated by the Ordinary Least Square (OLS), and it is assumed that the error term ε_{ij} is normally distributed.

Linnemann (1966) was the first including several additional variables to the basic gravity model, obtaining what has been successively called the 'augmented gravity model'. In fact, empirical estimations may add other variables like population, income per capita, exchange rates, and dummy variables for the presence of common language, colonial links or commercial agreements among the trading countries (Deardorff, 1995; Head, 2003).

At the empirical level, the gravity model gives very robust estimates and provides a good fit to the observed data. In fact, most of the estimations for bilateral trade volumes with respect to GDP, distance and other explanatory variables, measured values for the determination index (R^2) ranging from 0.65 to 0.95, depending upon the specification of the model (Harrigan, 2001).

Despite the success of the empirical analysis of trade patterns, the gravity model was extensively described as a theoretical orphan. However, in the last decade several authors have worked on reconciling international trade theories with the gravity model specification. Starting from the work of Anderson (1979), it has been shown that the formulation of the gravity model can be derived from different theoretical models such as Ricardian models, Heckscher-Ohlin (H-O) models and Increasing Returns to Scale (IRS) models of the New Trade Theory (Serlenga and Shin, 2004). As highlighted by Davis (2000), it is remarkable that in a short period of time, the gravity model has switched from being a theoretical orphan to a model for which many people are claiming its maternity.

In many studies, the estimation of gravity models refers to panel data. These are datasets formed by repeated observations of the same cross-sectional units over time. The use of panel data provides several advantages such as more variability in the data-set and the possibility of identifying the effects of time-varying variables such as, for example, the progressive reduction of trade barriers (Kennedy, 2003). More precisely, the use of panel data allows for the incorporation into the gravity model of a particular type of fixed effects, namely 'year-specific fixed effects', as indicated by the following notation:

$$\ln F_{ijt} = \alpha_0 + \alpha_t + \alpha \ln M_{it} + \beta \ln M_{jt} - \gamma \ln D_{ij} + \varepsilon_{ijt} \qquad (3)$$

Note that the intercept has two parts: one common to all years (α_0) and one specific to each year (α_t). This regression model is able to capture the relationship between relevant variables over time, as well as to identify the overall business cycle through the proper selection of dummy variables (t) for annual variations in trade flows.

4. Extended version of the gravity model for the analysis of Italian high quality wine exports

In this work, the investigation about the features of Italian QWPDR export flows is conducted in several steps.

Firstly, a basic gravity model was applied in order to test the assumption of the gravity theory for trade. The conceptual model is expressed by the following equation:

$$\ln Exp_{jt} = \alpha_0 + \alpha \ln QwProd_t + \beta \ln PcGDP_{jt} + \gamma \ln Pop_{jt} + \delta \ln Dist_j + \varepsilon_{jt} \qquad (4)$$

where:
Exp_{jt} = value of QWPDR exports from Italy to country j in the year t, expressed in Euro at constant prices;

α_0 = constant;
QwProd$_t$ = Italian QWPDR production in the year t, expressed in hectoliters;
PcGDP$_{jt}$ = GDP per capita of importing country j in the year t, expressed in U.S. dollars at constant prices;
Pop$_{jt}$ = population of importing country j in the year t, expressed in millions of inhabitants;
Dist$_j$ = distance between Italy and country j, expressed in kilometres.

The classic gravity model uses total GDP as a proxy for output capacity of the exporting country. Nonetheless, while total GDP is appropriate for more aggregated export data, in the case of a specific agro-food product such as quality wine, this variable could overestimate the country's output capacity. For this reason, the physical production of the analysed good (or alternatively its monetary value) is considered to be a good proxy for the output capacity for the exporting country, Italy in this case. This variable is expected to show a positive effect because, ceteris paribus, the higher the wine produced the higher the volume of its export, especially in the case of Italy where wine production exceeds internal consumption.

At the same time, the purchasing capacity for the importing countries is also considered by including total GDP in the standard gravity model. However, countries importing high quality wine from Italy are substantially different in terms of the size of their economy and their income per capita. Therefore, GDP per capita has been included in this model in order to explain the income effect in importing countries. A positive parameter for GDP per capita is expected, since the higher the income of the individuals, the higher the demand for quality wine. In addition, population of importing countries is also included in the model because, although GDP per capita controls for the income effect of one individual, it does not consider the size of the economy; by including population, total purchasing capacity of importing countries is captured. As regards the population, a positive coefficient is also expected because it is assumed that the larger the population, the more the country will import.

Finally, in the basic model, the distance between Italy and each importing country has been included, and it represents the proxy for transportation and transaction costs. Accordingly to the theory, a negative effect is expected because the longer the distance, the higher the costs; hence, the lower the trade flows.

In addition, the volume of trade between two countries might be influenced by historical, cultural, ethnic, political and geographical factors that might be difficult to observe and quantify. Because of this, in our analysis a list of dummy variables are included in the standard gravity model in order to control these factors. More precisely, each dummy variable identifies a group of countries that share a similar social-political-historical-geographical background. These groups of countries are illustrated below:
1. Anglo-Saxon countries (Australia, Canada, Ireland, Malta, New Zealand, the United Kingdom, the United States of America, South Africa): Anglophone nations sharing similar political and cultural characteristics derived from United Kingdom colonies.
2. Latin American countries (Argentina, Brazil, Columbia, Costa Rica, Dominican Republic, Ecuador, Guatemala, Mexico, Peru, Venezuela): Spanish and Portuguese are the main languages; the rich diversity of Latin American cultural expressions is the result of many diverse influences such as 'native' cultures, which populated the continent prior to the arrival of the Europeans, European cultures owing to the region's history of colonisation, and finally, African cultures whose presence derives from a long history of New World slavery.
3. Southeast Asian countries (Philippine, Japan, Malaysia, Southern Korea, and Thailand): group of countries nowadays involved in similar processes of industrialisation and westernisation.

4. Central and North European countries (Austria, Belgium, Luxembourg, Denmark, Finland, Germany, Norway, Netherlands, Sweden, Suisse): countries of this group share similar gastronomic tradition (Continental diet), plus, most of them are members of European Union, they have signed the Schengen agreements and have the same currency.
5. East-European countries (Czech Republic, Cyprus, Estonia, Russian Federation, Latvia, Lithuania, Poland, Slovak Republic, Slovenia, Ukraine, Hungary): these countries are linked by recent historical and political events; in particular, after Second World War these countries were under soviet influence and communist regime; since 1989, with the fall of the Iron Curtain, these countries have started a process of integration in the European Union.
6. Mediterranean countries (France, Greece, Portugal, Spain) have similar gastronomic tradition (Mediterranean diet) and they are, similarly to Italy, the most important wine producers of the Old World.
7. China has a large very heterogeneous population, in terms of ethnic groups, languages and cultures; usually, it is considered a group of different cultures with the same political system;
8. India: the same as China.

Thus, the model including country/culture specific effect is:

$$\ln \text{Exp}_{jt} = \alpha_0 + \alpha \ln \text{QwProd}_t + \beta \ln \text{PcGDP}_{jt} + \gamma \ln \text{Pop}_{jt} + \delta \ln \text{Dist}_j + \lambda_k \text{Group}_k + \varepsilon_{jt} \qquad (5)$$

where:
Group_k = dummy variable, it assumes value of 1 if the country j is included in the group k, 0 otherwise.

Finally, an additional complication for the analysis is the 'year-specific fixed effects' (one for each year), which has been included in the model in order to capture exports variations over time. In particular, this last specification allows evaluating the process of globalisation on Italian QWPDR exports volume. Thus, the definitive expression of the empirical model is the following:

$$\ln \text{Exp}_{jt} = \alpha_0 + \alpha_t + \alpha \ln \text{QwProd}_t + \beta \ln \text{PcGDP}_{jt} + \gamma \ln \text{Pop}_{jt} + \delta \ln \text{Dist}_j + \lambda_k \text{Group}_k + \varepsilon_{jt} \qquad (6)$$

where:
α_t = specific 'year-effect' for year t.

These regression models (4, 5 and 6) have been estimated by Ordinary Least Squares.

The data-set for this analysis has 605 observations over a period of 11 years (from 1995 to 2005). Data consist of 55 countries; they cover the largest importers of QWPDR from Italy. The volume of Italian high quality wine exported to these countries in 2005 accounts for more than the 92 percent of the total.

Data on Italian QWPDR exports (dependent variable) were extracted from the database of the Italian Institute of Statistics (ISTAT); exports are expressed in thousands of Euros at current prices. These data were deflated using Italian Consumer Price Indexes (CPI), furnished by ISTAT. Data for Italian QWPDR production were also obtained from the ISTAT database in thousands of hectoliters. Data for 'GDP per capita' were obtained from the World Economic Outlook Database of International Monetary Fund and they are expressed in U.S. dollars, which were also deflated using U.S. Consumer Price Indexes (CPI), from the U.S. Bureau of Labor Statistics. Finally, data for distance between Rome, the Italy's capital, and the first cities of the others countries were obtained using the Haversine formula that was applied on the coordinates from the CIA's The World Fact-book; distances are expressed in kilometres.

5. Estimation results

Results for the Equation 4 (basic gravity model), which includes the most important performance indicators, are reported in Table 2. In particular, the F-statistic is 389.946 and its *P*-value is less than 0.01, which means that the model shows a good overall significance. The R-squared measure is 0.735, which indicates a good fit to the observed data.

The variable measuring the size of Italian QWPDR production is significant (1% rejecting level) on Italian quality wine exports and its coefficient is positive, as expected. GDP per capita and population in importing countries also have a significant effect (at 1%) on quality wine imports from Italy; these variables are positively correlated with the demand in the importing countries. Finally, the variable for the distance is also statistically significant (at 1%) and it has a negative coefficient, as expected.

In the second step was estimated the Equation 5, which includes country/culture specific effects (Table 3). Similarly to the previous model, the F-statistic shows a *P*-value less than 0.01; both explanatory variables of the 'basic gravity model' and dummies for specific groups of countries are significant at a level of 5%; in addition, the value of the adjusted R-squared increased to 0.845 and the Akaike Information Criterion and Schwarz criterion lowered, relatively to the previous model. Concluding, the second model performs better than the first one.

The third model, Equation 6, includes the second model plus year-specific effects, in order to evaluate the temporal trend of Italian QWPDR exports. The results show that the model is significant, because the *P*-value of F-statistic is less than 0.01 (Table 4); the independent variables of the 'basic gravity model' and the country/culture specific effect remain statistically significant

Table 2. Regression results (basic gravity model).

Variable	Coefficient	Std Error	T-Statistic	P-value	Significant
Const	-20.152	5.561	-3.62	0.000	***
ln_QwProd	2.020	0.581	3.48	0.001	***
ln_PcGDP	1.616	0.044	36.35	0.000	***
ln_Pop	0.755	0.031	23.63	0.000	***
ln_Dist	-0.172	0.054	-3.14	0.002	***

Dependent variable = ln_Exp$_{jt}$.
Number of observations = 605.
F-Statistic (4, 600) = 389.946 (*P*-value < 0.00001).
R^2 = 0.735143.
Adjusted R^2 = 0.733377.
Log-likelihood = -981.296.
Akaike information criterion (AIC) = 1972.52.
Schwarz information criterion (BIC) = 1994.62.
Hannan-Quinn information criterion (HQC) = 1981.16.
Significant: *** at 1%; ** at 5%; * at 10%.

Table 3. Regression results (basic gravity model with dummies for groups of countries).

Variable	Coefficient	Std Error	T-Statistic	P-value	Significant
Const	-15.540	4.313	-3.60	0.000	***
ln_QwProd	2.099	0.439	4.78	0.000	***
ln_PcGDP	1.231	0.057	21.51	0.000	***
ln_Pop	0.846	0.029	28.82	0.000	***
ln_Dist	-0.491	0.069	-7.11	0.000	***
Dummies for groups of countries					
Anglo-Saxon	1.374	0.155	8.85	0.000	***
Latin American	0.739	0.195	3.79	0.000	***
South East Asiatic	0.634	0.188	3.38	0.001	***
East European	-0.413	0.208	-1.98	0.048	**
Mediterranean	-1.056	0.212	-4.97	0.000	***
Central North European	1.659	0.172	9.64	0.000	***
China	-1.251	0.312	-4.01	0.000	***
India	-2.054	0.360	-5.70	0.000	***

Dependent variable = ln_Exp_{jt}.
Number of observations = 605.
F-Statistic (12, 592) = 328.245 (*P*-value < 0.00001).
R^2 = 0.848448.
Adjusted R^2 = 0.845376.
Log-likelihood = -812.422.
Akaike information criterion (AIC) = 1650.84.
Schwarz information criterion (BIC) = 1708.11.
Hannan-Quinn information criterion (HQC) = 1673.13.
Significant: *** at 1%; ** at 5%; * at 10%.

(at 5%); moreover, the year-specific effects are positive and significant for 2000, 2001 and 2002. Compared to the second, this model shows a better fit, as demonstrated by higher value for the adjusted R-squared (0.847) and by the Information Criterions (lower values). Finally, it is important to underline that the estimates for the 'basic gravity model' were robust across the different model specification.

All the mentioned considerations encourage preferring the third model for explaining the size of QWPDR trade flows from Italy to its main importing countries.

Italian QWPDR production volume is a variable with a significant effect on Italian quality wine exports and its coefficient is positive, as expected. Considering the logarithmic form of the equation, coefficients are interpreted as elasticity. Therefore, a coefficient higher than one (1.69) means that an increase, or a decrease, in Italian quality wine production will lead to a more proportional increase, or decrease, in Italian quality wine exports. Consumption of high quality wines in Italy, in fact, regards only a small share of Italy's internal production. This fact has two important implications: first, Italy shows an export-oriented nature regarding the analysed good and, secondly, there is a real possibility that a strong increase in Italian quality

Table 4. Regression results (basic gravity model with dummies for groups of countries and year-specific effects).

Variable	Coefficient	Std Error	T-Statistic	P-value	Significant
Const	-12.100	7.303	-1.66	0.009	***
ln_QwProd	1.692	0.767	2.20	0.028	**
ln_PcGDP	1.250	0.059	21.18	0.000	***
ln_Pop	0.846	0.029	29.02	0.000	***
ln_Dist	-0.487	0.069	-7.06	0.000	***
Dummies for groups of countries					
Anglo-Saxon	1.366	0.154	8.84	0.000	***
Latin American	0.760	0.197	3.85	0.000	***
South East Asiatic	0.646	0.189	3.42	0.001	***
East European	-0.393	0.210	-1.87	0.063	**
Mediterranean	-1.062	0.209	-5.07	0.000	***
Central North European	1.641	0.170	9.63	0.000	***
China	-1.213	0.317	-3.82	0.000	***
India	-2.001	0.386	-5.18	0.000	***
Year-specific effects					
1996	0.094	0.188	0.50	0.617	
1997	0.067	0.158	0.43	0.671	
1998	0.162	0.153	1.06	0.290	
1999	0.155	0.163	0.95	0.341	
2000	0.368	0.161	2.29	0.022	**
2001	0.455	0.161	2.84	0.005	***
2002	0.484	0.155	3.13	0.002	***
2003	0.103	0.192	0.54	0.592	
2004	0.124	0.195	0.64	0.525	
2005	0.142	0.164	0.57	0.545	

Dependent variable = ln_Exp_{jt}.
Number of observations = 605.
F-Statistic (21, 583) = 192.698 (*P*-value < 0.00001).
R^2 = 0.852974.
Adjusted R^2 = 0.847678.
Log-likelihood = -803.35.
Akaike information criterion (AIC) = 1641.92.
Schwarz information criterion (BIC) = 1706.8.
Hannan-Quinn information criterion (HQC) = 1671.06.
Significant: *** at 1%; ** at 5%; * at 10%.

wine production could be absorbed by the international market. In other words, Italy should increase the proportion of high quality wine since there are favourable conditions in place which would increase exportation. In fact, although Italy exports high quality wine to more than fifty countries, a large share of these flows go to just a few large trading partners (the five largest importers absorb about 70 percent of Italian quality wine exports). On the other hand, the production of Italian high quality wine could easily be increased, from a production perspective. This is due to the fact that a large share of Italian wine production, especially in the southern regions, belongs to the 'table wine' category, despite the existence of favourable factors (land, climate, know-how, institutional context, etc.) which would allow for the production of a higher quality wine. Nevertheless, improvements in market performance must not be done without proper marketing and promotional activities. In order to expand its exportation of high quality wine, Italy must consider the increasing competition in the international arena and concentrate its communication and promotional efforts for the countries which indicate favourable market conditions. At the same time, together with focusing on expansion possibilities, the existing market shares must also be preserved.

GDP per capita in importing countries also has a significant effect on quality wine imports from Italy. This variable is a measure of demand in the importing countries and its effect is positive, as expected. More precisely, a one percent increase of GDP per capita in a given importing country could have as a consequence an increase of 1.25 percent in the value of quality wine imports from Italy, if other variables remain constant. According to these results, the value of Italian quality wine exports is income elastic. An income elasticity that is greater than one is an expected outcome for a processed good such as quality wine. A possible explanation is that the international market is larger if a bigger amount of product is available. Consequently, if Italian producers of high quality wine intend to expand their exportations, it is natural to look at those countries where income growth is constant and solid. In the same time, it is also important to observe that any decrease in income for the trade partners, in other words an economic recession, would have serious negative consequences on the volume of Italian quality wine exports. Looking at Table 5, which shows the IMF estimates for annual percent change of GDP per capita, it is interesting to highlight that among the countries with the highest income growth rates there are three very populous countries, China, Russia and India, for which expansion possibilities for Italian quality wine exports are very attractive. Currently, these countries import less than 1% of total exports of Italian high quality wine. However, this share could increase exponentially if Italian exporters succeed in penetrating these markets and in consolidating their presence. At the same time, it is important to highlight that the main countries importing Italian high quality wine (the United States, Germany, the United Kingdom, Switzerland, Canada, Japan and almost all western European countries) show a moderate but stable income growth (ranging between 1 and 2 percent). At this point, it would be advisable to defend and consolidate Italian market shares against any possible aggressions by the new wine producing countries.

The population of the importing countries has also a significant effect on quality wine trade flows from Italy. This variable is a measure of the purchasing capacity of importing countries and its effect is positive, as expected. According to the results, the demand for Italian quality wine exports is inelastic with respect to the variation of the population. More precisely, a one percent increase (or decrease) in population in a given importing country could result in an increase (or decrease) of 0.85 percent in the value of quality wine imports from Italy, ceteris paribus. Consequently, the variations of the population of the Italian trade partners would have small consequences on the volume of Italian quality wine exports.

The distance variable is also statistically significant and its coefficient is negative, as expected. This result suggests that an increase of 1% in physical distance could lead to a less proportional reduction (0.49%) of Italian high quality wine exports. In particular, the effect of the distance is lower than the averaged value reported in the literature of aggregated export data (Head, 2003).

Table 5. Annual percent change of Per capita GDP [1].

Countries	2005	2006	2007	Countries	2005	2006	2007
China	9.6	9.5	9.5	Jamaica	0.9	2.3	2.6
Latvia	10.9	11.6	9.4	Brazil	0.8	2.2	2.5
Estonia	10.1	9.8	8.3	Spain	2.8	3.0	2.5
Slovak Republic	6.1	6.5	7.0	Jordan	4.5	3.4	2.4
Russia	7.0	6.9	6.9	Finland	2.8	3.3	2.3
Lithuania	8.1	7.2	6.9	Colombia	3.5	3.0	2.3
Bulgaria	6.3	6.4	6.8	United Kingdom	1.2	2.2	2.2
Romania	4.4	5.9	5.9	Australia	1.3	1.8	2.2
India	7.2	6.7	5.6	Norway	2.0	1.7	2.2
Argentina	8.0	6.8	4.8	Israel	3.0	1.9	2.2
Czech Republic	5.9	5.9	4.7	Japan	2.6	2.7	2.1
Hong Kong	6.4	5.1	4.6	Austria	1.4	2.6	2.1
Poland	3.5	5.1	4.5	Denmark	3.0	2.4	2.1
Ireland	3.3	4.4	4.2	Canada	2.0	2.2	2.0
Malaysia	3.2	3.7	4.0	Belgium	1.5	2.7	2.0
Thailand	4.4	3.4	4.0	Mexico	1.5	2.5	2.0
Slovenia	3.6	4.2	3.8	United States	2.3	2.5	1.9
Hungary	4.3	4.7	3.7	Ecuador	3.3	3.0	1.8
Ukraine	3.4	5.8	3.6	Sweden	2.3	3.6	1.8
Peru	4.9	4.5	3.5	France	0.6	1.8	1.8
Dominican Rep.	7.7	4.0	3.5	Switzerland	1.7	2.9	1.7
Greece	3.7	3.7	3.5	Venezuela	7.2	5.4	1.6
Korea	3.5	4.2	3.4	Portugal	0.3	1.1	1.4
Kenya	3.7	3.6	3.4	Guatemala	0.6	1.5	1.4
Philippines	3.0	2.9	3.3	Netherlands	1.3	2.6	1.2
South Africa	3.9	3.0	3.0	Germany	0.9	2.0	1.2
Luxembourg	3.6	3.1	2.9	Malta	1.8	0.9	1.0
Cyprus	3.7	2.6	2.9	Italy	-1.0	1.1	1.0
Costa Rica	4.0	4.7	2.8	New Zealand	1.3	0.3	0.5
Singapore	3.7	5.1	2.7	United Arab Emirates	0.8	3.6	-1.7

[1] Data are IMF estimates.
Source: World Economic Outlook Database of International Monetary Fund (http://www.imf.org).

This difference could be explained by the nature of the product, bottled wine is a durable good and it is also characterised by a high unitary value, which involves a low incidence of shipping and transportation costs.

The analysis of year-specific fixed effects shows that, ceteris paribus, high quality wine exports increase over time. More precisely, the year-specific effects are positive and significant for years of the period 2000-2002, showing a regular increase at every year. Specifying, the dummy for the first year, 1995, has been omitted in order to avoid the dummy trap. Between 2000 and 2002, the export of Italian high quality wine increased by 62 percent ($e^{0.484} - 1 = 0.62$) independently to the variations of all the other variables. This could be considered a 'globalisation effect', taking into account that most of the WTO agreements are the result of the Uruguay Round Negotiations signed at the Marrakech ministerial meeting in April 1994. However, the high rate growth of Italian quality wine exports could also be explained by other factors like changes in consumers' preferences.

The estimation results of Equation 6 allow for evaluating the joint effects of historical, cultural, ethnic, political and geographical factors on Italian quality wine exports. More precisely, a positive or a negative value of the dummy coefficients indicates, respectively, an increase or a decrease of the Italian quality wine exports value, ceteris paribus. Firstly, an higher value of the coefficients has been registered by the dummy referred to Central and North European countries (1.64); obviously, historical and strong linkages based on political and economic agreements between Italy and the others European countries have facilitate conspicuously the trade. The dummy for Anglo-Saxon countries shows a positive and significant coefficient (1.37) probably due to the historical linkages between all countries with western culture; furthermore, the increasing diffusion of English as second language allowed exchanges and trades to become easier. A positive coefficient was estimated for the Southeast-Asian countries' dummy (0.65). These countries are facing an industrialisation process and, above all, westernisation; these countries seem to prefer Italian QWPDR, undoubtedly a symbol of western lifestyle and, more precisely, of *made in Italy*. The positive coefficient (0.76) for Latin American countries could be explained by the similarity between Italian and South American cultures and cuisines; moreover, it is important to consider that South America shows excess in demand, either for quantity and quality. Obviously, the negative coefficient estimated for Mediterranean countries (-1.06) enforces the hypothesis that the exceeding production of wine (also in qualitative terms) constrains Italian QWPDR exports, although these countries are consumers of big volumes of wine. The negative values of the dummies for China (-1.21) and India (-2.0) could be well explained by the large cultural differences, relatively to Italy and other western countries, in terms of life-style and gastronomic traditions. Income inequality for these countries is another plausible explanation. Finally, the moderate and negative value of the dummy concerning East-European countries (-0.39) emphasises the historical closure of these countries toward the western cultures: it is worth to mention the 'Iron Curtain', the boundary that symbolically, ideologically, and physically divided Europe into two separate blocks from the end of Second World War until the end of the Cold War. Nevertheless, it is important to highlight that at the end of the considered period, the European Union faced an historical enlargement: more precisely, on May 1 2004, ten new countries of Eastern Europe (Cyprus, the Czech Republic, Estonia, Latvia, Lithuania, Hungary, Poland, Slovenia, Slovakia and Malta) joined the fifteen existing member States. Later in January 2007, Bulgaria and Romania also became members of the EU. Moreover, the recent process of enlargement of the European Union could lead to an increase in Italian trade towards the East-European countries. The absence of customs barriers within the countries of the European Union, and the existence of common customs tariff applied to imports from non-EU countries, will surely facilitate this process. In addition, it is interesting to note that all the new members of EU and, in particular, the Baltic Republics (Latvia, Estonia, and Lithuania) show high income growth rates (ranging between about 4% and 9%). Therefore, these countries represent an interesting, and yet an unexploited market. Relatively to the New World competitors, the exporters of Italian quality wine could gain extra profit from the EU enlargement because of the cancellation of customs barriers.

6. Conclusions and final remarks

This work shows that the gravity model is a very useful analytical tool also for specific product's trade analysis. In particular, the model, once adapted to the specific research purposes, is able to explain with great accuracy the size of trade flows using easily disposable data. Moreover, the gravity model may also be used to predict potential trends in trade flows and to estimate the impact of various policy issues.

Examining the results of the analysis of Italian high quality wine exports, several points can be highlighted. The production of Italian high quality wine should be increased because of the advantageous opportunities in the international market. Considering that the analysed exports are income elastic, as shown in the empirical model, Italian producers should target their exports to specific markets/countries, accounting for their income growth, and developing a diversified portfolio strategy. Countries with high income growth rates, in fact, are advantageous targets because of the positive correlation between income growth and demand, even if the growth and the demand are both instable. In the same time, the inclusion of countries with moderate GDP growth in the portfolio will assure the maintaining of the existing market shares.

As mentioned before, the recent enlargement of the EU could represent a great opportunity for the exporters of high quality Italian wine. In fact, there are advantageous opportunities for penetrating the Eastern European markets, which are growing rapidly. Customs barriers will facilitate this step, even if further trade liberalisations would probably reduce this advantage.

At the same time, considering the effects connected to the WTO agreements, signed at the end of GATT Uruguay Round, we can observe the positive influence on the exportation of high quality Italian wines, even if concerned a short period of time. Hence, it is desirable that future WTO negotiations on agriculture in the Doha Development Agenda Round will rapidly conclude with an agreement. An opportune precaution is to refer these conclusive thoughts exclusively to the effects of a strong liberalisation on the exportation of only high quality Italian wines.

References

Anderson, P.S., 1979. A Theoretical foundation for the gravity equation. American Economic Review, 69: 106-116.

Brenton, P., F. Di Mauro and M. Lucke, 1999. Economic integration and FDI: An empirical analysis of foreign investment in the EU and in Central and Eastern Europe. Empirica, 26: 95-121.

Carbone, A., 2003. The role of Designation of Origin in the Italian food system. In: S. Gatti, E, Giraud-Héraud and S. Mili (Eds), Wine in the Old World – New risks and opportunities. Milano, Italy: Franco Angeli, pp. 29-41.

Davis, D.R., 2000. Understanding international trade patterns: advantages of the 1990s. Unpublished manuscript, Columbia University.

Dascal, D., K. Mattas and V. Tzouvelekas, 2002. An analysis of EU wine trade: a gravity model approach. International Advances in Economic Research, 8: 135-147.

Deardorff, A.V., 1995. Determinants of bilateral trade: does gravity work in a neoclassical world?. NBER Working Papers No. 5377. Cambridge, MA.

Harrigan, J., 2001. Specialization and the volume of trade: do the data obey the laws? FRB of New York Staff Report N. 140.

Head, K., 2003. Gravity for beginners. Working Paper. University of British Columbia.

Karemera, D., V.I. Oguledo and B. Davis, 2000. A Gravity Model analysis of international migration to North America. Applied Economics, 32: 1745-1755.

Kennedy, P. (Ed.), 2003. A guide to Econometrics. Cambridge University Press.

Linnemann, H., 1966. An econometric study of international trade flows. Amsterdam: North-Holland Pub. Co.

Martinez-Zarzoso, I. and F. Nowak-Lehmann, 2003. Augmented Gravity Model: an empirical application to Mercosur-European trade flows. Journal of Applied Economics, VI: 291-316.

Pöyhönen, P., 1963. A tentative model for volume in trade between countries. Weltwirtschaftliches Archiv, 90: 93-100.

Prentice, B.E., Z. Wang and H.J. Urbina, 1998. Derived demand for refrigerated truck transport: a Gravity Model analysis of Canadian pork exports to the United States. Canadian Journal of Agricultural Economics, 46: 317-328.

Rose, A.K., 2004. Do we really know that the WTO increases trade. The American Economic Review, 94: 98-114.

Serlenga, L. and Y. Shin, 2004. Gravity Models of intra-EU trade: application of the Hausman-Taylor estimation in heterogeneous panels with common time-specific factors. Edinburgh School of Economics Discussion Paper, n. 88, University of Edinburgh.

Ševela, M., 2002. Gravity type model of Czech agricultural export. Agricultural Economics [Zemedelská ekonomika], 48: 463-466.

Tinbergen, J., 1962. Shaping the World Economy: Suggestions for an international economic policy. New York: The Twentieth Century Fund.

Zanni, L. (Ed.), 2004. Leading Firms and Wine Cluster – Understanding the evolution of the Tuscan wine business through an international comparative analysis. Milano, Italy: Franco Angeli.

Evolution of trade flows for sheep milk cheese: an empirical model for Greece

G. Vlontzos and M.N. Duquenne

Abstract

This research examines trade flows for Feta cheese, where sheep's milk is the primary component. The findings of the gravity model demonstrate the significance of trade flows for Greek Feta worldwide, including the trading potential of Feta cheese, taking into account a positive outcome on the judicial and political level for Feta in the WTO negotiations. The results of the gravity model provide essential information for policy recommendations which follow, in order to provide viable solutions to the obstacles the sector faces. The suggested policy recommendations are focused on increasing the competitiveness of the sector and on providing it with all the essential quality and safety reassurances needed in a highly competitive market environment.

Keywords: trade flows, feta, quality, competitiveness, gravity model

1. Introduction

The cheese sector is one of the most important primary sectors of the Greek economy, due to the fact that the national sheep and goat herd is large, not only by national, but by EU standards as well. Sheep and goat's milk is the basic component for numerous types of cheese, the most popular being Feta. There has been a long standing dispute and litigation between Greece and its EU counterparts, Denmark, Germany, UK and France, regarding which country has the right to produce cheese named Feta. In 2002, Greece won the case and effective October 2007, has the exclusive right in the EU market, of producing white cheese named Feta from sheep and goat's milk (R1829/2002). While Greece now maintains the exclusive production rights for Feta in the EU, this is not true internationally. Thus, during the ongoing negotiations of the Doha Round, the EU's official request for increased protection of products with *Geographical Indications (GI's)* has much relevance for Greek Feta.

The purpose of this study is to examine Greek export flows of Feta cheese world wide. The findings of the gravity model used for this research were used as the basis for policy recommendations that address the most serious obstacles the Greek Feta sector faces and highlight new opportunities that recent political and commercial changes have created.

2. Background

One of the most contentious issues between the parties involved in the case of Feta *cheese* was that the northern European countries are using cow's milk in the production of their Feta. Greece's argument is that the name 'Feta' is not a generic name for cheese, but a name for a traditional Greek cheese produced by using a mixture of sheep and goat's milk only, with no cow's milk at all. This is the main reason for labelling the product as PDO (Product Designated of Origin) in 2002. The 2002 decision established a transitional period of five years for the EU countries to fully comply, that ended in October 2007. The outcome of this litigation is documented in the EU legislation under R1829/2002.

All the parties involved in Greek Feta production realised that the sector was characterised by increased competitiveness, with great risk to the growth of Greek Feta cheese producers when firms outside Greece use the name 'Feta' to increase their sales, without producing the product in the traditional way. Another important issue was that the consumers were incorrectly informed about the ingredients and the original flavour of the product. The white cheese from cow's milk is more yellowish colour and has a different taste from the original Feta cheese. This situation creates an unfavourable trading environment for any attempt from the Greek side to gain market shares in these countries as consumers' perceptions of Feta have already been shaped by their experience with the cow's milk-based product. It is not an easy task to change consumers' perceptions, to convince them of which is the original product and which is an imitation (Babcock, 2003).

The implementation of the gravity model for Greek export flows of Feta cheese appears as a crucial necessity for understanding the fundamentals of the product's trading environment. This trading environment has two tiers. The first one is the EU market, where there are no trade barriers and the second is the global market where there are numerous barriers to Greek Feta exports (tariffs, TRQs, technical barriers, etc.) maintained by individual countries. In order to reach feasible policy recommendations in a heterogeneous trading environment, it is necessary to present the most important export destinations and estimate the significance of each factor affecting trade dealings and performances.

3. Objectives

The first objective of this study is to analyse the main export flows of Greek Feta cheese worldwide and to assess the factors influencing the evolution of such flows, by applying a gravity model. Studying the impact of each factor affecting international trade is essential for the implementation of a new Greek Feta trade policy, given that the current WTO negotiations can be expected to introduce significant change. The EU market is an additional source of change, given that the name Feta can no longer be used by other countries, except Greece, as of October 2007.

The second objective is to measure the importance of each economic factor affecting Feta trade. The factors affecting Feta trade studied are distances, quantities exported, total trading quantities, value of exports and imports, evolution of GDP per capita, and population.

The third objective is to evaluate the importance of the non economic parameters affecting Feta trade. Such parameters are the existence or not of Greek immigrant societies, cultural similarities with Mediterranean diet and whether exporting destinations are EU countries or not.

The forth objective is to verify which policy recommendations are most effective for a rapid improvement of international trading performance of Greek Feta. This part of the research is crucial, because until now there has not been an aggregate trading and export policy for Feta cheese in Greece. The EU CAP is primarily responsible for this situation. The Greek Ministry of Agriculture gave priority to the expansion of arable crops in order to access CAP subsidies, rather than focusing on building competitive exports of feta.

4. Framework of analysis

4.1 Export flows of sheep's milk cheese

Until recent changes in the EU CAP, most Feta production was focused in the Greece with limited exports. In the last five years Greece produced an average of 115,000 tonnes of Feta cheese. The vast majority of it (85 per cent) is consumed internally within Greece, with only 15 per cent is being promoted abroad (National Institute of Statistics of Greece: www.statistics.gr). On the basis of FAO (FAOSTAT, http://faostat.fao.org) data, during the period 1990-2004, Greek exports of sheep's milk Feta[2] were subject to significant fluctuations (Figure 1). Until 1995, the volume of Feta exports increased reaching a total volume of 10,000 tonnes. Exports significantly decreased in 1996-1997, and have been increasing steadily since, to a total volume exceeding 13,500 tonnes in 2004. A dramatic export surge in 2000 was uncharacteristic and not sustained.

The most important market for Feta cheese is Germany, importing an average 67% of total Greek Feta exports. Figure 2 clearly indicates that fluctuations of total exports were until 2002, closely linked with German demand.

Table 1 indicates a new trend emerging after the year 2000, where the relative weight of the German market is progressively decreasing. By the end of 2004, this market represents 54% of

---

[2] As it has also been refereed, data related to exports are related to sheep milk and not Feta in itself. However, in actuality, exports of milk are insignificant, hence the data relates almost exclusively to Feta cheese.

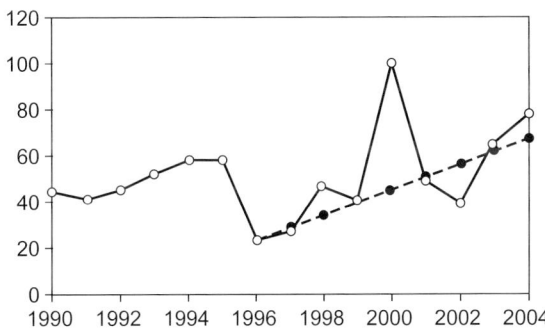

Figure 1. Greek exports of Feta cheese (tonnes, base 100 = 2000).

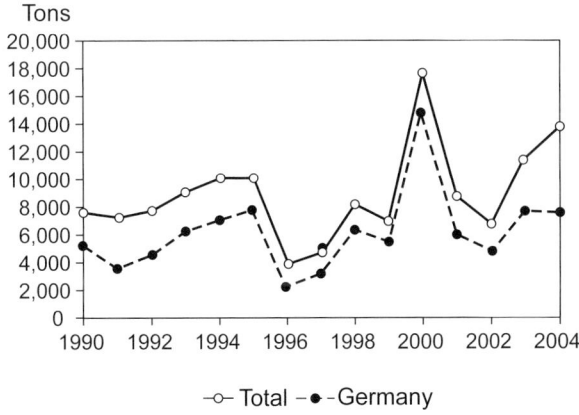

Figure 2. Greek exports of Feta cheese in Germany (in tonnes).

Table 1. The decreasing weight of German market.

Year	Number of trade partners	Exports in tonnes		
		Total	Germany	
1990	14	7,691	5,076	66%
1991	15	7,183	3,597	50%
1992	15	7,792	4,495	58%
1993	16	9,125	6,172	68%
1994	18	10,112	6,916	68%
1995	16	10,332	7,621	74%
1996	18	3,832	2,212	58%
1997	18	4,662	3,071	66%
1998	15	8,088	6,290	78%
1999	15	6,947	5,326	77%
2000	17	17,530	14,736	84%
2001	17	8,653	6,000	69%
2002	16	6,734	4,640	69%
2003	17	11,376	7,448	65%
2004	17	13,670	7,406	54%

total exports of Greek Feta compared to 77% in 1998-99. Table 1 also illustrates the continuous increasing demand for Greek Feta from other EU countries, mainly France, Italy, Netherlands and Sweden (Figure 3, Group B).

Outside the EU, significant export markets for Greek Feta are the US, Canada and Australia (Group C). In all these markets, demand for the product is due to immigrant Greek communities that consume Feta as an integral part of their diet. The remaining markets for Greek Feta exports are other EU countries as well as some Balkan countries (Albania, Bulgaria) and few African countries.

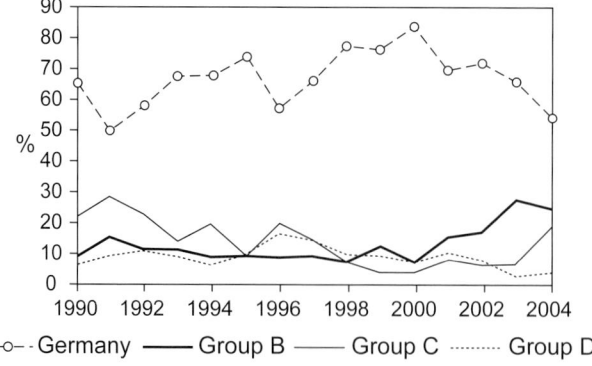

Figure 3. Structure of Feta exportations by group of countries (%).

The following conclusions stem from the analysis above:
- Despite annual fluctuations, exports of sheep's milk cheese and consequently, Feta, follow a positive trend during the examined period.
- These exports are mainly focused on countries with significant Greek communities that have migrated in a large part during the decades 1960-1980.
- New markets, especially in Europe, seem to emerge and pose a remarkable increase in demand for Greek Feta. Two factors could contribute to such a result: Greece is an important destination for European tourists who have the opportunity to become familiar with this product and secondly, there is an increasing interest in the Mediterranean diet for which Feta as well as olive oil are primary ingredients.

4.2 Factors influencing Greek Feta exports

Foundation of the model

In order to provide a comprehensive empirical analysis of Feta's trade flows world wide, a well-known gravity model was used. The model developed by Tinbergen (1962), Pöyhönen (1963) and Pulliainen (1963) is considered one of the most appropriate tools explaining bilateral trade flows. Since the early 1960's, the majority of empirical studies used the gravity model to estimate the total volume of bilateral trade (aggregated exports and imports). In 1979, Anderson affirms that 'the gravity model is probably the most successful empirical trade device of the last twenty years' while more recently Mátyás (1997) considers that the use of gravity type models is one of the most appropriate ways to formalise these flows. Numerous empirical models based on gravity equation have been formulated during the last two decades (Bun and Klaassen, 2002), providing evidence that this type of model can be applied not only to international economics but also to a wide range of problems such as migration flows, foreign direct investment, tourism flows etc.

According to the Law of Universal Gravitation discovered by Newton in 1687, the basic specification of bilateral trade flows between geographical entities can be expressed in the following way:

$$\text{EXP}_{i,j} = a_0 \frac{Y_i^{a_1} \cdot Y_j^{a_2}}{D_{i,j}^{a_3}} \varepsilon_{i,j} \tag{1}$$

where:
$\text{EXP}_{i,j}$ is the volume of exports from the country i to the country j, corresponding to the attractiveness of the market j for the country i

Y_i and Y_j are indexes of masses, reflecting the economic sizes of the two countries and more precisely the potential to participate in international trade such that trade flows between countries should be positively related to their economic size. In international economics, masses are usually measured with GDP.

$D_{i,j}$ measures the geographical distance between the two countries. The distance, as a measure of geographical proximity, reflects not only the difficulties that countries can encounter in developing exchanges because of transport cost and duration (Luo, 2001) but also difficulties in accessing appropriate information. In other terms, the geographical distance engulfs the resistance factors for two countries to develop bilateral trade. Even if technological progress contributes to reduce transport costs, recent research confirms that geographical distance still has a negative impact on trade flows[3].

Considering the extensive research and study of the past 20 years on the gravity model (Mátyás *et al.*, 2000), it appears that the econometric specifications of the gravity equation has often been improved by the introduction of different explanatory variables as well as dummy variables[4]. Linnemann (1966) in the H-O framework postulated that potential trade is dependent on differences in population size. After Linnemann, many empirical studies introduce variables relative to the population size of the countries. Moreover, geographic entities with high GDP are logically the most important commercial partners world wide while at the same time, exports are part of GDP such that the choice of GDP as an independent variable inflate 'quite artificially' the factor R^2 (Head, 2003). For this author, population size is mainly an instrumental variable for GDP in order to deal with the above specification problem. Tinbergen's model (1962) introduced dummy variables for neighbouring countries and membership to a preferential trade area. Thereafter, many studies introduced dummy variables in order to capture the effect of Regional Trade Agreements in bilateral trade and measure the role of the economic integration. With the addition of dummy variables, the authors attempt to account for different factors reflecting historical, cultural and institutional characteristics as a means to more effectively investigate trade flows. Even if dummy variables occasionally present a lack of theoretical justification[5], they improve appreciably the quality of the model. More specifically, cultural factors such as common language, seem to be highly correlated with trade. In other words, such variables contribute to estimating the effect of non-economic but cultural proxies.

Specification of the model

Despite the gravity equation being a generally popular tool to characterise the pattern of bilateral trade flows (aggregated or not), it has been rarely used in the investigation of the determinants of global agricultural trade (Paiva, 2005). The model we attempt to develop is effectively focused on a specific dairy product, Greek Feta cheese, which is also a PDO (Protected Designation of Origin) in the EU market. The specifications of the gravity equation must take into account (1) the specification of this product which cannot be considered as a current – basic consumption product (except for the Greek population) and (2) the geographic structure of Greek exports as described above which could be attributed to a cultural effect linked to the presence of immigrant Greek communities in some of the considered countries.

Finally, the equation used in the present paper is an augmented form of the basic gravity equation where the volume of Greek Feta exports ($EXP_{G,i}$) depends upon:

---

[3] See for example, Leamer and and Levinsohn (1995), Anderson and van Wincoop (2003) and especially Disdier and Head (2006). The last two authors investigate the sensitivity of the distance coefficient estimated within the gravity equation, taking into account the results of a sample of 103 papers corresponding to 1,467 estimations. They finally conclude that despite the technological progress that should reduce the importance of transport costs, the negative impact of distance has not disappeared and moreover, has slightly increased over the last century.

[4] An interesting review of the literature relative to the refinement of the explanatory variables and the addition of new ones is proposed by Martinez-Zarzoso and Nowak-Lehmann (2002).

[5] One criticism often directed at the analyses based on gravity models is that they are mainly empirical approaches without serious theoretical foundation. In fact, since the end of the 1970s, several authors as Anderson (1979), Bergstrand (1985, 1989, 1990), Helpman and Krugman (1985), Deardoff (1995) highlighted that the use of gravity equation can be in a large part justified from standard trade theories.

- Y_i: the size of the partners' respective economies which is measured through per capita GDP. This variable is usually chosen when estimating bilateral exports relative to specific products while total GDP is often used for the estimation of aggregated exports' (Martinez-Zarzoso and Nowak-Lehmann, 2002). Moreover, GDP per capita is based on purchasing-power-parity (PPP). If this variable is an alternative measure of economic size, it also reflects the per capita income which is a much more relevant variable regarding non-basic consumption products. We are expecting a positive correlation between the level of per capita GDP and the volume of trade flows (b_1 positive).
- Pop_i: the population of the trade partner i. As mentioned above, this variable has been introduced as an instrumental variable for GDP. We are also expecting a positive relationship (b_2 positive).
- $Dis_{G,i}$ indicates the geographical distance between Greece and its trade partners. The distance is estimated from capital to capital, although not the best measure of geographical proxy is at least the simplest one. Due to the fact that sheep's milk cheese is a perishable good, distance as a proxy for transport cost and duration cannot be neglected. Examining the flows for one product between Greece and its partners, absolute distance is a convenient measure which is not necessary the case for bilateral trade between pairs of countries where relative distance is more appropriate (Polak, 1996).
- Finally, D_{ik} corresponds to additional dummy variables introduced in order to evaluate the effects of institutional and cultural factors, more precisely:
  - Whether trade partners of Greece are or are not members of EU. This dummy variable can be interpreted as the existence of an institutional proximity (Frankel *et al.*, 1995), reinforced by the fact that Feta is a PDO classified product for the EU market.
  - Some countries have long-established immigrant Greek communities (the diaspora). The introduction of such a dummy variable should reflect the existence of an alternative form of proximity, not based on geographic criteria but on a relational one. Considering the structure of Greek Feta exports, the existence of Greek diaspora in partners' countries could produce a similar effect as the one of common language. Different authors (Gould, 1996; Head and Ries, 1998; Dunlevy and Hutchinson, 1999; Belair and Gauthier, 2004) have shown evidence of the relative importance of immigrant links to the home country in terms of preference for typical home country's products. Feta cheese fits well in a preference pattern for the Greek diaspora.

For estimation purposes, the Greek exports of Feta cheese are expressed in log-linear form:

$$lEXP_{G,i} = b_0 + b_1 \cdot lY_i + b_2 \cdot lPop_i + b_3 \cdot lDis_{G,i} + \sum_k c_k \cdot D_{i,k} + \varepsilon_{G,i} \qquad (2)$$

where:
l denotes that variables are expressed in natural logs.
$\sum_k c_k \cdot D_{i,k}$ is the sum of the added dummy variables taking the value one when the specific condition (belonging to UE or belonging to countries with significant Greek diaspora) is satisfied and zero otherwise. The coefficients relative to these dummy variables are expected to be all positive.
$\varepsilon_{G,i}$ is the error term associated to the Ordinary Least Square estimation.

The data set covers 23 trade partners (of which 13 are EU members) that imported Greek Feta during the period 1990-2004. Data relative to the volume of exports were collected from the database of FAOSTAT, a detailed trade matrix. For per capita GDP and population, data was sourced from the World Economic Outlook Database of International Monetary Fund. Per capita GDP based on PPP is expressed in international dollars while population is measured in millions of persons. Since Feta exports are subjected to serious annual variations during the examined period, the variables have been expressed in 3-year period averages in order to reduce the effect of irregular fluctuations. Consequently, five successive periods are examined, namely the periods 1990-92, 1993-95, 1996-98, 1999-2001 and 2002-2004.

5. Results

Estimation with SPSS software was completed via Ordinary Least Squares (OLS). The results for the 5 successive periods are presented in the Table 2. As it appears, the model has a relatively low explanatory power with adjusted R^2, varying between 55% and 68%. Compared to other empirical models of trade flows, the above power is actually not that satisfactory. Moreover, taking into account that Feta cheese for non-Greek consumers is an uncommon product and could be considered as an 'exotic' good, different non-economic factors not easily measurable can be determinant in measuring such specific trade flows.

More precisely, it appears that the coefficients of the basic gravity equation have the expected sign, with one exception relative to EU trading block:
- Exports of Feta to the trade partners of Greece are positively affected by their economic size while demographic size has no significant impact. Export flows of Feta cheese increase in a greater proportion compared to the change in GDP per capita. Except for the second period, consumers' purchasing power is particularly high. In other words, a 1% increase in purchasing power could result in an increase of at least 2% in Feta exports from Greece. If we accept that Feta cheese is not a common product for the majority of countries, this result confirms that in countries with higher standards of living, consumers have the economic ability to splurge on 'ethnic' or non-typical products in their diet.
- The distance variable appears as a 'resistance factor': Feta exports are negatively affected by geographic distance and the magnitude of the coefficient is quite high in absolute terms. It can be suggested that the negative impact of geographic distance is not only due to transport costs – even if this factor is truly important for fresh products such as Feta – but also indirectly reflects a 'cultural distance' and a lack of information. However, it appears that the impact of geographic distance is decreasing over the last three periods. This may suggest that the development of markets for ethnic products, especially in developed countries, contributes in narrowing the cultural gap between the partner countries involved in this trade.
- Regarding the selected dummy variables, capturing institutional and geopolitical characteristics, it appears that the EU does not have a significant impact on the exports of Feta. In other words, whether a trading partner is a member of EU docs not seem to be determinant in contributing to greater exports. Moreover, it is observed that during the three first periods, the sign of the coefficient is unexpectedly negative. The fact that a change of sign is observed for the last two periods could confirm the emergence of new European markets as mentioned above. This result requires further analysis taking into account the expected role of tourism.
- Conversely, the existence of an important Greek 'diaspora' – as the case in Germany, Australia or USA – has a positive and significant impact on trade flows and the effect of this factor remains determinant for all the considered periods. This is particularly interesting because it shows that even if Greek emigration is a relatively old phenomenon, the attachment and links that the 'diaspora' maintains with its country of origin remain quite strong and likely contributes to the promotion and consumption of 'typical' Greek products.

Table 2. OLS results for the gravity equation.

Independent variables	Coefficients				
	1990-92	1993-95	1996-98	1999-01	2002-04
Constant	-18.723	-9.933	-7.397	-11.241	-9.321
Per capita GDP (in PPP) of importer	2.348	1.712	2.356	2.298	2.231
countries	(2.696)**	(1.487)	(2.677)**	(2.166)**	(2.019)**
Population of importer countries	0.428	0.964	0.538	0.751	0.544
	(0.901)	(2.013)	(1.443)	(1.826)	(1.249)
Distance between Greece and	-1.268	-2.603	-2.182	-1.573	-1.476
importer countries	(-1.693)	(-3.042)***	(-3.298)***	(-2.387)**	(-2.104)**
Importer countries with significant	3.477	3.527	2.772	3.102	3.599
Greek community	(2.035)**	(2.127)**	(2.148)**	(2.010)**	(2.541)**
EU member countries	-1.336	-1.047	-1.299	1.220	2.188
	(-0.853)	(-0.619)	(-0.224)	(0.850)	(1.437)
Adjusted R^2	0.615	0.549	0.681	0.682	0.681
F-test	5.755***	4.384***	7.687***	7.688***	7.686***
Durbin-Watson, d	1.846	1.878	2.094	1.998	1.958

All variables except dummies are expressed in natural logarithms.

Estimations use White's heteroskedasticity-consistent covariance matrix estimator.

t- Statistics are in parentheses, ** denotes significance at 5% and *** significance at 1%.

6. Final remarks

The results of the application of the gravity model mentioned above are quite supportive of spcific policy recommendations which can improve the export performance of Greek Feta. The value of per capita GDP is significant, providing considerable hints regarding which markets should be first priority targets for exports. Such markets are in developed countries, as well as developing country markets such as the Far East or Russia, where high income consumers generate considerable demand for *non usual* or *luxury* products. By focusing on such target groups the risks of attempting export expansion into new markets can be substantially reduced.

Although the result of the model confirms that while the distance factor is significant, when it is combined with the significance of the Greek Diaspora, regardless of distance from Greece, the most essential factor is the cultural distance between consumers and not the actual geographic distance between markets. According to this assumption, export markets with significant Greek communities must be chosen for expansion in the short term, due to pre-existing awareness and acceptance of the product, and in the long run, gradually increasing sales.

Finally the finding that the EU market, when considered as dummy variable is not significant, was unexpected. The establishment of the EU was based upon the idea of providing free access for products being produced within a larger unified market, compared with the individual country markets. According to the findings of the model, this free access is not significant for improving the export performance of Greek Feta. Other parameters which have previously been underestimated, create far greater positive preconditions for strengthening Feta cheese exports than access to the

EU market. It is obvious that the product now requires a differentiated marketing strategy, due to dramatic changes in the internal and external environments of the sector.

References

Aitken, N.D., 1973. The effect of the EEC and EFTA on European trade: a temporal cross-section analysis, American Economic Review, 63: 881-892.

Anderson, J.E., 1979. Error components and seemingly unrelated regressions. Econometrica, 45: 199-209.

Anderson, J.E. and E. van Wincoop, 2003. Gravity with gravitas: a solution to the border puzzle. American Economic Review, 93: 170-192.

Babcock, B., 2003. Geographical indications, property rights and value added agriculture. Review paper (IAR 9:4:1-3). Centre for agricultural and rural development. Iowa State University Ames. Iowa.

Bélair M. and B. Gauthier, 2004. Les effets de l'immigration sur le commerce bilatéral Australien. Cahier de recherche de l'IEA, no. IEA-2004-12, HEC Montréal, Canada.

Bergstrand, J.H., 1985. The gravity equation in international trade: some microeconomic foundations and empirical evidence. Review of Economics and Statistics, 67: 474-481.

Bergstrand, J.H., 1989. The generalized gravity equation, monopolistic competition, and the factor-proportions theory in international trade. The Review of Economics and Statistics, 71: 143-153.

Bergstrand, J.H., 1990. The Heckscher-Ohlin-Samuelson model, the Linder hypothesis and the determinants of bilateral intra-industry trade. Economic Journal, 100: 1216-1229.

Bun, M. and F. Klaassen, 2002. The importance of dynamics in panel gravity models of trade. University of Amsterdam, Faculty of Economics and Econometrics.

Deardoff, A.V., 1995. Determinants of bilateral trade: does gravity work in a neoclassical world. In: J.A. Frankel (Ed.) The regionalization of the world economy. University of Chicago Press.

Disdier, A.C. and K. Head, 2006. The puzzling persistence of the distance effect on bilateral trade. Available at: http://strategy.sauder.ubc.ca/head/

Dunlevy J.A. and W.K. Hutchinson, 1999. The impact of immigration on American import trade in the late nineteenth and early twentieth centuries. Journal of Economic History, 59: 1043-1062.

Frankel, J.E., E. Stein and S. Wei, 1995. Trading blocks and the Americas: the natural, the unnatural and the super-natural. Journal of Development Economics, 47: 61-95.

Gould, D., 1996. Immigrant links to the home country: implications for trade, welfare and factor returns. New York: Garland Publishing, 111p.

Head, K., 2003. Gravity for beginners. Available at: www.economics.ca/keith/gravity.pdf

Head, K. and J. Ries, 1998. Immigration and trade creation: econometric evidence from Canada. Canadian Journal of Economics, 31: 47-62.

Helpman, E. and P. Krugman, 1985. Market structure and foreign trade. Increasing returns, imperfect competition, and the international economy. Cambridge, MA: MIT press.

Leamer, E.E. and J. Levinsohn, 1995. International trade theory: the evidence. In: G. Grossman and K. Rogoff (Eds.), Handbook of international economics, Volume 3. the Netherlands: Elsevier.

Linnemann, H., 1966. An econometric study of international trade flows, Amsterdam.

Martinez-Zarzoso, I. and F. Nowak-Lehmann, 2003. Augmented gravity model: an empirical application to Mercosur-European Union trade flows. Journal of Applied Economics, VI: 291-316.

Mátyás L., 1997. Proper econometric specification of the gravity model. The World Economy, 20: 363-368.

Mátyás, L., L. Kónya and M.N. Harris, 2000. Modelling export activity of eleven APEC countries. Melbourne Institute Working Paper, No 5/00, ISSN 1328-4991, ISBN 0 7340 1485 6.

Luo, X., 2001. La mesure de la distance dans le modèle de gravite: une application au commerce des provinces Chinoises avec le Japon. Revue Région et Développement, 13: 163-180.

Paiva, C., 2005. Assessing protectionism and subsidies in agriculture: a gravity approach. IMF Working Paper, WP/05/21.

Polak, J.J., 1996. Is APEC a natural regional trading block? A critique of the gravity model of international trade. The World Economy, 19: 533-543.

Pöyhönen, P., 1963. A tentative model for the volume of trade between countries. Weltwirtschaftliches Archiv., 90: 93-99.

Pulliainen, K., 1963. A world trade study: an econometric model of the pattern of the commodity flows in international trade in 1948-1960. Ekonomiska Samfundets Tidskrift, 16: 69-77.

Tinbergen, J., 1962. Shaping the world economy. Suggestions for an international economic policy. New York: Twentieth Century Fund.

Effectiveness of Appellations of Origin on international wine market

G. Malorgio, L. Camanzi and C. Grazia

Abstract

The objective of this chapter is to evaluate the role of the Appellation of Origin (AO) system on the international wine market, given (1) the GIs international legal protection system, (2) the main aspects of world wine demand evolution and (3) the strategic choices of firms on the international market. On the demand side, we show through descriptive statistics and economic literature review, the increasing wine consumer appreciation of reputation and origin attributes. On the supply side, we identify the main quality strategies implemented on the international wine market and show an increasing role of origin in firms strategic choices worldwide. Finally, a direct survey on Italian Appellations of Origin concerning the registration on the international market shows an increasing risk of an imperfect use of geographical place names. The main consequences are identified for both producers and consumers. Firstly, a misperception of product quality attributes can arise and menace the effectiveness of AO as informative tool. Secondly, the free riding phenomenon may arise and affect the AO collective reputation with a consequently long term demand drop.

Keywords: appellation of origin, consumer information, intellectual property right protection, international trade agreements, wine market.

1. Introduction

International wine markets are subject to an increasing competition. As traditional wine producing countries in the EU-25 address the domestic challenges of increasing stocks and stagnating per-capita consumption, the emergence of the so-called 'New World' producers has animated extensive international discussions on the issues of labelling, brand protection and Geographic Indications of Origin (GIs) (Camanzi *et al.*, 2008).

In this environment, differentiated products can offer the hope of maintaining profitability. As a sensory experience good, wine differentiation hinges primarily on the transmission and perception of information on product quality.

Consumers face the problem of asymmetric information, with the potential that the average quality in the market will be less than optimal. Conversely, producers need to find ways to efficiently transmit information on their product quality, so as to maximise the potential price premium.

In the traditional European approach, producers tend to organise through consortia, which centre around the AO designation. This mechanism is much more than a simple geographic delineation. The consortium can be governed by history, tradition, culture, *terroir*, and even by tight controls over production decisions, irrigation, plant varieties, etc. Product quality is embodied in everything the Appellation stands for. The AOs play also an important role in EU exports. In fact they provide a tool for product differentiation in order to better fit demand segmentation as to create higher added value for producers.

With growth in international trade, subtle national differences in regulatory and legal frameworks can become major irritants between exporting and importing countries. Pragmatically, there is a need to find common ground so that trade can continue to flow. Discussions aimed to find that common ground have been taking place for a number of years in different fora. From the 1891 Madrid Agreement to the more recent talks taking place within the TRIPS framework.

At present an important debate is taking place about the meaning of the notification and protection system. According to the US and other 'New World' producing countries the GIs should be based on a voluntary registration system as an identification tool. Therefore GIs should be considered as a form of territorial right and their utilisation should be discussed in national legislation. On the other hand according to EU the GIs should enter in a multilateral register that should be enforced in all countries.

The aim of this paper is to discuss the efficiency of AO system on the international wine market as an instrument that can satisfy both producers and consumers needs, and then to give some suggestions to improve the market performance in the future.

2. Background

2.1 The protection of the Appellations of Origin on the international market

The issue of international protection of GIs goes back to the Paris Convention for the Protection of Industrial Property in 1883 ('Paris Convention'), which included 'indications of source or appellations of origin' as objects for protection by national industrial property laws[6]. Nowadays, this Agreement has more adherents than any other treaty addressing the protection of geographical indications. Namely, the Paris Convention currently requires member countries to prohibit the importation of goods bearing false indications. Hence, the importation of goods marked with a GI that might be liable to mislead, but does not rise to the level of being false, need not be prohibited under the Paris convention (Bendekgey and Mead, 1992).

However, a more comprehensive form of regulation is provided by the Madrid Agreement for the Repression of False or Deceptive Indications of Source on Good (1891), which prohibits the importation of goods bearing a false *or misleading* indication to signatory countries or to a place in that country. Nevertheless, as highlighted by Blakeney (2001), this Agreement failed to attract the accession of significant trading nations such as the United States, Germany or Italy.

The Lisbon Agreement for the Protection of Appellations of Origin and their International registration (1958) established an international system for registration and protection of appellations of origin by adopting the French definition of appellation of origin, which defines the appellation of origin as '*the geographical name of a country, region, or locality, which serves to designate a product originating therein, the quality and the characteristics of which are due exclusively or essentially to the geographical environment (milieu géographique), including natural and human factors*'. As underlined by Romain-Prot (1995), this Agreement failed to attract support from more than only a few nations. At first, the accession was confined to those nations which protected appellations of origin 'as such'. Hence, as highlighted by Geuze (2007), the international registration of an appellation of origin provides its protection as long as the appellation is protected in its country of origin. Secondly, no exception was made for GIs, which had already become generic in MS.

---

[6] The term 'indication of source' can be defined as an indication referring to a country, or to a place in that country, as being the country or place of origin of a product. This definition does not imply any special quality or characteristic of the product on which an indication of source is used.

The Trade Related Intellectual Property (TRIPS) agreement defines GIs as '...*indications which identify a good as originating in the territory of a Member, or a region or locality in that territory, where a given quality, reputation or other characteristic of the good is essentially attributable to its geographical origin*'[7]. As highlighted by Romain-Prot (1995), this definition expands (but weakens) the Lisbon Agreement concept of appellation of origin. Firstly, the criteria (quality, reputation, other characteristics) are alternative and independent; secondly, the link between natural and human factors disappears. For example, goods which have merely a certain reputation, but not a specific quality due to their place of origin, are covered by the definition provided by the TRIPS, but not by that provided b y the Lisbon Agreement. Finally, under the TRIPS Agreement, a GI has to be an indication in order to be protected, but not necessarily the name of a geographical place on earth. Hence, as highlighted by Höpperger (2007), non-geographical place names or emblems would fall into the category of signs that could constitute GIs under the TRIPS Agreement.

The TRIPS Agreement provides an additional protection for wines and spirits[8]. Firstly, the Agreement specifies that each Member shall provide legal protection for wine GIs even where the true origin of the goods is indicated or the geographical indication is used in translation or accompanied by expressions such as 'kind', 'type', 'style', 'imitation' or the like. As no mention is made of misleading the public or unfairly competing, the presumption is that no such conditions are required for GI protection for wines and spirits (Josling, 2006)[9]. Hence, the level of protection for wines and spirits is enhanced beyond that provided for GIs, under which protection is limited to cases where the public is mislead as to the true geographical origin of a product or where the use of the GI constitutes an act of unfair competition. Secondly, even if the Agreement does not set out the registration requirements for a geographical indication, it addresses the issue negatively by permitting Members to legislate to provide 'an interested party' to request the refusal or invalidation of the registration of a trademark which contains a GI identifying wines or spirits, which contains or consists of a GI which does not have the indicated origin. Thirdly, a protection for GIs for wines in the case of homonymous indications is provided. Conflicts typically arise where products on which homonymous GIs are used and are sold into the same market. Concurrent use of homonymous GIs in the same territory may be problematic where the products on which a geographical indication is used have specific qualities and characteristics which are absent from the products on which the homonym of that geographical indication is used. In this case, the use of the homonymous geographical indication would be misleading, since expectations concerning the quality of the products on which the homonymous geographical indication is used are not met (Blakeney, 2001).

However, important exceptions limit the effectiveness of this additional protection. Firstly, a Member is not obligated to protect a GI of another Member where that GI has become the generic ('customary') name for products and services or in respect of products of the vine for which the name is identical to the grape variety. For example, US. Bureau of Alcohol, Tobacco and Firearms (BATF) permit the use of 'semi-generic names' such as 'Champagne', 'Burgundy' and 'Chablis' if 'the correct place of origin is directly conjoined to the name' (Brody, 1994). The main exception relates to the so-called prior trademarks. Hence, when a trademark has

---

[7] See Article 22 of the TRIPS Agreement.

[8] See Articles 23-24 of the TRIPS Agreement In the current debate, some Countries consider this additional protection as an unacceptable discrimination against all other products and they have agitated for an extension of that protection to all kinds of geographical indications (Blakeney, 2001).

[9] Article 23.1 permits each Member to 'provide the legal means to interested parties to prevent the use of a geographical indication' identifying wines or spirits which do not originate in the place indicated by the geographical indication in question.

been acquired or registered in good faith before the date of application of the Agreement in that Member, or before the GI was protected in its country of origin, eligibility for or the validity of the registration of a trademark or the right to use a trademark shall not be prejudiced, on the basis that such trademark is identical with or similar to, a geographical indication. Finally, the third main exception concerns the so-called grand fathered uses, that is, the continued uses of a GI identifying wines or spirits for goods or services prior to the conclusion of the Uruguay Round, even when the GI has not become generic and there is no pre-existing trademark right. Namely, uses must have taken place in good faith or for at least ten years before 15 April 1994.

Parallel to, but distinct from the TRIPS Agreement, there are a number of bilateral and multilateral (including regional) agreements, which contain provisions modifying the TRIPS provisions dealing with geographical indications. For example, in 1994, the EU negotiated an agreement with Australia which included the phasing-out of European wine names used by Australian wine-makers that had entered into generic use. The Agreement also provided for mutual recognition of oenological practices of each party and improved European market-access conditions for Australian products, by removing a number of technical barriers to trade between both parties. On March 10, 2006, the US-EU wine trade Agreement has been signed. The Agreement covers wines with an actual alcohol content of not less than 7% and not more than 22%. It addresses several key issues, sets a framework to facilitate future wine trade between the United States and Europe and provides for mutual acceptance of existing oenological (wine making) practices (with the mutual acceptance of wine making practices the US will exempt EU wine from new US certification requirements for imported wine), certification (the EU will simplify its import certification requirements for US wine) and labelling (the Protocol on Wine Labelling, sets specific conditions for the use of names of vines, vintage characteristics, production methods, product types and variety names).

Moreover, the US and the EU agree to recognise certain of each other's names of origin in specific ways (article 7) and the US agrees to seek legislative changes to limit the use of 16 semi-generic names. The 'traditional expressions' that the U.S. will be allowed to use under specified conditions are: chateau, classic, clos, cream, crusted/crusting, fine, late bottled vintage, noble, ruby, superior, sur lie, tawny, vintage and vintage character. These terms may only be used if they have been approved for use on wine labels in the U.S. on a Certificate of Label Approval (COLA). Current US laws permit these names to be used on non-European wine. The new rules will prohibit new brands from using these names on non-European wine, but will grandfather existing uses of these semi-generic names.

2.2 The effectiveness of Appellations of Origin for producers and consumers

According to the economic theory, the creation of a brand has important effects on social welfare. Firstly, when quality is not adequately signaled to consumers, a decrease in the average quality provided on the market is expected to arise. In this sense, the brand acts as informative tool and can increase consumer utility. Secondly, the brand creation increases quality differentiation and thus let producers gain positive profits in the short-term, according to the degree of products substitutability (Dixit and Stiglitz, 1977). Finally, as far as the brand corresponds to an actual quality differentiation, the Intellectual Property Right acts as a tool to protect both consumers and producers interests.

In the specific case of Appellations of Origin, we can consider that an AO has an important role for both producers and consumers. On the demand side, the Appellation of Origin represents a quality signal, which provides information about the region of origin and the wine's average quality. On the supply side, the Appellation of Origin represents a long-term commitment constraining a firm's strategy in terms of quantity and quality; in exchange, producers have access to a collective reputation.

On the one hand, Appellations of Origin represent a way to solve of the asymmetric information problem (Laporte, 2001). In a context where the wine's quality is not directly observable to consumers, AO represents an important quality sign concerning the wine characteristics by providing information about the wine geographical origin and its average quality. In fact, wine market is characterised by a very heterogeneous supply and the impossibility to observe the product quality before purchase. This leads to relevant asymmetric information between producers and consumers and consequently implies strong promotional and information research costs (Nelson, 1970, Darby and Karni, 1973). The major consequence of the inefficiency of quality signals as regards consumer expectations on quality and typicality is the risk of a decrease in the average quality level supplied on the market, which can imply a long-term demand drop (Akerlof, 1970). In this context, the AO aims at reducing consumer information costs.

On the other hand, Appellations of Origin have important consequences on the 'characteristics space' (Lancaster, 1966). The *delimited production area* and the existence of *specific production requirements* (the maximum yield of wine from grapes, the minimum density of rootstocks per hectare, the minimum natural alcohol level by volume, the minimum total acidity, etc) confer to wine *specific quality characteristics and substantially differentiate each Appellation of Origin from the other ones*. As a result, the construction of an AO provides an increase in the *inter-appellation* quality differentiation and a decrease in the *intra-appellation* quality differentiation, by conferring specific quality characteristics to the wines belonging to the same AO. The quality differentiation is thus based on the *specific production requirements* to which producers commit. In exchange of quantity restrictions (delimited production area and maximum yield per hectare), which limit producer strategic flexibility in the long term, producers have access to a collective reputation, which may increase consumer willingness to pay for the AO (Chambolle and Giraud-Héraud, 2003).

3. Objectives and methodology

The objective of this paper is to discuss the performance of the Appellation of Origin system on the international wine market, with respect to some relevant context factors, such as: (1) the GIs international legal protection system; (2) the world wine demand trend and size; (3) the strategic choices of the competitors on the international market. In particular, the analysis aims at identifying the key factors that determine the effectiveness of Appellations of Origin to provide both profitability for producers and satisfaction for consumers. We will show AO importance for producers by describing how it helps to build and give access to a collective reputation, making it profitable for them to undertake relevant investments for quality. As regards consumers we intend to point out that AO are an effective tool for them to recognise the quality attributes they look for, especially when they are seeking an actual link with *terroir*: this means that the AO represent a valid solution to the asymmetric information problem. All this considered, and given the importance of the AO to protect and incentive intangible investments such as *terroir*, tradition and social history, we suggest that AOs should be accorded a more extensive recognition on the international market.

As for the methodology, the study is conducted in three steps. First, we carry out a demand analysis in order to evaluate consumers appreciation of origin attributes. The demand analysis is conducted through descriptive statistics and a critical review of the related economic literature.

The second step of our methodology consists in a supply analysis carried out in order to describe and evaluate alternative market strategies adopted by the main wine producing countries and in order to identify the role of origin in firm's strategic choices. Thirdly, through two empirical analyses we intend to show the risks that arise for both consumers and producers as a consequence of the coexistence of the brand names and AOs on the international markets.

The first investigation is conducted the United States Patent and Trademark Office (USPTO) trademark register database in order to illustrate some cases of imperfect use of quality signals on the international market. This analysis is aimed at quantifying the actual risks of altering of consumer quality perceptions and of weakening of AO reputation on the international markets. Further, we conducted a direct survey on the Italian AO Consortiums Association (Federdoc) in order to give some insights into producer efforts to register the collective brand on the international markets.

4. Results

4.1 Consumer appreciation of quality and origin attributes

This section of the paper aims at evaluating the role of quality in consumer behaviour through descriptive statistics and literature review. The demand analysis through descriptive statistics shows that wine's quality seems to be a fundamental factor behind consumption trends. In fact, if we consider the demand for wine from 1984 to 2003, we observe that the two categories 'quality wine' and 'table wine' have been moving in different directions. In particular, there has been a substantial fall in consumption of 'table wines'. Over the same period there has been a growth in consumption of 'quality wines', but not sufficiently large to compensate for the reduction in the first category. If we consider the traditionally producing and consuming countries (France, Italy and Spain), the gross human consumption per-capita of total wine has decreased about 40% from 1984 to 2004, whereas the opposite trend is registered in the case of quality wines PSR. Figure 1 shows the role of quality wines produced in specific regions (PSR) on the total GHC per capita in France, Italy, Spain and Portugal.

European consumers appear to be more quality-oriented than quantity-oriented. The raising importance of occasionally wine consumption is confirmed by several socio-economic surveys. In 2003, about 67% of Italian wine consumers consume wine each day, while about 33% consumes wine occasionally. The 75% of occasional consumers is identified as 'wine-passionate' consumers, which also have a 'wine-culture'. As for France, the INRA-ONIVINS survey 2005 confirms the increasing role of occasionally consumption.

As for the Italian market, a recent ISMEA's survey (ISMEA, 2005) examines the role of the designation of origin in consumer purchase choices. According to this survey, Italian consumers recognise the Appellations of Origin as high quality products from the point of view of (1) taste and (2) food safety (due to the existence of production system's control mechanisms). Moreover, an increasing knowledge concerning AO is registered, which highlights an increasing interest in these categories of products.

Further, the demand analysis through the review of economic literature shows, an increasing relevance of objective characteristics (as region of origin, the reputation and other objective characteristics) on consumer willingness to pay for wine. Hence, when a product has a high proportion of attributes that can only be assessed during consumption (experience attributes) as with wine (Chaney, 2000), then the consumers will fall back on extrinsic cues in the assessment

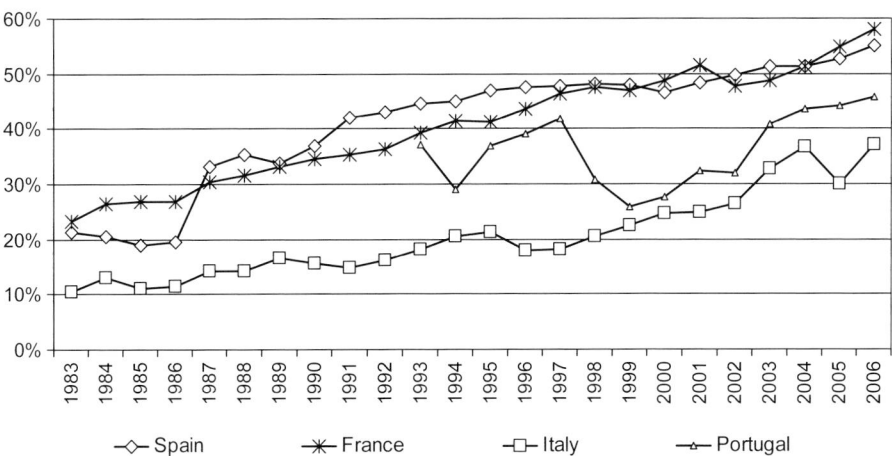

Figure 1. Role of quality wines PSR on per-capita gross human consumption in the traditionally wine producing and consuming countries (1983-2006). Source: Eurostat data, Wine Balance Sheet, 2006 (http://epp.eurostat.ec.europa.eu).

of quality (Speed, 1998). Several papers show the impact of objective characteristics on price differentials. This category includes the vintage's year, the Appellation, the region, the grape variety, which usually appears on the label and are therefore easy to identify by consumers. Combris *et al.* (1997, 2000) use data for Bordeaux and Burgundy wine to estimate a hedonic price function. In both studies, price is strongly explained by objective attributes appearing on the label of the bottle. The authors conclude that consumers may decide to vary their willingness to pay for wine primarily according to observable attributes. See also Nerlove (1995) and Gergaud (1998) for an analysis carried out using the data for Champagne. The relevance of objective traits is also underlined in Oczkowski (1994). Landon and Smith (1998), use an unbalanced panel of 196 red wines from the five Bordeaux vintages from 1987 to 1991 and estimate two hedonic price equations. The authors confirm the relevance of the objective traits and show that long term reputation explains much more variation in the consumer willingness to pay than does short term quality changes. This finding has been confirmed by focusing only on a balanced panel of 151 wines for the 1989 and 1990 vintages (Landon and Smith, 1998). Subsequent applications to premium wines from North America, Australia, South Africa and Chile by Schamel (2000) and to Australian premium wines by Oczkowski (2001) support the presence of significant reputation effects. Schamel (2003) estimates a hedonic pricing model of premium wines sold in the U.S in order to analyse the factors behind price differentials based on regional origin and points out that the domestic regions command higher prices than wines imported from other New World sources.

As for the Italian market, Benfratello *et al.* (2004) estimate a hedonic model using a dataset on two premium quality wines (Barolo and Barbaresco) covering the 1995-1998 vintages and show that the reputation acquired by wines and producers during the years is more important than taste in driving market prices.

Other papers, dealing with experimental studies, point out that the AO can improve consumer's WTP (Bazoche *et al.*, 2005).

Mtimet and Albisu (2006) examine Spanish AO wine consumer behaviour by the use of a choice experiment technique. Empirical results indicate the importance of the designation of origin and the wine aging attributes on wine selection. The grape variety variable, although it has a lower utility values, is also found to be significant (especially a foreign variety), thus confirming the emergence in the Spanish wine market of the 'New World' marketing strategies based on well-known varietal wines.

4.2 Quality strategies on the international wine market: the role of origin

Two main production–marketing systems coexist on the international wine market. Behind these systems two main strategies can be identified: the private brand strategy and the Appellation of Origin system. These two strategies can be distinguished through the degree of commitment-flexibility, which characterises producer strategic choices.

The *private brand strategy* is advantageous for the firms, because it allows speedier adjustments to market conditions, particularly changing in this field of the agrifood consumption. Let us consider as an example the large firms of 'New World' producing countries (Jacob's Creek, Gallo, Southcorp, etc.). These firms develop a whole series of brands, easily identified by consumers, thanks to a great market volumes and notoriety. Considerable investments in promotion are associated with these brands. The firm efficiency is based on its capacity for scale economies, which allows it to meet market volume requirements and to develop strategies of price promotion. For example, around 66% of Australian wine is sold on price or multi-buy promotion on the UK market.

On the other hand, the *Appellation of Origin system* requires the producer's commitment to specific production requirements, which constraint the producers in terms of quantity. In exchange, the producer benefits from a collective reputation related to the Appellation. The quantity constraints may result in a loss of strategic flexibility (Giraud-Héraud and Grazia, 2008). The loss of strategic flexibility can constitute a limitation of firm's expansion in the markets which are characterised by an increasing wine consumption trend (especially the Anglo-Saxons countries) and thus by a great level of competition between Appellations of Origin and 'New World' wines. Indeed, whereas the wine consumption is nowadays stagnating in the countries with the highest wine production (and consumption) as France, Italy or Spain, on the other hand, it is not the same in the U.S.A, in the United Kingdom and in the Asian countries, as China or Japan, where the competition between the AO and the private brands is very strong and leads to several strategic difficulties for the producers.

The importance of wine origin for traditionally producing countries can be appreciated from Figure 2, showing the trend of wine *production* and *exports* in the leading trio of producing and exporting countries (France, Italy and Spain)[10].

Table wines still make up more than half of Community wine production (98 million hl in the 2004/05 wine year) but their share is declining in favour of quality wines. The increase in the share of quality wines on the total wine production is manly resulting from conversion of lands and reclassification on some table wines in response to changing demand. The analysis of the trend of volume of *exports* by category of wine (for France, Italy and Spain), points out that the growth in exports of quality wines has been slower but more constant than for table wines. This points out a relatively stability of quality wine image on the exports markets. The conjoint analysis of

---

[10] France, Italy and Spain together account for 50% of world production and 60% of world exports.

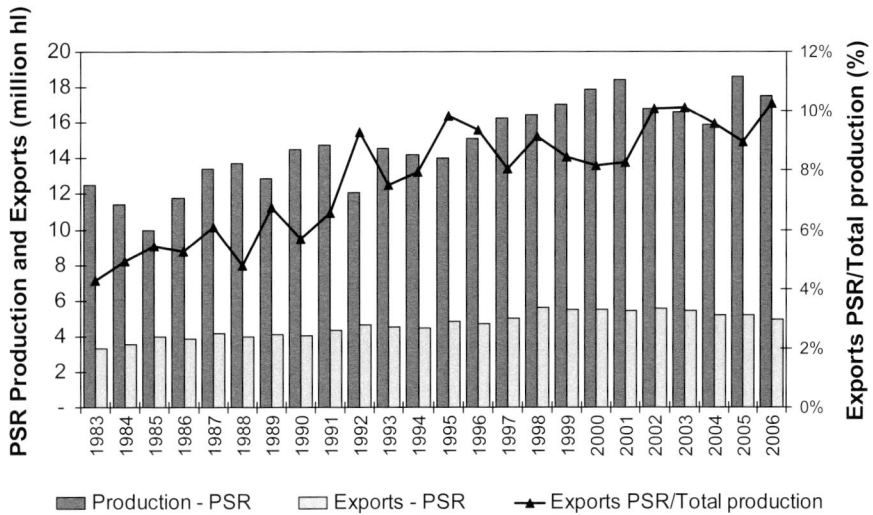

Figure 2. The relative importance of quality wines PSR exports on total wine production in the trio of leading world producing and exporting countries. Source: elaborations on Eurostat wine balance sheet; 1983-2006 (http://epp.eurostat.ec.europa.eu).

the trend in production and exports points out that the relative importance of exported volumes of quality wines with respect to the total production has increased in the period 1983-2006 from 4% to 10%. This points out an increasing importance of quality wines strategy for the traditionally producing countries with respect to exports markets.

The competition between the two systems mentioned above (*private brand* vs. *Appellation of Origin*) is particularly tight in those markets characterised by increasing consumption. Nevertheless, we observe that many producers around the world started to use Geographical Indications to differentiate their product (Hobbs, Kerr, Phillips, 2001): the increasing competition by foreign wines and the evolution of consumer behaviour towards an increasing appreciation of quality, implies the implementation of origin-oriented strategies.

In this perspective is worth noticing the development of the American Viticultural Areas (AVAs) in California and in particular in Oregon and Washington (Rousset, Traversac, 2006): over 160 American Viticultural Areas are nowadays approved. An American Viticultural Area (AVA) is a delimited grape-growing region distinguishable by geographic features, with boundaries defined by the United States government's Alcohol and Tobacco Tax and Trade Bureau (TTB). The TTB defines these areas at the request of wineries and other petitioners. An AVA specifies a location. Once an AVA is established, at least 85% of the grapes used to make a wine must be grown in the specified area if an AVA is referenced on its label. Current regulations impose the following additional requirements on an AVA: (1) evidence that the name of the proposed new AVA is locally or nationally known as referring to the area, (2) historical or current evidence that the boundaries are legitimate and (3) evidence that growing conditions such as climate, soil, elevation, and physical features are distinctive. It can be noticed that the AVA implies a lower level of commitment as compared to the European AO. In fact, it does not limit the type of grapes grown, the method of vinification, or the yield, for example. Some of those factors may, however, be used by the petitioner when defining an AVA's boundaries.

The use of Geographical Indications in Australia started in 1993 when the Australian Wine and Brandy Corporation Act (1980) was updated to enable Australia to fulfill its Agreements with the European Community on Trade in Wine and the Agreement on Trade-Related Aspects of Intellectual Property Rights (TRIPS). The use of GIs is aimed at 'providing the legal means for interested parties to prevent use of a geographical indication identifying wines for wines not originating in the place indicated by the geographical indication in question'. With respect to the European AO system, it is much less restrictive in terms of viticultural and winemaking practices. In fact the only restriction is that wine which carries the regional name must consist of a minimum of 85% of fruit from that region. This protects the integrity of the label and safeguards the consumer.

4.3 The assessment of the effectiveness of Geographical Indications on the international market: empirical results

In order to assess the risk of consumer misperception of the link between the geographical place name and the actual region of origin, we carried out an analysis on the USPTO database with respect to the 17 *semi-generic names* concerned by the EU-US Wine Agreement. This analysis points out some examples of trademarks, which explicitly refer to European Appellations of Origin, but have been registered by firms located outside the delimited production area. The main results of the analysis are the following (see Table 1):

- Several semi-generic names appear in non-wine related products. In this case the level of consumer misperception is relatively low. See for example, 'The Champagne of Tea', 'Pink Champagne' (Beauty products), 'The Champagne of Water (Drinking Water), 'Champagne Honey mustard splash' (salad dressing) or 'Marsala' (Fresh olives and grapes), Porto's (Bakery goods).
- Some of the semi-generic names are explicitly mentioned in trademarks referring to wine (relatively high risk of misperception), which have been registered by producers located outside the delimited production area. See examples in the Table below.
- A few semi-generic names are not registered as trademarks, neither from producers located in the delimited production area, nor from US firms (Haut Sauterne, Hock, Moselle, Retsina, Sauterne). 'Porto' and 'Malaga' do not appear in trademarks registered from producers located outside the delimited production area.
- The most 'used' geographical place names (both in non-wine and wine related sectors) are likely to be those with the highest notoriety on the international market; thus, in addition to the risk of consumers misperception, an opportunistic behaviour may take place, when producers located outside the original production area may take advantage of the Appellations of Origin collective reputation.
- In particular some multinational firms seem to develop a sort of strategy based on an explicit mention to European Appellations of Origin (Arbor Valley).

As a second step of the investigation, we searched the USTPO database for names similar to the Italian Controlled and Guaranteed Denominations of Origin (DOCG). The results are illustrated by Table 2.

Eleven out of the thirty-four DOCG names are not registered at all (neither from the Consortium nor from other firms not related with the actual product's origin or with the wine sector): Albana

Table 1. Registration on the US market of semi-generic names. Source: elaboration on United States Patent and Trademark Office (USPTO); http://tess2.uspto.gov/bin/gate. exe?f=tess&state=lmi2nf.1.1.

Burgundy	Arbor Valley American Burgundy
	Inglenook Classic Burgundy
	Taylor California Cellars Burgundy
Chablis	Arbor Valley American Chablis
	Inglenook Chablis
Champagne	Chamblue (sparkling wine)
Claret	Bearitage California Claret, Vanderbilt Claret, Crown Claret
Madeira	Arbor Valley American Madeira
Marsala	Arbor Valley American Marsala
Rhine	Taylor New York Rhine Wine
Sherry	Arbor Valley American Sherry, Arbor Valley American Cream Sherry
Tokay	Y-Tokay

di Romagna, Bardolino, Carmingnano, Ghemme, Soave Superiore, Taurasi, Torgiano Rosso Riserva, Valtellina Superiore, Vermentino di Gallura, Vernaccia di San Gimignano, Gattinara.

More interestingly, we found that some DOCG are not registered by the Consortium, but their geographical place name has been registered as trademark or service mark by non-wine related firms (Barbaresco, Barolo, Chianti, Gavi o Cortese di Gavi). In this case the risk of misperception is relatively high, in particular for the DOCG Chianti, which has not been registered by the Consortium. In fact, its geographical place name appears in wine-related trademarks (Arbor Valley American Chianti, Inglenook Chianti, Good Chianti, Chianti Station).

The risk of misperception can arise in spite of the registration from the Consorzio di Tutela. For example, the DOCG Asti has been registered by the Consortium, but the geographical place name 'Asti' appears in trademarks registered by non-wine related firms.

An effective intervention of the Consortium is registered for Brachetto d'Acqui, Brunello di Montalcino, Chianti Classico, Franciacorta Spumante, Gattinara, Ramandolo, Recioto di Soave and Vino Nobile di Montepulciano.

A relatively important action is that of individual firms, which register their individual brand (containing the geographical place name of the AO): Marchesi di Barolo, Primore Casa Vinicola in Gattinara, Gavi La Scolca, Martini & Rossi Asti Spumanti Martini, The Bosca Millennium Collection Asti, Poggio Rosso Chianti Classico, Barone Pizzini Franciacorta DOCG brut. In some cases the individual registration strategy allow the firm to protect its brand (and indirectly the geographical place name of the concerned AO), in spite of a lacking intervention of Consorzio di Tutela (Ruffino Chianti 2004 dal 1877 DOCG, Chianti DOCG 2001 Piccini, Chianti Vino Pasolini).

In order to give an insight into traditional wine producing countries attitude towards brand registration in international markets we conducted a direct survey among the most representative Consortia in Italy. Consortia were chosen from the National Confederation for Voluntary Consortia for the Oversight of the Denominations of Origin (Federdoc). As shown by Table 3, preliminary

Table 2. Registration on the US market of Italian DOCG. Source: elaboration on United States Patent and Trademark Office (USPTO); http://tess2.uspto.gov/bin/gate.exe?f=tess&state=lmi2nf.1.1.

	Registered by the Consorzio di Tutela	Risk of misperception	
		Registered in wine sector (high risk)	Registered in non-wine sectors (low risk)
Asti spumante – Moscato d'Asti	Consorzio dell'Asti (Trademark), Asti (Certification Mark)		Astipure, Asti, Asti aircraft safety technology, Asti magnetics corp.,
Barbaresco			Barbaresco (Service Mark)
Barolo			Villa Barolo Ristorante and Wine Bar (Service Mark), Barolo, Barolo Tuscan Grill (Service Mark), Barolo (watches), Barolo (shoes)
Brachetto d'Acqui	Brachetto d'Acqui (Certification Mark)		
Brunello di Montalcino	Brunello di Montalcino (Certification Mark)		
Chianti		Arbor Valley American Chianti, Inglenook Chianti (Constellation Brands), Good Chianti, Chianti Station	Chianti
Chianti Classico	Chianti Classico (Trademark), Chianti Classico dal 1716 (Trademark), Consorzio Vino Chianti Classico (Collective Trademark)		
Franciacorta Spumante	Franciacorta DOCG (Trademark)		
Gavi o Cortese di Gavi			Gavi, Gavi Fund, Piazza Gavi
Montefalco Sagrantino	Montefalco Sagrantino (Certification Mark)		Sagrantino di Montefalco (Service Mark), registered by an italian firm
Ramandolo	Ramandolo (Trademark)		
Recioto di Soave	Recioto di Soave (Certification Mark)		
Vino Nobile di Montepulciano	Vino Nobile di Montepulciano (Certification Mark)		

Table 3. Sample representativity. Source: direct survey on Appellations of Origin associated to Federdoc.

		Italy	Sample	
AO production 2004	(0,000 hl)	12,740	5,140	40.3%
AOCG	(n)	24	9	37.5%

results refer to 21 Consortia that account for 40.3% of Italian production with Appellation of Origin and that include 9 out of 24 Appellations of Origin Controlled and Guaranteed (37.5%).

As detailed in Table 4, the survey shows that almost one out of two Consortia interviewed (48%) have not taken any action yet to register their Appellation of Origin as a brand, neither on the national or EU market, neither on the international market. At least two of them are presently evaluating the cost of registration in few countries which are their main importers.

Among those Consortia that already have registered a mark we notice that quite a few (19%) have taken this action only to protect their Appellation on the National or European market. Therefore only one third of the Consortia considered makes use of international marks, in the form of individual trademarks, collective marks and international marks (according to the Madrid Agreement).

The most used tool for Appellation protection on the international markets is the Individual trademark, chosen by 24% of Consortia of our sample, followed by the collective mark which is used by a smaller percentage of Producers Associations – 19%. Only in one case (5%) we recorded the use of the Madrid Agreement through which the Appellation is protected in 31 countries.

Table 5 illustrates the main countries in which Italia Consorzia register their marks. As regards the countries in which Appellations seek for protection, Canada and the US are leading the list (71% of cases), followed by Japan (57%), Argentina, Australia, Chile and South Africa (43%). Another

Table 4. Attitude of Italian consortia towards marks. Source: direct survey on Appellations of Origin associated to Federdoc.

Strategy	Consortia	
	(n)	(%)
No action	10	48%
Presently evaluating costs of registration	2	10%
Registered international marks	7	33%
Trademark	5	24%
Collective mark	3	14%
Madrid Agreement	1	5%
National or European collective mark	4	19%
Total	21	100%

Table 5. Countries in which marks are registered by Italian Consortia. Source: direct survey on Appellations of Origin associated to Federdoc.

Countries	Registered marks	
	(n.)	(%)
Canada, USA	5	71%
Japan	4	57%
Argentina, Australia, Cile, South Africa	3	43%
Brazil, Phillippines, Mexico, New Zealand, Venezuela	2	29%
India, Indonesia, North Korea, Paraguay, Perù, South Korea, Switzerland, Taiwan, Thailand, Uruguay	1	14%
Total	7	100%

relevant group of countries includes Brazil, Phillippines, Mexico, New Zealand, Venezuela, in which 29% of our sample Consortia registered their marks. Finally there are several countries such as India, Indonesia, North Korea, Paraguay, Peru, South Korea, Switzerland, Taiwan, Thailand and Uruguay, where only one Appellation is registered as mark.

In the last three years the overall registration process cost was about 126,000 euro and it has been more expensive for trademarks (almost 89,000 euro) than for collective marks (37,300 euro), but this is due to the greater use of the former as compared to the latter.

At present two important Consortia are pursuing registration of both trademarks and international marks in many other countries such as Albania, Algeria, Bulgaria, Croatia, Cuba, Malta, Morocco, Romania, Singapore, Tunisia, Turkey, Vietnam.

The main difficulties come up in the registration process relate to refusals, in particular in Australia, Canada Russia and Switzerland. Other issues arose because of the bureaucratic burden, the excessive time length and costs (consultants and personnel) required by the procedure.

As for the legal actions in protection of the Appellation or the mark the survey shows that Consortia had to spend even more than for the registration process (164,000 versus 126,100 euro). However we notice that in most cases they are oriented at protecting the Appellation of Origin, with a cost up to 114,000 euro, while the protection of the trademark / collective mark occurred more rarely with a lower overall cost.

5. Final remarks

The higher competitiveness on the international wine market, in the last years, has increased the implementation of strategies to differentiate production and at the same time, it has increased the demand for a protection system apt to guarantee high investments and commitments by producers.

The study conducted aimed at assessing the effectiveness of the AOs on the international wine market. The results of the analysis conducted are both positive and negative.

As for the positive aspects, AOs are a key to ensure fair competition and consumer information. According to market surveys, they are perceived both as origin and quality indicators. They assist consumers in making the right choice, be it whether to buy an AO product or not. Consumers show willingness to pay a premium price if the origin of the product is guaranteed.

As for the negative factors, we observe a weak performance of AOs on international markets. In fact, given the present IPR system, in some cases we observe a double registration of brand and double costs for producers: one for the AO registration and one for the industrial brand registration. Further, we observe a weak recognition of specific investments and quantity and quality commitment for AO producers and some risk of altering of consumer quality perceptions.

The debate at national and international level, concerning industrial brand and AO brand, is also linked to distribution of monopole rent derived from monopolistic competition by the brand. In the case of industrial brands, since these are property of a firm, the firm will directly benefit from them. In the case of AOs, the beneficiaries are all the producers of the area, who may be considered as a club. In fact the management of AO is always a collective concern, with many difficulties because of the different interests and behaviours of the beneficiaries.

This is the reason why, in order to develop its potential benefits, the AOs need a strong economic regulation and specific controls to adapt, by one hand, supply to demand to avoid short term opportunist behaviours and stabilise product's quality in the long term, and, by the other hand, to increase its notoriety and information guarantee and trust to consumers. Individual and collective brand should coexist, with differentiated and specific dynamic to fit in wider segmented wine markets.

Moreover the AO implies specific techniques, a traditional competence linked to territory, a collective patrimony, with an economic value and also a strong social and cultural dimension that constitute determinant factors of quality policy for European producers.

The Agreement on Trade-Related Aspects of Intellectual Property Rights represents an important step toward the universal recognition of GIs protection. While previous agreements, including the Madrid and the Lisbon Agreements, have already regulated related legal figures such as indications of source and appellations of origin, the TRIPS Agreement is today the standard subscribed by all Members of the World Trade Organization and therefore the one with widest international recognition.

The TRIPS Agreements not only sets some minimum standards but according to Article 23.4, calls for negotiations for the '*establishment of a multilateral system of notification and registration of geographical indications for wines and spirits eligible for protection in those Members participating in the system*'. Negotiations on a multilateral system of notification and registration of GIs for wines and spirits are currently underway in the special session of the TRIPS Council. At the heart of the debate are a number of key questions: when a geographical indication is registered in the system, what legal effect, if any, would that need to have within member countries and to what extent should the effect apply to countries choosing not to participate in the system. WTO Members remain divided over whether countries should be obliged to protect the GIs to be covered through the multilateral system – as advocated by the EU and Eastern European countries – or whether it should be left to each country to decide at the national level – as favored by Australia, Canada, Japan and the United States[11].

The former countries propose that when a geographical indication is registered, this would mean that the term is to be protected in other WTO members[12]. The latter group of countries proposes a decision by the TRIPS Council to set up a voluntary system where notified geographical indications would be registered in a database. Those governments choosing to participate in the system would have to consult the database when taking decisions on protection in their own countries. Non-participating members would be 'encouraged' but 'not obliged' to consult the database.

At present the debate is very heated and a solution does not seem to be at hand, nevertheless it is necessary to set out common rules as soon as possible so as firms and consumers will dispose of all the relevant information to make their choices. Preventing misleading information and opportunist behaviours means preventing a market failure and helping the development of a more diversified, profit-oriented agriculture.

References

Akerlof, G.A., 1970. The market for 'lemons': quality uncertainty and the market mechanism. Quarterly Journal of Economics, 84: 488-500.

Bazoche, P., P. Combris and E. Giraud-Héraud, 2005. Willingness to pay for Appellation of Origin in the world chardonnay's war: an experimental study. XIIth Œnometrics, Macerata, Italy, 27-28 Mai 2005.

Bendekgey, L. and C.H. Mead, 1992. International protection of Appellations of Origin and other Geographical Indications. The Trademark Reporter 82: 765-792.

Benfratello, L., M. Piacenza and S. Sacchetto, 2004. What drives market prices in the wine industry? Estimation of a hedonic model for Italian premium wines. Ceris-CNR Working Paper n.11.

Blakeney, M., 2001. Proposals for the international regulation of Geographical Indications. The Journal of World Intellectual Property, 4: 629-652.

Brody, P., 1994. Protection of Geographical Indications in the wake of Trips: existing United States laws and the administration's proposed legislation. The Tradmark Reporter, 84: 520.

Camanzi, L., C. Grazia, D. Leishman, G. Malorgio and A. Menghini, 2008. Strategies toward quality and brand protection. In: R. Fanfani, E. Ball, L. Gutierrez and E. Ricci Maccarini (Eds.), Competitiveness in agriculture and food industry: US and EU perspectives. Bononia University Press, pp. 173-190.

Chambolle, C. and E. Giraud-Héraud, 2003. Certification de la qualité par une AOC: un modèle d'analyse. Économie et Prévision, n. 159, 2003-3.

Chaney, I.M., 2000. External search effort for wine. International Journal of Wine marketing, 12: 5-21.

Combris, P., S. Lecocq and M. Visser, 1997, Estimation of a hedonic price equation for Bordeaux wine: does quality matter? The Economic Journal, 107: 390-402.

Combris, P., S. Lecocq and M. Visser, 2000. Estimation of a hedonic price equation for Burgundy wine. Applied Economics, 32: 961-967.

Darby, M. and E. Karni, 1973. Free competition and the optimal amount of fraud. Journal of Law and Economics, 16: 67-88.

Dixit, A.K. and J.E. Stiglitz, 1977. Monopolistic competition and optimum product diversity. American Economic Review, 67: 297-308.

---

[11] These proposals have been laid out side by side so that they can be compared easily, in a Secretariat paper (document TN/IP/W/12 of 14 September 2005), available on Documents Online http://docsonline.wto.org.

[12] Except in a country that has lodged a reservation within a specified period: if it does not make a reservation, a country would not be able to refuse protection on these grounds after the term has been registered.

Gergaud, O., 1998. Estimation d'une fonction de prix hédonistiques pour le vin de Champagne. Économie et Prévision, 136: 93-105.

Geuze, M., 2007. Let's have another look at the Lisbon Agreement: its terms in their context and in the light of its objects and purpose. International Symposium on geographical indications, Beijing, june 26-28.

Giraud-Héraud, E. and C. Grazia, 2008. Certification of quality, demand uncertainty and supply commitment in agriculture: a formal analysis. In: R. Fanfani, E. Ball, L. Gutierrez and E. Ricci Maccarini (Eds.), Competitiveness in agriculture and food industry: US and EU perspectives. Bononia University Press, pp. 119-136.

Hobbs, J.E., W.A. Kerr and P.W.B. Phillips, 2005. Identity preservation and international trade: signaling quality across national boundaries. Canadian Journal of Agricultural Economics / Revue Canadienne d'Économie, 49: 567-579.

Hopperger, M., 2007. Geographical Indications in the international arena. The current situation. International Symposium on geographical indications, Beijing, june 26-28.

ISMEA, 2005. I prodotti agroalimentari protetti in Italia. Le tendenze della produzione e del mercato e la situazione a livello comunitario.

Josling, T., 2006. The war on *terroir*: Geographical Indications as a transatlantic trade conflict. Journal of Agricultural Economics, 57: 337-363.

Lancaster, K., 1966. A new approach to consumer theory. Journal of Political Economy, 74: 132-157.

Landon S. and C.E. Smith, 1998. Quality expectations, reputations, and price. Southern Economic Journal, 64: 628-647.

Laporte, C., 2001. L'Appellation d'Origine Contrôlée: une solution efficiente pour résoudre le problème de l'asymétrie d'information sur les marchés des vins de qualité? INRA-ESESAD working paper No 5.

Mtimet, N. And L.M. Albisu, 2006. Spanish wine consumer behavior: a choice experiment approach. Agribusiness, 22: 343-362.

Nelson, P., 1970. Information and consumer behaviour. Journal of Political Economy, 78: 311-329.

Nerlove, M., 1995. Hedonic price functions and the measurement of preferences: the case of Swedish wine consumers. European Economic Review, 39: 697-716.

Oczkowski, E., 1994. A hedonic price function for Australian premium table wine. Australian Journal of Agricultural Economics, 38: 93-110.

Oczkowski, E., 2001. Hedonic wine price functions and measurement error. The Economic Record, 77: 374-382.

Romain-Prot, V., 1995. Origine géographique et signes de qualité: protection internationale. Revue de Droit Rural, 236: 432-438.

Rousset, S. and J.-B. Traversac, 2006. Dai vini ordinari ai nuovi standard di qualità: strategie di identificazione delle winery californiane. In: G.P. Cesaretti, R. Green, A. Mariani and E. Pomarici (Eds.), Il Mercato del vino - Tendenze strutturali e strategie dei concorrenti. Milano: Franco Angeli.

Schamel, G., 2000. Individual and collective reputation indicators of wine quality. Technical report, Centre for International Economic Studies, Adelaide University.

Schamel, G., 2003. International wine market: analyzing the value of reputation and quality signals. Paper prepared for the 2003 AAEA Annual Meeting, Montreal, Canada, July 27-30.

Speed, R., 1998. Choosing between line extensions and second brands: the case of the Australian and New Zeland wine industries. Journal of Product and Brand Management, 7: 519-536.

International marketing and trade of protected designation of origin products

C. Mora and D. Menozzi

Abstract

This paper aims to analyse the international marketing and trade strategies implemented by Italian quality food producers with special attention to the export and trade issues for two important Italian PDO products: Prosciutto di Parma PDO and Parmigiano-Reggiano PDO. The international marketing and trade strategies will be analysed according to the four Ps approach, widely applied in national marketing studies especially for food industrial products. In particular, PDO products and other EU quality labelled products, has been analysed in international trade contexts using other methodologies (such as sector analysis). The paper reports theories and case studies of the supply chain players strategies, focusing on the rapid change in trading and distribution channels. The future of PDO and PGI products is not only connected to their positioning, promotion and international protection but also to the strategies of the retail leaders and to the overall strategy of the producers, ranging from small to international companies. The latter are more and more present in the market of EU quality labelled product which represents a diversification and a good investment to improve market positioning.

Keywords: PDO products, international marketing, Parmigiano-Reggiano cheese, Prosciutto di Parma, consolidation in traditional food sector

1. Background

The PDO/PGI product sector accounts today for about 8% of national consumption of food products. Cheese and prepared meats account for 95% of the value of PDO/PGI. In terms of export, these products are above the national average of the food industry; 18% of output value compared to 13% average. Many products that are frequently exported are of high value; they include 'Toscano' PGI olive oil (67% product exported), 'Pecorino Romano' PDO cheese (63%) or niche products such as 'Terre di Siena' PDO olive oil (55%), 'Garda' PDO olive oil (50%) and balsamic vinegars (45%). The proportion of export within the segment of protected dairy products, shows big differences (Table 1).

Table 1. Production and export of some Italian PDO cheeses (2005, tonnes).

	Production	Export	Share export / production
Grana Padano PDO	159,607	32,718	20.5
Parmigiano-Reggiano PDO	118,979	17,617	14.8
Gorgonzola PDO	48,480	14,027	28.9
Asiago PDO	23,621	1,444	6.11
Provolone Valpadana PDO	12,745	3,911	30.7

Note: Pecorino Romano PDO overall production, which is not shown in the table, is rapidly declining. In 2000 less than 34,000 tonnes were produced. This decrease in output is explained by the pessimistic forecasts for export, partly due to its low competitiveness on the US market, given the unfavourable exchange rate euro/dollar.

Overall, cheese accounted for two thirds of total exports (650 million euros), followed by cured ham at 270 million euros (2005). Almost 70% of the PDO/PGI product exports were delivered to EU markets.

The analysis of agri-food exports to European markets shows that the most fragmented segments (cheese, deli meats and wine) are 'weaker' in terms of volume compared to products bearing strong brand labels (tomato conserves, past and olive oil.) rather than geographical indications. The same is true for the USA where 70% of imported olive oil is 'Italian', 35% of pasta, 28% of wine, 23% of cheese, 17% of tomato products, while the percentage for deli meats is only 3% (INDICOD and Nomisma, 2005).

As known, the image of Italian products is positive, as proved by the way they are cloned in all corners of the earth. A survey conducted by INDICOD and Nomisma (2005) showed that sales of Italian sounding products on the USA market accounted for ten times the retail value of 'real' Italian products and that this level is growing.

2. Objectives

This paper aims to analyse the international marketing and trade strategies implemented by Italian quality food exporters with special attention to the export and trade aspects for two of the most important Italian PDO products.

The study focuses on the rapid change in trading and distribution channel, that will affect the exports of high quality food products, by discussing the international marketing and trade strategies of the quality food exporters.

Italy has increasing difficulties in unravelling the knot of international market competition, due to the size of the export companies which, on the whole, is inadequate as regards to global market standards and due to Italy's distribution system which has failed to expand abroad, leaving the advantage to the foreign chains to take, preferentially, their national products to the countries they set up. However recent market concentration process is rapidly change the competitive arena for main PDO and PGI products.

3. Data and methodology

The international marketing and trade strategies of Italian PDO and PGI exporters are analysed considering their marketing mix strategy. As Van der Lans *et al.* (2001) stated, it is important for PDO and PGI labelling to be effective, to be inserted into an appropriate and well-articulate marketing strategy.

In this paper, the four Ps approach is used to evaluate the selection and development of the *product*, the determination of *price*, the selection and design of distribution channels (*place*), and all aspects of generating or enhancing demand for the product, including advertising (*promotion*).

Data are collected from internal data of institutional Consortium, companies data (producers and retailers), ISTAT data on trade movements and other quoted sources.

The approach of the marketing mix in studying the strategic position of a product in the marketplace is widely used in the literature of the agri-food product (Padberg *et al.*, 1997). In particular, some papers apply this methodology, as well as the specification of the four Ps, in analysing the role of the PDO and PGI labelling in the international trade of quality products.

For instance, Magni and Santuccio (1999) give an overview of the possible role and perspectives of the marketing policy for geographical indications in Italy. In this context, the authors show the strength and weak points for the international positioning of such products. The four Ps approach has been used by Bonetti (2004) in analysing the organisational and branding strategies adopted for traditional Italian production 'Mozzarella di Bufala Campana' PDO. Cardinali (1998) gives an example of the marketing mix for the national trade of regional apples ('Mela della Val di Non' PDO) with the collective brand 'Melinda'. The strategic position, according to the four Ps scheme, within the national and international market of the main Sicilian wines with Geographical Indications are studied and exposed in the paper by Bellia (2005), while Rossetto (2002) studied the ones from Veneto region. Marette and Zago (2003) provide some economics of the strategies needed to compete on foreign markets and/or to enter into new markets for wines. In Crescimanno *et al.* (2002), the trade marketing of organic producers of olive oil is examined with regards to both national and international markets.

Finally, the role of PDO and PGI in the international trade is generally described in Boccaletti (1999), Sodano (2001), Bureau and Valceschini (2003); in these papers, the denomination of origin is mostly considered as a marketing tool per se, without paying particular attention to the marketing policies behind (price, product, distribution, etc.).

Finally, the case of 'Prosciutto di Parma' and 'Parmigiano-Reggiano' has been studied in previous researches like, for instance, O'Reilly and Haines (2004) and Arfini (1999); however, in these papers, the cases are exposed with reference to the local marketing networks and strategies without a particular attention the foreign channels.

3.1 The selection and development of the product

The whole set of decisions relating to selecting a product for export can be considered as the international product policy (Valdani and Bertoli, 2006; Pellicelli, 2007). Traditional products themselves are already extremely differentiated by their very nature and thus enjoy competitive advantage, and price positioning absorbs transport costs and potential import barriers.

Competitive pressure may appear between one EU quality labelled product and another, or as is often the case for Italian products, between a protected geographical indication and an imitation. The extent of this depends on how correctly consumers perceive the product.

Through prism or amplification effect, products with guaranteed origin are usually perceived overseas to be at a higher level than they are in the country of origin (Valdani and Bertoli, 2006). One of the causes of the 'prism' is the country of origin effect, or consumers' previous experience of a country's products and country attributes. This may be direct experience as an immigrant or tourist, or it may stem from interpersonal information or mass communications, or generalised opinions about the country (Valdani and Bertoli, 2006).

It is not by chance that the top export-oriented products (PGI Tuscan olive oil, Pecorino Romano PDO cheese and Terre di Siena PDO olive oil) come from Tuscany which is one of the regions most popular with tourists or from the area of Lake Garda (Garda PDO olive oil). The main consumption areas overseas are traditionally those with a large proportion of Italian immigrants. But today global wide competition means that the country of origin effect is not enough; new requirements of the international consumer have to be taken into account, even for traditional and regional products. In China for example, extra virgin olive oil is promoted as a gastronomic speciality, but operators say that its success depends on its nutritional and health characteristics.

In the eyes of East Asian consumers, Italy represents biodiversity. The most active regions, Emilia Romagna and Lombardy carry out local marketing which means promoting an image of the country, not just organising a stand with regional finance. The two main producer's consortia of Emilia Romagna Region (Parmigiano-Reggiano and Prosciutto di Parma) decided to focus on this country image when in 2005 they halted EU financed joint promotions with a French cheese, to focus on 'Parma' products. The importance of the 'country of origin' cue justifies the EU regulations on traceability and labelling.

Adaptation and communication of product

The requirements of different overseas markets has led consortia of various quality labelled products, in their institutional function of promotion, to modify product image or rather enhance the country of origin effect (Jaffe and Nebenzahl, 2001) for the culture or environment of the new market. Important or desirable features in one country are not necessarily effective in another[13].

Given that intrinsically a traditional product is not easy to adapt, it is the service content rather than the product itself that has to be modified. For instance, the 'Grana Padano' PDO cheese is exported in different seasoning patterns according to the destination country to meet consumers' perception and need.

International promotion of traditional and regional products having a real or potential overseas market has mainly been carried out by institutional bodies such as producer consortia. It has strengthened the country of origin effect by playing on the geographic origins of the product.

Only recently commercial publicity, such as publications, cultural initiatives etc., has been developed to improve the relationship with clients. Commercial advertising by institutions tends to adapt the message taking cultural differences into account. It also depends on finance being available and local regulations on advertising. The EU makes available funding for international promotion of geographical indications, either directly or financing companies or institutions[14]. Another important method of communication overseas are specialised trade fairs, which can be opportunities for sales or method of direct communication to potentially interested operators. They are an efficacious form of communication.

In 2006 Italian companies spent about 7,000 million euros on information, promotion, publicity and support for PDO and PGI products, trade fairs and export in general. The funds came from

---

[13] Market research is expensive especially on new and distant markets, so that international and local support is essential to overcome the language barrier and other difficulties.

[14] Recent press advertising campaigns in the USA by the Consortia of Prosciutto di Parma and Parmigiano-Reggiano focus on product imitation and use pictures showing 'photocopies' or the 'DNA' of the products. The EU has spent 4 million euros over three years for this press advertising.

the EU, government, regional authorities, the National Institution for Foreign Trade (ICE – Istituto nazionale per il Commercio estero), Buonitalia Spa[15] and Chambers of commerce.

There have also been recent initiatives by big Italian companies to consolidate market position overseas. An example is 'Italia del Gusto – Taste of Italy'. 'Taste of Italy (Great Food Good Living)' is a consortium composed by the most important Italian companies with high quality products in the food sector[16]. This consortium undertakes activities in marketing, promotion and communication on foreign markets to assist the development of international sales for its members. The aim is to spread the taste and the Italian way of eating all over the world with the products of consortium companies. It develops trade marketing and initiatives in the distribution and the Horeca sector[17]. 'Taste of Italy' also carries out public relations with the media and other institutions. In particular, it provides advice and opinions in the areas of marketing and global communications, research, events and sales promotions. It also participates in international trade fairs, and it can help create commercial and logistical synergies and partnerships amongst its members.

Intellectual property rights

International product policy involves ascertaining the legal system of intellectual property rights, which in some countries may give grounds for uncertainty. This is particularly important for definition, communication and promotion of quality labelled products, which tend to be more complex because of the collective ownership. Indeed, the protection of geographical indications is usually guaranteed by producers' associations or consortia that, in general, are unlikely to register brands or patents overseas.

The protection given to geographical indications at international level is considerably enhanced by the TRIPs (Trade-Related Aspects of Intellectual Property Rights) Agreement. Approved as part of the Final Act of the Uruguay Round, it lays down minimum standards of protection for several categories of intellectual property but the negotiating process has not as yet achieved the hoped-for results. The situation within the EU is of course different; Council Regulations (EC) No 510/2006 and No 509/2006 aim to protect the names of products whose specific character is determined by their geographical origin or by a particular method of production. More information can be found in the literature on the protection at international level of geographical indications (Evans and Blakeney, 2006; Germanò, 2005; Handler, 2006).

The experience of the 'Prosciutto di Parma' Consortium in the US market is an example of intellectual property rights protection 'problems'. The brand was purchased by the Consortium on the US market while on Canadian, Japanese and Mexican markets this was not possible and trademark remain owned by local firms. Italian companies may not therefore sell prosciutto di Parma ham under the name 'prosciutto di Parma' in Canada. The geographical indications are not guaranteed in spite of Art. 22 of TRIP. In Japan the trademark Parma was accepted for a Canadian company but not for Prosciutto di Parma, because it was considered generic. Mexico

---

[15] Buonitalia Spa is a private company 'for the promotion, valorisation and internationalisation of the Italian food industry, established by the Ministry for Agricultural and Forestry Policies in July 2003'.

[16] Members are Amica Chips, Auricchio, Barilla, Conserve Italia, Cremonini, Granarolo, Illy Caffè, Italia Zuccheri, Noberasco, Orogel, Parmacotto, Parmalat, Parmareggio Unigrana, Pastificio Rana, Regnoli, Riso Gallo, Salov, Sammontana, Acqua San Benedetto e Aia (Gruppo Veronesi).

[17] The Horeca sector includes mainly: hotels, restaurants, cafés, pubs and bars, camping sites, canteens and catering services.

does not accept the registration of Prosciutto di Parma as a trademark because Parma is already a registered trademark; but in this case the authorities allow both products to be sold on the market.

In practice, promotion by group of producers (consortia or producers' associations) is carried out on markets where there are imitation products only where legal protection is guaranteed. As long as legal protection is assured, it is possible to implement effectively promotion and communication; where it is absent or pending, only private brands are protected and promoted.

In the case of products with geographical indications strongly imitated abroad, the international protection is so important to become the fifth P!

3.2 Price policy

Price has critical role in defining the international marketing policy. Several variables prevent producers from formulating a true pricing policy for quality labelled products, especially in the long and fragmented chains. The increasing number of channels, the changing role of intermediaries and modern distribution leaves little margin for manoeuvre by producers. The more fragmented and distant from end markets is the production, the more critical these aspects may become.

The analysis of 'Parmigiano-Reggiano' cheese market shows that retail price trends are not correlated so much with production costs, but with more important factors such as distribution costs (Mora and Arfini, 1997) and, for overseas markets, the exchange rates, import duties and taxes. The growth of large scale modern distribution will further reduce producers' control over the price variable. Supermarket price policies will probably make product positioning more transparent than it is today, and the product will finally be positioned and perceived in the same way as it is in the origin country.

On the other hand, in order to reduce the effects of product imitation, producers and their associations are trying to maintain high levels of differentiation with particular attention to 'speciality' distribution channels such as high class restaurants, delicatessens etc. which assure higher selling prices.

3.3 Placement: selection and design of international distribution channels

A survey performed on a sample of 800 firms accounting for 15% of total turnover in the food industry, reveals that different channels are used for overseas markets and that the same company often uses multi-channel distribution (INDICOD and Nomisma, 2005).

The most frequent channels are importers (Table 2), overseas distributors and a sales network overseas. Follow the direct sale to overseas supermarket chains (6% of respondents) and traditional import–export companies (4%). 15% of respondents (Multiple reply) intend to strengthen the channel of direct selling to overseas supermarkets. Importers are the most frequent channel chosen to export PDO and PGI products towards US and Northern European markets. In the US market, New York is the main destination for many reasons: for its geographical and cultural closeness to Italy, for the higher standard of living compared to the rest of the country and for the logistic advantages since the main importers are located along the East coast.

Table 2. Channels to overseas markets. Source: INDICOD and Nomisma, 2005.

	First answer (%)	Multiple reply (%)
Foreign importers	45.6	60.8
Foreign distributors	14.0	38.5
Import-export companies	4.0	14.5
Trading companies	2.9	5.5
Other traders	5.0	12.1
Branches	2.4	5.5
Own sales network overseas	12.3	21.1
Direct sales to overseas supermarkets	5.9	15.7
E-commerce	1.2	2.4
Total	100.0	

Traditionally, the export of a quality product starts with an importer who distributes the product to top quality restaurants. It then moved to specialised retail and, finally, to modern distribution. This process is typical for traditional products with smaller producers, or for new markets. On consolidated or larger markets, producers tend to set up overseas branches or invest directly in local production[18]. They may also have direct contact with the supermarkets, although specialised importers may have stronger competitive advantage.

Choosing suppliers, language translations and, often, merchandising are all services offered to the distributor by the importer; in this way transaction and commercial costs, quality risks connected to the product are lower for the supply chain, at least in the initial stages.

According to National Association for the Speciality Food Trade[19], the key factors in overseas supermarket buyer decisions are in order of importance: standardised product quality, consumer demand for product category, newness, (certified) reliability of producer, price, ethnic characteristics of shoppers, investment in promotions and advertising, exclusive rights. Trade marketing and very different actions from entering and staying in traditional channels is essential for firms selling to overseas supermarkets.

From the commercial point of view, the Italian quality labelled products have shown a wide range of operational solutions. For hard grated cheeses, only the company Zanetti (22% of total export of hard cheeses) have direct export agencies in USA, Germany and France; in general, the other exporters have foreign agents. Auricchio, leader in provolone cheese, exports to 50 countries worldwide, historically to countries with a large number of Italian emigrants and descendents; exports account for 20% of turnover which was 104 million euros in 2004. The most important destination is North America; Auricchio has an exclusive distributor, The Ambriola Company, with a network covering all channels in USA and Canada. Another example is the company

---

[18] There are many direct investments overseas outside the EU in the sectors of cured pork and packaging.

[19] *The National Association for the Specialty Food Trade* is a not-for-profit business trade association established in 1952 to foster trade, commerce and interest in the specialty food industry. The NASFT is an international organisation composed of domestic and foreign manufacturers, importers, distributors, brokers, retailers, restaurateurs, caterers and others in the specialty foods business. The organisation has more than 2,300 current member companies throughout the U.S. and overseas.

IGOR. IGOR Srl accounted, in 2004, for more than 30% of the national market of Gorgonzola cheese, the 35% of which was exported especially through modern retail (2005 turnover was 66 million euros).

International clients include the retailers chains Carrefour, Auchan, Metro, Marks & Spencer, Wal Mart, Aldi, Lidl, Rewe, Penny Mark, Netto and Kaufland[20].

Due to the territorial link of PDO and PGI products, international investments cover traditional products not protected by European quality regulations; indeed, direct overseas investments are frequent in the un-protected deli meat sector and packaging.

Direct investment and trading agreements in agriculture and food are very low compare to the total Italian figures (2% in 2005), although there have been recent promising developments in the cured pork sector on Asian markets. Grandi Salumifici Italiani for example accounted for 15% of the total export of cured pork in 2005. It is present in China in a joint venture with an important Chinese company (third largest world player), by controlling Shanghai Yihua Food Co and with four facilities for production and curing of traditional Italian deli products. Grandi Salumifici Italiani (GSI) has a trading company as a sales and distribution platform in the main chains across Asia. GSI has made also direct investment in Brazil producing quality brands such as Casa Modena and Senfter as well as a more competitive one, Sino Sul.

4. Trends in international large scale distribution

PDO and PGI international distribution is more complex than simple product placement. Current trends in international distribution (Fornari, 2005), and their impact on producers, selection and design of distribution channels are summarised below.
1. Segmentation of private label supply
    In 2006 in the UK retailers private labels market share (in value) for grocery products were about 42%, in Germany 35%, in Spain 26%, in France 25%, in the Netherlands 21%, while in Italy it accounted 12.5% (Giacomini, 2007).
    The market share in value of the private labels in the European supermarket chains shows the following figures: Aldi (D, 95%), Tesco (UK, 42%), Carrefour (F, 30%), Casino (F, 25%), Intermarchè (F, 25%), Ahold (NL, 22%), Auchan (F, 21%), Rewe (D, 20%), Metro (D, 12%) (Giacomini, 2007).
2. Internationalisation of private label management
    Retailers operating on different international markets tend to select co-packers on a world-wide basis and stipulate transversal supply contracts for private label products for all markets. Today this is happening for those quality products where the firm has critical mass. There are often 'category management' projects involving fancy food or gourmet with these suppliers. This allows to lower transaction costs and improve quality standards[21].
3. Internationalisation of distribution
    Many large retail companies perform a large share of their business in food and non-food outside their country of origin (cross-border distribution).

---

[20] Strategic assets of IGOR Srl range from great attention to health and hygiene standards, high levels of product standardisation and process and product innovation leading to variety and completeness of product range.

[21] This trend is forcing many producers to adopt BRC or IFS standard certification. This is a further barrier to entry and leads to exporter companies becoming more specialised.

4.1 Effect on PDO products

The factors considered to explain the trends in international distribution, have important effects on national and international end markets for PDO and PGI labelled products.

Internationalisation of distribution

The last considered factor was the internationalisation of distribution; this issue is interesting for export products made in Italy, which are sold from long term in overseas markets and promoted as 'Italian' or 'traditional'. This can enlarge end markets and also have positive effects in terms of 'product transparency'.

An example of this approach comes from Carrefour, which 'company vision' includes enhancement of local tradition and develops own brand lines of regional products of small and medium enterprises. In Italy Carrefour promotes the brand 'Terre d'Italia' (see Table 3), in France 'Souvenirs du Terroir', 'Reflets de France', 'Destination Saveurs'[22], etc., and in Spain 'De Nuestra Tierra'. In Switzerland and other countries 'national' product lines enrich other own brand lines such as 'Filiera Qualità Carrefour' ('Filieres Qualité'[23]), which is present in all countries. In this way Italian products are exported to Japan, Switzerland, Belgium, Argentina and Columbia and, with special promotions, to France and Spain.

Internationalisation of private labels management

Today, co-packers range from specialised transnational companies to local producers for particular specialities. The relative absence of industrial brands has helped retailers to develop autonomously their private label in EU quality labels business. The absence of big companies in the sector and known brands, sometimes leads supermarkets to create upwards vertical integration of quality products marketing; this allows the retailer to control and lead the whole supply chain. In the case of Carrefour, for example, buyers of retailer's own quality brand in France frequently bypass traditional intermediaries allowing a direct contact between producers and supermarket. In other countries, however, products were bought through more traditional channels such as importers, wholesalers and trading companies. In some cases, there might be a horizontal partnership with cooperation agreements; this is the case of the agreement between Conad and Leclerc, which commits the former to share its know-how in selecting producers of Italian products, particularly traditional foods.

---

[22] This line was recently replaced by Traiteur Charcuterie.

[23] The values associated to this product line are: taste, food safety, continuity (development lasting over time), authenticity (enhancement of local products), price/quality ratio. For example, in Switzerland, Gorgonzola, Parmigiano-Reggiano and Prosciutto di Parma are sold through the 'Filieres Qualité' line.

Table 3. Retailers' quality product lines in Italy.

Chain	Line	Description
SMA Auchan – 'I sapori delle regioni'		Original and quality regional products at medium-low prices. Packaging shows the supplier name, method and place of production for purposes of transparency. More than 120 products are included in the line.
Carrefour - 'Terre d'Italia'		It was introduced in 2000 and is now present in all Italian Carrefour stores. More than 160 products are included, authenticity deriving from suppliers selected from different Italian regions and regional foods, some of which are PDO/PGI labelled.
Conad – 'Sapori e dintorni Conad'		This product line offers a wide range of quality products (over 100) based on strong gastronomic traditions: cheeses, preserves, pasta, biscuits and cakes, deli meats and oil. The line is constantly developed.
Coop – 'Fior Fiore Coop'		128 quality products representing the best of Italian gastronomic tradition are included in this line. PDO labelled cheeses in the line include Parmigiano-Reggiano 30 months seasoned.
CRAI – 'Piaceri Italiani'		The line includes products obtained through regional and traditional methods guaranteed by PDO and PGI labelling and other certifications. Cheeses include Parmigiano-Reggiano PDO 'Selezione di Montagna' (Mountain selection).
SISA		Some product lines has been developed with regional indications and imaginative names. Today there is the brand 'I sapori dell'antica locanda' (Old Inn flavours) for deli meats and 'Antiche bontà di Sardegna' (Traditional goodness from Sardinia).
Esselunga – 'Naturama'		Naturama is a brand owned by Esselunga which includes various deli meats, as well as 'Prosciutto di Parma' PDO. Key benefits offered under the scheme are full traceability, animal welfare provision, conformity to PDO guidelines and further inspection by Esselunga veterinary staff, careful processing of pork in selected curing houses. Moreover, the Esselunga line for organic product (Esselunga Bio) includes 'Parmigiano-Reggiano' PDO produced from organically farmed milk in the area defined by the product specification.

Segmentation of own brand supply

International large scale retailers have felt the need to escape from price pressure by adding their own brand to the assortment of products offered. Today's fourth generation own brands correspond to the development of a different own brand for each segment, from the lowest to

premium price. Intense horizontal competition has led modern distribution chains to attempt gaining the competitive edge by means of two strategies (Fornari, 2005):
- by selling products able to generate higher profitability compared to the leading brand names;
- by generating consumer loyalty to the point of sale.

The large scale retailers have recently developed their own umbrella labels for a range of products including fresh food (fruit, vegetables and beef) with a specific quality and safety content, and for traditional and organic products.

In some cases, the retailer acts in a vertically cooperative way, by cooperating with the other players along the food-chain (in particular industry), but also horizontally, cooperating with other retailers (competitors), acting in Europe or on global markets. Examples of horizontal cooperation are the development of food safety standards and, more recently, integrated traceability systems, with the example of the international Global Food Safety Initiative and the national action of INDICOD-ECR[24]. Other vertical cooperative strategies have been developed throughout the supply chain (with agricultural producers, processors, transporters, and so on), allowing retailers to gain a greater market strength. The main examples are the retailers' brands for EU quality labelled products (premium price products), for organic foods, for baby food, for allergen free line, as well as for other voluntary certifications. These are quality chains which confirm the high positioning of private label in consumer perception, increasing profits and market share. For commercial brand products this means increasing competition and intensifying promotions. But for EU quality labelled products the general absence of commercial brands means less competition, even though it does not stop the continuous drop of producers' contractual power.

The Carrefour's initiative 'Terre d'Italia', the first quality product line developed in year 2000 by a retailer in Italy, was imitated by other supermarket chains. Esselunga initiatives was to propose weekly gastronomic specialities from different regions; Coop Italia's 'In viaggio tra i sapori d'Italia' or MDO's 'Scopri i sapori d'Italia'. After this starting period, the largest supermarkets adopted strategies like Carrefour, with lines based on traditional and regional products (Table 3). In this way the retailer's own name appeared alongside the consortium or producers' association name guaranteeing the respect of production standards and the complete consumer protection.

Coopernic was the first attempt at a European level to join cooperative such as Conad, Leclerc, leader in France, Coop Suisse, the second Swiss group, Rewe, number two in Germany and Colruyt, third in Belgium. Coopernic is thus overall the largest European distributor with over 90 billion euros total sales in 17 countries. It aims to gain more contractual power and to develop private label products. Early declarations by its members say that each partner will retain its own logo and private labels, but will also become leading seller for some product categories. Conad in its international alliance with Leclerc has thus created a centre for synergic product management. An example is Conad's range of quality products, 'Sapori e Dintorni'. The line accounts overall for about 10% of the portfolio of own brands, with a constantly increasing turnover since its launch in 2001 (Cristini, 2006). Conad own branded products account for about 17% of the overall turnover (Cristini, 2006). In the French Leclerc supermarkets about 54 Conad labelled products are offered (Cristini, 2006) and, during the last three years, promotions entitled 'Vive l'Italie' have been developed involving mainly the 'Sapori & Dintorni Conad' line. In 2004, Conad signed another agreement with Rewe, the German cooperative group active in Italy through

---

[24] The reader may find further information in the web sites http://www.indicod-ecr.it/index.php. and http://www. ciesnet.com/2-wwedo/2.2-programmes/2.2.foodsafety.gfsi.asp.

Billa and Standa supermarkets. This will offer new opportunities for Italian producers. Smaller but dynamic chains have recently started internationalising. One example is SISA, which offers Italian and other foreign quality labelled products, especially Greek. Overseas SISA Hellas and Sisa Malta supermarkets sell SISA branded products made in Italy.

Co-marketing and co-promotion

Retailers' strategy of developing brands in order to achieve competitive differentiation may also lead to lower transaction costs and long term advantages with producers and their associations or Consortia. This happens in particular for communications, when co-marketing takes place with Consortia, and for price and product policy, in this case with producer and trading companies. Consortia or producers' associations managing PDO and PGI labelled products, consider certain types of intervention particularly important: institutional promotion and communication of labels, protection of denominations on international markets, chain agreements and sales contracts with supermarkets, catering firms and restaurants.

An example of this strategy comes from the promotions run by the Parmigiano-Reggiano Consortium across Spain in 2005, that involved 75 'El Cortes Inglés' supermarkets and hypermarkets, and 125 Carrefour stores. At the same time, the Consortium has supported also Italian restaurants belonging to the association Arris Gourmet of Barcellona and other more specific initiatives for catering and restaurants in Poland and the UK.

Viceversa, Conad in developing the line 'Sapori e Dintorni' cooperates closely with suppliers to maximise the efficacy of often joint marketing. Through the supplier, Conad contacts the producers' association or Consortium which may be involved in order to sell out a particular local product.

Finally, in 2006, many joint promotions were made on the Russian market by the Consortia of 'Prosciutto di Parma' and 'Parmigiano-Reggiano' with international chains such as Auchan and Metro, as well as local chains.

5. Case studies: products and players

This section will focus on the supply chain structure and export channel and actors for two PDO products: Parmigiano-Reggiano and Prosciutto di Parma[25]. Given the internal market difficulties perceived in the last few years, export has become more and more important for these products and it is believed to be even more important in the next future. However export depends on different factors such as hygienic and sanitary regulations, trade marks rules, tariff and quota existence and general market conditions (exchange rate, country risk, financial availability, etc.).

---

[25] In this sections different sources were used: institutional data published in the websites of Parmigiano-Reggiano and Prosciutto di Parma Consortia, companies data collected trough direct interviews with representatives of Consortia and other supply chain operators. Other specific sources, such as the CLAL company website and Montanari and De Roest 2006 paper, are specified in the text.

5.1 Parmigiano-Reggiano PDO

The production structure of Parmigiano-Reggiano cheese is based on a networks of farms located in the production area[26] (some figures can be cound in Table 4). The origin of milk for the production of Parmigiano-Reggiano covers the provinces of Parma, Reggio Emilia, Modena and parts of Bologna and Mantova; the first four provinces are in Emilia-Romagna region, while Mantova is in Lombardy. Thus, the link between the production zone of Parmigiano-Reggiano and the source of raw material is very strong in Emilia-Romagna region; about 80% of the milk produced in Emilia Romagna and 18% of the overall Italian milk is processed into Parmigiano-Reggiano (www.clal. it); in the mentioned four provinces (representing 80% of the farms and cows of the region) almost the whole milk production is directed to Parmigiano-Reggiano cheese making. Recent studies pointed out remarkable processes of reorganisation, with a progressive concentration of cows in the largest farms especially in plain areas. The average productivity per dairy expressed in tons of processed milk moved from 699 tons of milk in 1970, to 2,713 in 1998 and to 3,824 tons of milk in 2005 (Montanari and De Roest, 2006). The overall number of dairies in the last fifty years has diminished by 28.5% passing from 733 in 1993 to 524 in 2003 and to 461 in 2006. Despite the great decrease in the number of dairies recorded (at least of 20 dairies/year), the Parmigiano-Reggiano production increased according to a constantly growing trend (www.clal.it). The decrease has affected both dairies in the plains and those in the mountains in a very similar way, even if the most remarkable concentration, over 42%, has involved cooperative dairies, the number of which diminished to 384 units in 2003 with respect to the 606 existing ten years before (Montanari and De Roest, 2006). During the last decades, cooperatives have been the fundamental organisational structure of Parmigiano-Reggiano production and even today the cooperative model still prevails with respect to the private one; however, the trend shows a progressive erosion of the cooperative dairies production share and an increase in private production.

If we consider the strategies of the cheese dairies up to now, their main efforts have been the re-organisation to a bigger scale; however, the most common dimension is still small, especially if compared to the producers of Grana Padano cheese.

---

[26] National Law of 10 April 1954, n.125 on 'Tutela delle denominazioni di origine e tipiche dei formaggi' (G.U. n. 99, 30 April 1954).

Table 4. Parmigiano-Reggiano cheese in figures (2006). Source: Parmigiano-Reggiano Consortium website: www.parmigiano-reggiano.it.

---

12 months of minimum ageing

16 litres to make 1 kg of cheese

20-24 average ageing of the wheels (in months)

38 average weight of a wheel (in kg)

461 number of dairies

600 litres to make one wheel

4,750 producers of milk

251,000 cows

3,089,837 number of wheels produced in the 2006

800 million euros value of production in 2006

The most part of Parmigiano-Reggiano is distributed through a 'long channel'; with the intermediation of the wholesaler-ripener between cheese dairy and distribution channel (retailers or traditional shops). A small quota is sold in farm shops annexed to the cheese dairies (direct channel). In this way, the ripening stage of a great part of the production is mostly entrusted to farms external to the processing and of quite different types. They range from wholesalers who ripen the product directly before selling it, to the complex structures of consortia associating dairy cooperatives. The latter, through the link between production and ripening, aim to avoid or at least reduce the price speculations, that cause the cyclical crises typical of Parmigiano-Reggiano market. In 1995, 80% of the production of Parmigiano-Reggiano was seasoned in 'private' structures, while today cooperative system presents the most important seasoning and trading player on the market (Unigrana Spa). This represent an example of vertical integration. Apart from cooperatives, there are more than 200 wholesalers, but 1/4 of them controlled 90% of the private companies market share.

In 2006 the export of Parmigiano-Reggiano (and those of the main competitor, Grana Padano) has increased by 4.5% compared to year 2005 (www.clal.it). The growth within the EU was less important compared to the rest of the world (respectively 2.2% and 7.7%). The exports decrease in the two main outlet markets, France (-9.1%) and Germany (-3.1%), has been more than compensated by the increase in the other EU countries, such as United Kingdom, Austria and Greece. On the other way, as also shown in Table 5, exports are increasing rapidly towards the other extra-EU destinations, especially in North America (+11% in the United States and +4.5% in Canada).

Table 5. Export of Parmigiano-Reggiano and Grana Padano (tonnes, 2001-2005). Source: Parmigiano-Reggiano Consortium website: www.parmigiano-reggiano.it.

	2001	2002	2003	2004	2005
Volume	35,444	38,736	43,594	46,423	50,335
Value[1]	276,712	301,037	348,941	388,063	385,083
Europe	24,513	26,733	30,174	32,207	34,876
France	3,503	4,189	5,510	4,366	5,030
Belgium, Luxemburg	1,314	1,562	1,818	1,764	1,604
Germany	5,754	6,604	7,229	8,205	9,688
United Kingdom	2,808	2,916	2,942	3,675	4,026
Danmark	442	492	588	764	856
Sweden	403	531	639	677	774
Switzerland	6,550	5,600	6,141	5,753	5,298
Austria	976	1,268	1,432	1,428	1,438
Spain	938	1,288	1,385	1,976	2,017
Greece	748	943	947	1,281	1,379
Africa	98	119	94	161	210
USA	6,128	6,600	8,328	8,232	9,164
Canada	1,547	1,779	1,647	1,727	2,102
Brasil	147	198	69	108	95
Asia	1,594	1,812	1,997	2,354	2,297
Japan	1,346	1,477	1,595	1,857	1,617
Australia	973	1,076	892	1,146	1,195

[1] Million euros.

The main channel is represented by agents or importing societies in consolidated markets, or trading companies, importer and broker in new and far markets.

The old way to open a new market passed from contact with restaurants, gourmet shops and after retailer (usually large scale ones). Now the strategies changed and for new market the contact with restaurant and international retailer is simultaneous (Berti *et al.*, 2005).

The first important thing to point out is that the two PDO hard cheeses are increasingly marketed together because wholesalers are widening their product portfolio and diversifying their supply in order to offer retailers a more complete cheese product mix.

They implemented also a set of product differentiation strategies (different ages, improving packaging, snacks, grated, etc.), as well as specific retail strategies selling differentiated product in different retail channels, regions and countries. The packed products are normally branded (in total 40% of the production) and the quota of private label is increasingly high in this segment.

The main players are private companies or cooperative groups (Table 6). On the one hand, the process has been characterised by the enlargement of the corporate structure and vertical integration in the great cooperative groups while, on the other hand, the growth by horizontal integration (with partial diversification) and vertical integration (in particular for seasoning and packaging) has been applied by the big companies producing Grana Padano which, as known, are characterised by a less fragmented and more marketing-oriented industrial structure (Menozzi, 2005).

The first ten exporters of hard 'grana' cheese count less than 50% of the overall figure (Table 6). The data are aggregated (Grana Padano, Parmigiano-Reggiano plus other hard cheeses), and it has to be considered that the first two players, Zanetti and Agriform, are specialised in Grana Padano, Consorzio Latterie Virgilio associates dairies facilities producing both PDOs, Colla

Table 6. *Eexport of Grana Padano, Parmigiano-Reggiano and other hard cheeses (cow milk) in value (million euros) and quantity (tonnes), 2004 data. Source: European Commission, Directorate-General JRC (2006).*

Company	Export in value	%	Export in quantity	%
Zanetti Spa	98.0	19.6	15,200	22.4
Agriform	25.5	5.1	3,482	5.1
Consorzio Latterie Virgilio	19.9	4.0	2,350	3.5
Colla	17.4	3.5	2,042	3.0
Unigrana	16.2	3.2	1,859	2.7
Saviola	14.2	2.8	2,100	3.1
Ambrosi	14.0	2.8	2,000	2.9
Parmareggio	13.3	2.7	1,080	1.6
Casearia Brazzale	12.5	2.5	2,196	3.2
Nuova Sala	12.3	2.5	2,000	2.9
Zarpellon	10.5	2.1	1,470	2.2
Others	*247.4*	*49.4*	*31,292*	*46.1*
Total	501.2	100.0	67,850	100.0

produces Grana Padano and commercialises Parmigiano-Reggiano with its own brand or with retailer and catering brands. The first player uniquely specialised in Parmigiano-Reggiano is Unigrana Spa.

Zanetti Spa is specialised in the production of Grana Padano and maturing of Grana Padano, Parmigiano-Reggiano and other traditional Italian cheeses. It sells its products on the domestic market and abroad, both in wheels, pre-packed portions and grated. It produces every year about 130.000 wheels seasons 410.000 wheels of Grana Padano. It's a very diversified and innovative firm, with facilities for portioning and grating. It exports in Europe, Japan, USA, Australia, Canada, China and South Africa. With more than 50% of external trade turnover, Zanetti represents the biggest hard cheeses exporter.

In association with ten member cooperative dairies, Agriform has been producing, maturing and selling quality cheeses since 1980. In 2004 Agriform had a turnover of over 76 million euros, 35% of which was made abroad (Mora, 2005). The main destination countries were Germany, USA, Australia, Canada, France and United Kingdom. About 75% of the overall turnover comes from Grana Padano sells (300,000 wheels sold in 2004). With more than 1.500 breeders associates, its overall production reaches every year about 220,000 wheels produced, making Agriform the main producer in the Grana Padano Consortium (Mora, 2005).

The company 'Consorzio Latterie Sociali Mantovane Virgilio' is a second degree cooperative founded in 1966, accounting today 110 associated companies (producers of Grana Padano and Parmigiano-Reggiano), grouping together over 2.500 breeders. The company works in the dairy - cheese and meat sectors. In the dairy sector it produces milk, butter, cooking cream, mascarpone cream and quality cheeses such as Parmigiano-Reggiano PDO, Grana Padano PDO and Provolone[27]. Virgilio belongs to the business group Virgilio-Ghinzelli which controls Bertana Spa, Brendolan Prosciutti Spa and Castelcarni Spa too. The overall turnover in 2006 amounts at 340 million euros, and 620 million euros considering the whole group. The company is the head of a group which includes facilities involved in the pork meat sector. The vertical integration of cutters in companies within the group, leads to the marketing of fresh meat and most prestigious Italian PDO deli meats. The group is the Italian leader in the pork slaughter sector accounting approximately 20% of the national production of heavy pig, assigned to the deli meats and fresh meat production (including the recent PDO label 'Gran Suino Padano'). The group works in the deli meats sector with a company that is part of Brendolan Spa, producing four PDOs: 'Prosciutto di San Daniele', 'Prosciutto di Parma', 'Prosciutto di Carpegna', 'Prosciutto Veneto'.

The Granterre Consortium was founded in 1959. It guarantees full control over the supply chain, from livestock to market. The Group includes Parmareggio Spa, which processes 170,000 quintals of milk to product Parmigiano-Reggiano cheese; 3,600 heads of livestock divided into four farms and 76 cheese factories represent approximately 1,600 farms; Unigrana started its activity in 1991 as a trading company of the cooperative Granterre Consortium. Unigrana holds the leadership in Europe in the Parmigiano-Reggiano market and is among the first operators non manufacturers for Grana Padano. Other industrial activities are butter production, pig breeding, warehousing, ripening and packaging of cheese.

---

[27] This is the second company with regards to total sale of Grana Padano PDO and the 8th company with regards to total sale of Parmigiano-Reggiano PDO in value.

Among the other important companies, Dalter has to be mentioned, with over the 50% of turnover made abroad (exports). Some companies (i.e. Ambrosi and Boni) have branches in EU countries such as France, or are connected with local importers.

5.2 Prosciutto di Parma PDO

The cured ham market includes the seven recognised PDO hams as well as the unbranded product, the raw material of which can be both Italian or foreign. The Prosciutto di Parma is the most important product in this market: it represents the 40% of the overall ham production in Italy and the 75% of the PDO ham production. About 82% of the entire Prosciutto di Parma production is sold in the Italian market (7.7 million hams, total value at the consumption stage 1.8 billion euros), while the remaining quota (18%), is exported (1.6 million hams) (Table 7).

The companies producing Prosciutto di Parma were 201 in 1993 and 171 in 2005, according to Consortium data. It's important to point out that 60% are joint-stock companies (Spa and Srl). According to the Consortium data, the Prosciutto di Parma is exported in over 60 countries, with more than 1.8 million hams exported every year. Normally on the whole 170 firms, only 70 are exporters, of which 40 habitual exporters.

The European Union is the principal market, accounting for approximately 74% of the total exports. The three leading non-EU destinations of Prosciutto di Parma PDO are the United States, Japan and Canada. The Consortium has recently opened new markets in Singapore, New Zealand, Lithuania, South Korea and Australia.

The pre-sliced Prosciutto di Parma is the packaging format with the best growing rate; it almost doubled its export quota in the last five years, representing now the 20% of total exports (20 million packets produced in 15 authorised laboratory of slicing and packing). In 2006, for example, the sliced Parma ham registered +33.2% in Italy and +27.6% abroad.

The Italian sector of deli meats is characterised by a substantial fragmentation that embraces both production, with many small and medium local firms next to few big multi-specialist national groups, and distribution, where the traditional channel still retains the upper hand. The market

Table 7. Prosciutto di Parma in figures (2006). Source: Prosciutto di Parma Consortium website: www.prosciuttodiparma.com.

Production: 171 companies (so called 'prosciuttifici')
9,839,000 hams produced in 2005
5,386 pig breeding farms in 10 Italian regions
139 slaughterhouses
3,000 workers directly involved in the production
500 workers indirectly involved
885 million euros the wholesale market value
1,800 million euros the consumer market value
Market: national 82%; foreign 18%
1,800,000 hams exported in 2005 (+ 8.2% compared to 2004)
8.6 million pre-sliced packets (173,000 hams) sold in Italy (+19.5% from 2004)
23.1 million pre-sliced packets (463,000 hams) exported (+15.7% from 2004)

conditions have forced leading companies to diversified the product portfolio with quality products and through technological innovation, targeting goods with a higher service content.

All big producers are exporters, mainly direct to foreign big retailer, while some of the medium-size producers export preferentially with specialised operators. All the leaders of the pig meat market are operating in the Prosciutto di Parma district. These groups, usually completed vertically integrated from rearing to cutting and slicing, have in their product mix many EU quality labels as well as substitute products of Prosciutto di Parma, such as unbranded ham or other deli meats.

In the case of deli meat market the growth process is complex. On the one hand, it is characterised by a strong chain integration process from slaughterhouses merging processing/seasoning plants[28]; on the other hand, it shows an horizontal integration and diversification from companies operating in other food sectors[29].

At the moment there are strong interests from large industrial groups for local production firms and a growing interest of foreign capital for Italian leading brands.

'Brand industries' account for 1/3 of the total Italian ham market (PDO and not); these are multi-specialist companies, usually with foreign branches, direct investments and joint ventures with local leaders in Europe and overseas (USA, Japan, China). They are specialised on convenience foods and are able to supply international retailers.

The deli meats market in 2005 is shown in Table 8. Unfortunately, it was not possible to consider only PDO cured ham market share and export data. Apart from Galbani, all the other companies operate in the production and trade, both national and international, of Prosciutto di Parma and other important PDO products.

Recent market operations modified the competitive arena. In 2005, the leader has become 'Grandi Salumifici Italiani Spa', including Unibon Salumi (selling brand 'Casa Modena' and 'Unibon') and Senfter. The group acts as a big cooperative pig meat plant controlling all the supply chain, and as a company specialised in the quality deli products.

Fiorucci is still the leader brand of Italian deli products in the world market. It operates mostly with the retail and food service sector. The turnover made abroad is one fourth of the total one. It has eight plants, seven of which in Italy and one in the US. The 55% of its capital is owned by the fund Vestar Capital Partners. Fiorucci family owns 25% of the company capital.

The Beretta Group is third in the Italian deli products market, but leader in the take away retail market. It has 13 plants in Italy and one in New Jersey (USA) specialised in deli meats and used as the company's logistic centre for the North America.

---

[28] An example is given by the 'Industrie di Macellazione Marino Ghinzelli Spa' which has controlling interests in Brendolan Prosciutti, Macellerie Bertana Spa and Castelcarni Spa (a firm specialised in the processed meat products, generally sold to large scale retailers). Brendolan operates in the markets of Prosciutto di Parma PDO, Prosciutto di San Daniele PDO and Prosciutto Veneto-Berico Euganeo PDO. It has the only plant producing Prosciutto di Carpegna PDO. It is a leader in the Italian PDO ham market with a production of about seven million kilograms produced in its six production sites.

[29] The Cremonini Group, leader of the beef market, is also operating in the deli products market using the brand 'Montana'. Its 2005 turnover was 153.3 million euros, with 16% made abroad.

Table 8. Players in the deli products segment.

Company	Turnover 2005 (million euros)	Deli meats market share (estimation on volume, 2005)
Grandi Salumifici Italiani	434	6
Cesare Fiorucci	375 (consolidated)	5.1
Gruppo Galbani Spa	1,125 (consolidated)	3.3
Fratelli Beretta	335	3.3
Rovagnati	200	2.7
Ferrarini	250	2.4
Gruppo Veronesi	1,700 (consolidated) 570 (deli products)	3.6
Citterio	350	1.4

Rovagnati was founded in 1941 as a food distribution company; in 1967, it began producing cooked hams becoming the sector leader in the 90's. In 1994, Rovagnati continued its industrial growth with the acquisition of a company specialised in the production of Mortadella di Bologna IGP. In addition, with the set up of a new production plant, has recently entered in the Prosciutto di Parma PDO production.

The agro-industrial group Ferrarini is very diversified one. Indeed, it includes Ferrarini Spa, specialised in the Prosciutto di Parma PDO production, Fattorie Ferrarini Srl, specialised in the Parmigiano-Reggiano PDO cheese and wine production, and Vismara Spa operating in other deli meats segments.

The Veronesi Group is the fourth Italian agro-food group, with a turnover of 1.7 billion euros (2006), 94 million of which made abroad. AIA-Veronesi Group has recently rationalised its activity transferring to Negroni Spa the management of the most important deli products brands, representing 300 million euros turnover: Negroni, Montorsi, Fini Salumi and Daniel. The Group has more than 1000 employees, six plants in Italy and several sales offices in France, Germany, Switzerland and USA.

Citterio, has eight plants in Italy plus a facility in the United States. It operates in other European countries such as Switzerland and France with sales offices and direct sales organisation. It is also present in other countries with area managers, key accounts and distributors.

6. Final remarks

Some final conclusions can be drawn regarding the marketing and positioning of PDO and PGI products. First of all, the necessity of quality standardisation will be more and more stressed in the future especially for those exporters connected with international large scale retailers. This does not mean that the same standardise quality must be sold worldwide, since consumers from different countries have heterogeneous perceptions about the same product; thus, product quality have to be segmented according to the final market, as well as the service content.

Another central aspect regarding the EU quality labelled products' future worldwide diffusion is the protection at international level, as it will be designed within the WTO context. This factor may be defined, given its importance for EU geographical indications export, the 'fifth P' of producers' marketing mix.

The development of producers associations (second level cooperatives or consortia) acting as trading companies, autonomous with respect to large scale retailers and foreign importers, may represents a success factor in the definition of the distribution strategies.

Promotion and communication campaigns will depend on the other marketing mix's Ps, in particular on the product protection level in the targeted country. Moreover, promotion strategies defined in collaboration and synergy with other PDOs and PGIs producers will likely be more efficient and effective if able to improve the overall country image.

Finally, in the price definition the role of the different supply chain stages have necessarily to be recognised, especially if both farmers and producers are involved in the respect of stricter specifications rules defined for exported products. The price positioning, in particular within international large scale retailers, will have to consider the higher product quality differentiation, trying to prevent risky price competition.

The future of the EU quality labelled products is not only connected to their positioning, promotion and international protection but also to the strategies of the retail leaders and to the overall strategy of the brand industry. The latter is more and more present in these that once were defined as 'niche markets', while now represent a diversification and a good investment to improve company's market positioning.

References

Arfini, F., 1999. The value of typical products: the case of 'Prosciutto di Parma' and 'Parmigiano-Reggiano Cheese', 67th EAAE Seminar EAAE, October 28th, 1999, Le Mans (France).

Bellia, C., 2005. Il ruolo del marketing mix nelle imprese vitivinicole siciliane marketing-oriented, XLII Convegno SIDEA: 'Biodiversità e tipicità. Paradigmi economici e strategie competitive'.

Berti, A., M. Canavari and R.P. King, 2005. The Supply Chain for Parmigiano-Reggiano Cheese in the United States. In: Defrancesco, E., L. Galletto and M. Thiene (eds.) Agriculture and the Environment. Economic Issues, Milano: Franco Angeli, pp. 117-133.

Boccaletti, S., 1999. Signalling quality of food products with Designations of Origin: advantages and limitations, World Food and Agribusiness Congress, Florence, June 15th, 1999.

Bonetti, E., 2004. The effectiveness of meta-brands in the typical product industry: mozzarella cheese. British Food Journal, 106: 746-766.

Bureau, J.C. and E. Valceschini, 2003. European food-labeling policy: successes and limitations. Journal of Food Distribution Research, 34: 70-76.

Cardinali, M.G., 1998. Le strategie di marketing dei prodotti agro-alimentari tipici. Trade Marketing, 22: 43-66.

Crescimanno, M., S. Di Marco and G. Gruccione, 2002. Production and trade marketing policies regarding organic olive oil in Sicily. British Food Journal, 104: 175-186.

Cristini, G., 2006. Marketing d'insegna e marca privata. Strategie e implicazioni operative per distributori e copackers. Il Sole 24 Ore, Milano, Italy.

European Commission, Directorate-General JRC, 2006. Food quality schemes project, case study: Parmigiano-Reggiano, (F. Arfini, S. Boccaletti, C. Giacomini, D. Moro, P. Sckokai), Bruxelles.

Evans, G.E. and M. Blakeney, 2006. The protection of geographical indications after Doha: quo vadis? Journal of International Economic Law, 9: 575-614.

Fornari, D., 2005. La rivoluzione del supermercato, Il modello del Gruppo SISA, Milano, Egea.

Germanò, A., 2005. Il panel WTO sulla compatibilità del regolamento comunitario sulle indicazioni geografiche con l'accordo Trip's. Agricoltura – Istituzioni – Mercati, 2: 279-290.

Giacomini, C., 2007. Crescita e prospettive dei prodotti a marca commerciale nella GDO europea, Unione Parmense degli Industriali Seminar: 'Standard di certificazione e GDO europea', October 3rd, 2007, Parma (Italy).

Handler, M., 2006. The WTO geographical indications dispute. The Modern Law Review, 69: 70-91.

Jaffe, E.D. and I.D. Nebenzahl, 2001. National image and competitive advantage. The theory and practice of country of origin effects, Copenaghen Business School Press.

Magni, C. and F. Santuccio, 1999. La competitività dei prodotti agro-alimentari tipici italiani fra localismo e globalizzazione. Rivista di Economia Agraria, LIV: 299-324.

Marette, S. and A. Zago, 2003. Quality and international trade: what strategies for EU AOC system. International Conference: Agricultural policy reform and the WTO: where are we heading?, Capri (Italy), June 23-26, 2003.

Menozzi, D., 2005. I formaggi veneti a denominazione di origine protetta (DOP), in AA.VV.: 'Analisi Economica del Comparto Lattiero-Caseario nel Veneto', Ed. VenetoAgricoltura, Padova.

Montanari, C. and K. De Roest, 2006. I cambiamenti strutturali dei caseifici del comprensorio del Parmigiano-Reggiano dal 1993 al 2005, C.R.P.A., Reggio Emilia.

Mora, C., 2005. La specificità del mercato Veneto e le strategie degli operatori, in AA.VV.: 'Analisi Economica del Comparto Lattiero-Caseario nel Veneto', Ed. VenetoAgricoltura, Padova.

Mora, C. and F. Arfini, 1997. Typical products and local development: the case of Parma area, 52nd EAAE Seminar on 'EU Typical and Traditional Productions: Rural Effect and Agro-Industrial Problems', Parma, 19-21 June 1997.

O'Reilly, S. and M. Haines, 2004. Marketing quality food products - A comparison of two SME marketing networks. Food Economics, 1: 137-150.

Padberg, D.I., C. Ritson and L.M. Albisu, 1997. Agro-food marketing, CIHEAM, Cab International.

Pellicelli, G., 2007. Il marketing internazionale, ETAS, Milano.

Rossetto, L., 2002. Marketing strategies for organic wine growers in the Veneto Region, 8th Joint Conference on Food, Agriculture and the Environment, August 25-28, 2002, Red Cedar Lake, Wisconsin, U.S.

Sodano, V., 2001. Competitiveness of regional products in the international food market, In 77th EAAE Seminar, August 17-18, 2001, Helsinki.

Valdani, E. and G. Bertoli, 2006. Marketing dei mercati internazionali, Il Sole 24 Ore, Università Egea, Milano.

Van der Lans, I.A., K. van Ittersum, A. De Cicco and M. Loseby, 2001. The role of the region of origin and EU certificates of origin in consumer evaluation of food products. European Review of Agricultural Economics, 28: 451-477.

Food safety in international trade: the Spanish experience in Mediterranean products

J. Briz, M. Garcia and I. de Felipe

Abstract

International agricultural and food trade between developed and developing countries is increasing due to a worldwide liberalisation movement. Tariffs and technical barriers are decreasing and many exporters in developing countries are ready to supply developed markets. However, there are still non-tariff barriers. The objective of this paper is to analyse how some food safety and quality control measures may be overcome by developing countries through a benchmarking exercise. The identification of 'best practice' in operating firms may be useful to others. We describe the evolution of international trade, the situation at destination and origin markets through the international channel. It is important to identify the role of food safety and traceability in food international relations. A case study is presented with the analysis of the fresh food export supply in Spain.

Keywords: international trade, food safety, benchmarking, Spanish fresh food export supply chain, traceability

1. Introduction

Increasingly international agricultural trade between the European Union (EU) and developing countries is the first step towards long-term economic and social integration. In addition to marketing strategies oriented to meet consumer demands and other elements such as the liberalisation of commercial barriers, there are still important factors that limit the development of efficient import-export relations.

The objective of this paper is to highlight the importance of safety and quality requirements and traceability as potential non-tariff barriers discriminating against developing countries. Exporters in developing countries wanting to export their products to the EU market should be aware of these increasing demands and implement systems to demonstrate compliance.

The paper focuses on the food safety and quality control systems in the Spanish fresh food exporting sector as a reference for the benchmarking analysis. As a leading exporter of fruit and vegetable products, it is of interest to identify some of the main activities in the Spanish fresh produce export supply chain. The study focuses on the identification of key non-compliance areas in food safety and quality.

2. Characteristics of international food trade between developed countries (DC) and less developed countries (LDC)

The evolution of international trade in the last decades shows how agricultural products are less dynamic than manufactures or energy (Figure 1) with yearly variations that are generally lower (Figure 2). Although agricultural trade accounts for about 10 per cent of total international trade, it has a strategic position and WTO negotiations are quite often dependent upon agreement on international agricultural trade.

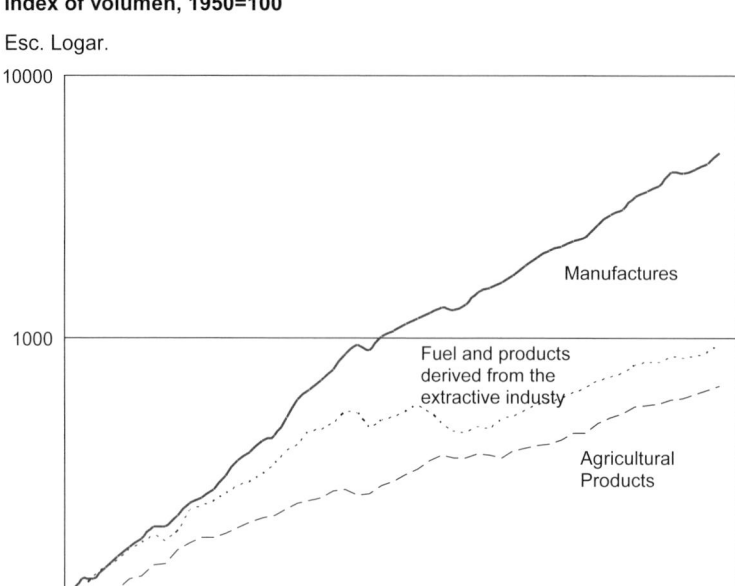

Figure 1. Evolution of international trade. Source: WTO.

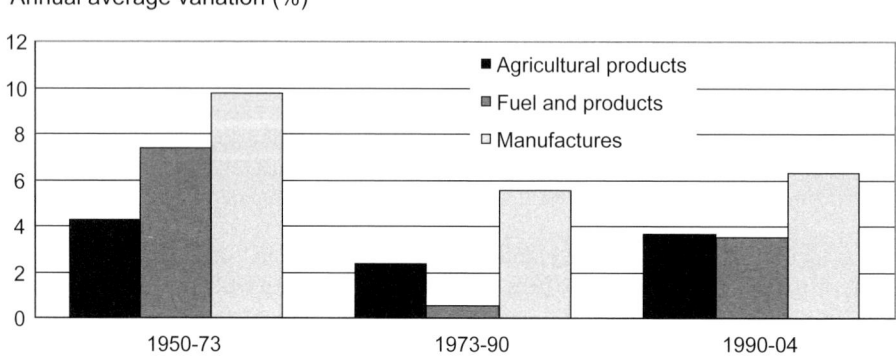

Figure 2. Annual variation of international trade. Source: WTO.

Less Developed Countries (LDC) and the non-Governmental Organisations on Development (NGOD) are placing a special emphasis on opening the borders of the rich countries and allowing the entrance of products coming from them, with the conviction that international trade will impact positively on economic and regional development.

The efficiency of the commercial chain which goes from the producer of underdeveloped areas to the consumer of rich countries needs to be taken into account (Camps, 2004). Any failure along the chain can imply a discontinuity in the supply or an abuse of dominant position of brokers who take great part of the benefit. Thus, experience shows that even in Developed Countries a collapse in the prices perceived by farmers is not automatically reflected in the equivalent fall in consumers' price. Hence, we must consider the scenario as a whole and consider that the added value generated by the possible liberalisation of markets in rich countries must be addressed essentially to help marginal areas and needed population. Otherwise, with a mere agreement on reduction of tariffs we have only covered the first stage.

Three dimensions are considered in this study: the destination market of the product, the origin market and the union channel of both (Briz and Trueba, 2006).

2.1 Destination markets

In a market liberalised of quotas and tariffs, the products from LDC will find several problems, (e.g., Non Tariff Barriers, NTB) derived from exigencies in quality and food safety required by consumers. These types of barriers cannot be considered discriminating since the requirements are the same for national producers and foreigners. This is the case for the traceability requirements that came into effect in January 2005 in the European Union (EU). The EU provides support on equipment and education of experts, so if there are qualified experts in the LDC they can fulfil the controls of quality and traceability in the exports directed to the EU. A different matter is the regulation of bio terrorism in the U.S.A., where requirements apply only to imported products.

Another aspect to consider is the distribution system used. Until now, some products from LDC are sold through Fair Trade logo. This is a good initiative that is consolidating, oriented to a sensible public within the Third World. However, if the imports with Fair Trade Logo increase in volume, it would be necessary to think about the use of regular trade channels able to absorb those amounts in a good relation quality/price. Multinational companies related to retailers may develop a good role, given their commercial agility. It is necessary to consider the participation in traditional channels of wholesalers and retailers to support this initiative.

2.2 Origin market of products from LDC

Since the goal is to help small farmers to improve their income, they must be the receivers of a great part of the added value obtained by the liberalisation of markets in DC. Trade systems in underdeveloped areas are usually inefficient, with lack of transparency and abuse of dominant position of certain economic agents. Consequently, a good proportion of the benefits may be lost through waste or through fraudulent practices or corruption, which would produce frustration and discrimination.

In addition, farmers oriented to export products are usually the richest ones, whereas the marginal ones are centred in self-consumption or, in the best case scenario, in the domestic market. Hence, it is necessary to evaluate the opportunity of those poor ones to receive the benefit, for which there are to design support policies, among them: proper distribution of land and other productive factors, the search of market windows, infrastructure improvement in origin markets, transports, storage, in addition to the traditional agricultural extension services.

2.3 International trade channels

We may identify some general elements and factors that condition commercial relation. The logistic, financial and administrative complexity of agricultural export channels limits the participation of Small and Medium Enterprises (SME´s) that in the best of the cases are centred in very specific fields where they have evident comparative advantages. For that reason, this commercial link could perceive great part of the added value, without being perceived by the link of the chain, the farmer.

Another aspect to consider in marketing channels is the customs bureaucracy which faces the international trade relations. The discretion of the Administration, slowness of the operations, bureaucratic overlapping and lack of coordination between civil employees and industrialists is added to the existing lack of transparency.

As an example and according to the 'United Nations Conference of Tariff and Trade' (the UNCTAD) (WTO, December 2005) a normal transaction in border requires the participation of about 25 agents, 40 documents, 200 data items (a third of which are repeated 30 times).The temporary delay, cost and errors made, are a big weight to carry for the companies, forced to give the merchandise in a determine place and time. According to the WTO, the agility in this transaction could generate up to 0.26% of the GIP of the countries, which almost counts double of the benefits derived from tariff liberalisation. The situation is still more serious in the SME's that operate with small volumes, where unitary effects are greater. It all affects to competitiveness in the international scene, mainly to the LDC.

It is important to distinguish between food security, food safety and integral food security that it includes both. The *Food Security* responds to the need to have the available food amount at any moment and place for the survival of people. In this area, organisms such as FAO have a relevant role. *Food Safety* has complementary problems, especially concerning developed countries that have preoccupation in health and hygienic (Briz, 2003).

Market economy shows that, in a certain way, its operation acts as a safeguard to maintain a certain level of food safety, since companies need to have the confidence of their clients (consumers). Nevertheless, the market does not guarantee enough levels of food safety. On one hand, consumers cannot define exactly their needs and have difficulties in identifying the degree of food safety they demand. On the other hand they do not respond to the enterprise's efforts in food safety, by paying higher prices.

Some firms in Food Industry face unfair competition from other companies that do not respect safety discipline, so they have lower costs. This problem repels to all the companies supplying similar products. It is interesting to highlight that food safety has a wider range, concerning possible negligence. There are social costs derived from the attention of public services, loss of working hours, minor labour yields and global distrust that may cause asymmetric movements in the markets and deviation of natural resources.

Public powers must be involved through regulations, inspection and actions, that guarantee innocuous foods to their citizens. Within these regulations those of the foreign trade are fitted.

In the international field the concerns for the Sanitary and Phytosanitary measures (SFS) has a greater dimension and displays greater problems. On one hand there is a greater heterogeneity in food supply and the regulations vary depending on the country; on the other hand consumers have different socioeconomic characteristics and different levels of communication and information.

That is why it is not uncommon to face problems related with food safety. There are several solutions, from the total cease of trade flows until the problem is solved, direct negotiations between the affected countries, the intervention of international organisms like the WTO and the World Health Organization (WHO), or the improvement of productive and elaboration processes that have caused the conflict.

The process of increasing globalisation forces the companies to get involved in the markets of other countries, taking care of the sanitary – hygienic - legislations and of quality in order to attend different consumers segments. Besides the international regulations, companies oriented to foreign countries usually have their own quality regulations, frequently more demanding, if they are positioned with known marks, since any problem is quickly transmitted to other countries.

The operating system and consequent sanitary quality controls in the international relations (Mitchell, 2003) can follow diverse modalities according to the participation degree and enterprise responsibility. There is the possibility of exporting directly the finished product to its destiny markets or establishing their own factories in other countries.

When an investment is made in facilities and manufactured in foreign countries, companies have the advantage that products are already within the destiny market and do not face commercial barriers. However, in addition to their own hygienic and quality regulations, must fulfil the regulations established by the corresponding governments with respect to processing and manufacturing. In other words, food trade have to accomplish all the requirements of International Food Security, which include adequate supply in innocuous conditions.

2.4 Food safety requirements in developing countries

International trade implies an extension of the commercial chain, supplying distant markets at more competitive prices and greater variety of products, with more opportunities for the LDC. Great efforts are being made to open the borders of rich countries through tariff disarmament, without paying attention to the subject of food safety and control.

It is important to consider that globalisation implies safety and quality risks that can interfere in any of the links of the chain. In order to avoid these risks, the opportune controls are required, which means a greater operative cost. Normally, the disagreements in this field are usually solved through the World Trade Organization (WTO) by a group of experts on sanitary and phytosanitary measures without greater consequences. The regulations on quality and food safety are considered to be trade barriers. However, controls and certifications may be an intensification of exchange flows because of the greater confidence of consumers.

Therefore, it is necessary to strengthen international cooperation, where rich countries facilitate to the LDC the economic and human resources that allow the improvement of their control equipment and systems. The EU is aware of this subject and especially in these last years when it has established cooperation projects and studies to identify the main problems and search solutions.

Quality and food safety controls are going to cause discrimination in the market, being fomented by that companies which obtain products adapted for consumption. The effects in the LDC may be important, since companies oriented to exportable products will constitute groups, providing the entrance of foreign currencies, and advantages on those oriented exclusively towards the domestic market. The distortion can be harmful if food production diminishes for basic feeding and affects imports. Therefore a degree of food sovereignty would be lost, depending on the international market with the inherent risks.

In this case, the public powers and the private sector must reach an agreement to obtain the balance between the self-supply and exports. Each country has its own peculiarities and, in any case, a frame must be contemplated including broad, viable and sustainable geo-economic areas that surpass national borders. Also, in destination markets of developed countries there is an increasing preoccupation for food safety, trying to improve control and information systems (Buzby and Unneveh, 2003). Previously, food scandals had local dimensions and passed unnoticed for mass media. Nowadays, the complexity of food chains and its amplitude, multiply the risks at world level, quickly spread by the New Technologies of Information and Communication (NTIC).

3. The role of food safety and traceability in agro food international trade

Food safety regulation has undergone significant changes in many developed economies during the last decade. Due to the recent developments, some countries have increased their national efforts in maintaining high quality standards and ensuring the safety of food supply for both domestic consumption and export. However, it is recognised that Developing Countries have difficulties in meeting certain requirements associated with the implementation of sanitary or phytosanitary measures and which come in connection with technical regulations, standards and conformity tests (IMF/World Bank, 2002). As more sophisticated governments and industry introduce regulations, there is the risk that new regulatory barriers will be erected. This is of particular concern for Developing Countries, where existing infrastructure may not allow for the adjustments needed to meet new requirements.

In exporting countries with established and organised supply lines, the coordination of safety and quality through private retailer supply relationships or through a centralised organisation is possible. Traceability systems for food safety may represent a technological barrier to exporting firms in Less Developed Countries. The process is much more problematic where there are fragmented supply chains, less direct multi-producer relationships with exporters, and less vertical integration in the supply chain. Food systems in developing countries are not always as well organised and developed as in the industrialised world, and moreover, knowledge of standards is often lacking.

Food safety is more likely to be a concern in fresh food product international trade than in other types of agricultural trade (Unnevehr, 2000). Firstly, since fresh products are transported and consumed in fresh form, handling throughout the entire supply chain can influence food safety and quality (Zepp *et al.*, 1998). In addition, it is the relatively high perishability of fresh produce and the susceptibility to damage and disease pre- and post harvest that imposes high requirement levels for quality assurance. Secondly, standards in Developed Countries tend to be significantly higher than those in developing countries; hence compliance with those standards may require greater initial investment in quality control and health system in Developing Countries. Thirdly, fresh commodities are subject to increasing scrutiny and regulation in Developed Countries where food safety hazards are better understood and more often traced to their sources.

The long-term solution for Developing Countries to sustain an international demand for their products lies in building up the trust and confidence of importers in the quality and safety of their food supply systems. This requires improvements within national food control systems (Alvarez, 2005) and within industry food quality and safety programmes (FAO, 1999). 'Farm to table' process control to manage both quality and safety is increasingly in demand in developed countries, and new institutions are evolving to certify production practices (Unnevehr and Jensen, 1999). Hence, there are market incentives for developing exporters' countries to adapt these management practices, and to coordinate safety and quality management more closely with importers (Busch *et al.*, 2000).

A key to product quality and safety management throughout the fresh produce supply chain is *traceability*, enabling product tracking and accountability at each stage. Nowadays, the facility to trace fresh produce production and handling practices is required by the importer/retailer complex, and all major operations, from planting to export, must be documented. This approach ensures a better understanding of the steps and conditions to which fresh produce have been subject (Ait-Oubahou and El-Otmani, 2000). Traceability requires the identification of all physical entities (locations) from where fresh produce originates and where it is packed and stored.

Due to the globalisation of the fresh produce supply chain and because of the diversity of international produce supply chain practices, the fresh produce sector in March 2001 agreed upon *Fresh Produce Traceability Guidelines* (FPT guidelines). The FPT Guidelines were developed together with the Euro-Handels-Institut (EHI), the European Association of Fresh Produce Importers (CIMO), the Euro Retailer Produce Working Group (EUREP), the European Union of the Fruit and Vegetable Wholesale, Import and Export Trade (EUCOFEL) and the Southern Hemisphere Association of Fresh Fruit Exporters (SHAFFE) to provide a common approach to tracking and tracing of fresh produce by means of an internationally accepted numbering and bar coding system – the EAN•UCC system (EAN International, 2001). The adoption of the guidelines is voluntary and the degree to which companies will implement them may vary because of differences in commercial operations. However, the use of common identification and communication standards will significantly improve the accuracy and speed of access to information about the provenance of fresh produce. Therefore, it is likely that this traceability model will be a requirement for fresh produce exporters in the near future.

As a final word in this section, it must not be assumed that there is an easier commercial option in domestic markets for firms who do not wish to meet the challenges of more sophisticated export markets. There are many reasons besides the ethical and moral imperatives why firms must strive to achieve high levels of performance in respect of safety and quality. Social and economic losses due to poor food safety and quality are probably as serious, if not more serious in developing economies where standards and systems are lower, than are losses in advanced economies (Poole *et al.*, 2002). Improvements in the health and safety of poor people are fundamental to international efforts to achieve the Millennium Development Goals for poverty reduction, as is the development of vibrant food systems where the sector is, or has the potential to be, a major source of employment, export earnings and other macro-multiplier effects.

4. Case study: the Spanish fresh agro food export supply chain

Spain is the second largest European fresh produce producer after Italy and the largest world exporter. The fresh produce sector represents around 45% of total agricultural production in Spain. It generates over 450,000 jobs, and in several Autonomic Communities the activity of the sector is fundamental to rural employment (Garcia, 2003).

During the last years, production and market performance has been very positive, but producers are facing internal and external problems. The competitive advantages of Spanish vegetable production are focused on low production costs and out of season production that effectively out-competed nearby community competitors (France, the Netherlands and Italy).

However, these competitive factors have shown themselves to be fragile, in that low prices are important but are not the only determinant element of the consumer's choice and the European markets have been opened increasingly to competition with third countries. In the last decade, the Spanish horticultural sector has been losing competitiveness through an improper modernisation of the production and marketing structures.

One of the major problems for the fresh produce sector is the fragmented and small scale production. Within the context of third countries' competition and the concentration of demand (with increasing requirements) there is an urgent need for producer organisations and vertical integration. With this objective the European Common Agricultural Policy (CAP) in fresh produce has imposed the formation of Producer Organisations (OPFH) in order to receive subsidies. In the year 2000 there were 675 OPFH in Spain, with 582 having processing plants, and representing half of the total production value. 100% of banana and tomato firms are organised in OPFH, with 30% in Citrus, 20% in fruits and 7% in vegetables (MAPA, 2005). As a consequence, the analysis of the marketing channel and traceability it's important

The Spanish export chain with a long tradition and socio-economic importance may be of interest for entrepreneurs either in other exporting sectors or developing countries. In order to know the 'best practice' for fresh produce safety and quality in food export chain, it is necessary to implement inspections and certifications.

Generally, there is high efficiency in the organisation of Spanish exports. Direct connection of exporters with distribution companies give more flexibility and better information through the marketing chain than is seen in other countries. This is one of the arguments that places an integrated marketing process in a better position than wholesale or 'veiled marketing' where there is not a continuous producer-retailer experience.

In the last few years in Spain there has been an increasingly concentrated demand in the food distribution chain. Along with the fragmented nature of supply, this causes serious imbalances in the markets, giving exporting companies a weak position in the market. The consequences of fragmented supply to Spanish exporting companies are multiple: it prevents the enterprises from incorporating added value, inadequate design and promotion; it does not permit the exporting enterprises to acquire large commitments; and marketing costs are high which weakens the position of the professional associations and limits their capacity to press the national and EU administration in order to defend their interest and position in the CAP.

In relation to the structure of the Spanish exporter companies, strong fragmentation and spatial concentration are important factors. Thus, enterprises dealing with more than 25,000 tonnes of exports are concentrated in Andalusia, Murcia, Valencia and the Canary Islands. In this last region, the cooperative system has major participation in export.

The cooperative sector in Spain is responsible for about 30% of exports and involvement in the domestic market is probably greater. The cooperative sector includes Sociedades Agricolas de Transformación (SATs) as well as traditional cooperative organisations. SATs differ from traditional cooperatives in that their membership and business is not restricted to a specific geographical area. Figure 3 summarises the main relationships in the fresh produce supply chain for exports.

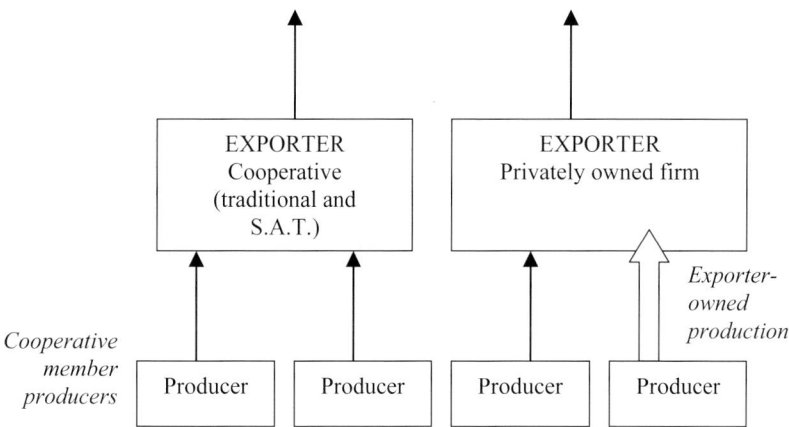

Figure 3. Spanish fresh produce export supply chain (García, 2003).

If we consider the bigger exporter enterprises, some of their characteristics are:
• Great concentration on marketing activities. There are group of enterprises that have joined activities in the marketing channels through some joint-ventures, either in origin or in destination. There are also groups of family enterprises that maintain a joined marketing management's functions. Thus, export companies have a great capacity to adapt to market conditions, either to the slow growth of consumption in some markets or to big changes in production and export of new products, such as Iceberg lettuce, broccoli, and Galia melon.
• In general, many enterprises are simultaneously producers and exporters, regardless of the juridical structures. Perhaps, the only exception is Almeria. This kind of integration is one of the most relevant strengths in the foreign trade.
• Strong participation in the regional production of export enterprises, especially in the Canary Islands, Alicante, Huelva, Seville, Cadiz, Almeria and Murcia, in products like tomatoes, lettuces, melon, peaches and strawberry.
• Many export enterprises are trying to orient their activity to greater distribution enterprises. For that reason, they pay special attention to homogeneity, quality, and regularity in the quantities supply, during a long period of time. Simultaneously, they have been able to create alliances with wholesalers and importers in destination countries.
• In the future, we foresee a continuation in the concentration process of the producer-exporter companies, especially in the marketing area, trying to provide adequate logistic and commercial services.

5. Best practice and benchmarking in a more efficient export supply chain

An export supply chain showing 'best practice' for fresh produce safety and quality involves many inspections and certifications. Some of these controls will be carried out by government authorities, both at exporting and importing countries, based on public standards and regulations, while others will be undertaken by private organisations (i.e., third party certification) on behalf of importers/retailers and based on private specifications.

In exporting countries with more established and organised supply lines, the coordination of safety and quality through private retailer-supplier relationships, or through a centralised organisation is made possible. The process is much problematic in developing countries when there are fragmented supply chains, more indirect multi-producer relationships with exporters, and less vertical integration in the supply chain.

We have to outline the different stages of the fresh produce export supply chain. Given the scope of this report, the analysis will focus on the control activities carried out in exporting countries, leaving out of the study those activities undertaken by importing authorities in country destinations.

Exporters will receive the produce from their suppliers in an unsorted or partially sorted conditions and requiring further processing (i.e. washing, grading, selection, etc.), and/or packaging. It is essential at this stage that the raw material is safe, legal and meets the standards laid down by the packer/exporter/importer/retailer.

At this stage a number of quality and safety checks will be carried out:
- produce quality, weight and labels checked for conformance with specifications;
- produce inspected for physical contaminants and mechanical damage, including chill damage;
- need for ripening assessed;
- produce sampling for quality testing specific to product (e.g. sugar content in citrus);
- produce sampling for phytosanitary purposes;
- produce sampling for pesticide residue checks.

Traceability depends basically upon accurate and timely record keeping. The EAN system includes the transmission of traceability data by electronic means, a technology that is not available to all the firms interviewed in this study. The EAN standard bar-coding system allows the identification of all locations where the fresh produce originates from and where it has been packed and stored. Some firms in this study are able to track produce units around the pack house itself using barcode recognition apparatus. Hence a more simplified scheme for best practice could be that followed by ANECOOP, the largest second-tier cooperative in Spain in the fresh produce sector.

Carrying out a benchmarking exercise will enable the comparison of the level of implementation of food safety and quality practices, across countries, across sectors and across different sizes and ages of firms, with the identification of key non-compliance areas for exporting companies in the countries under study. The process involves working with those operators considered to be examples of 'best practice' in the industry and those firms with less market share.

In UK industry, a study by the Food and Drink National Training Organisation also used a scoring system in a benchmarking process for the food and drink manufacturing industry. The objectives of this research were to set a benchmark for UK food and drink manufacturing companies to identify and promote world class manufacturing activities, to establish a set of benchmarking criteria founded on international best practice for UK companies to measure themselves and identify areas for continuous improvement and to produce an industry action plan. The key areas looked at were:
- business measures;
- personnel and training measures (statistical data);
- skills profiles.

The benchmarking process involved a questionnaire on Business Measures and Personnel and Training Techniques, explored with senior management team during a visit to company and then assessing the skills of personnel on visits to manufacturing operations (García, 2003).

Some of the key strengths of the benchmarking process in previous studies, has been the bringing together of participants from companies in various sectors and of various sizes, providing a forum for exchanging information and experiences to help resolve problems (e.g. Andersen *et al.*, 1999). In this study, the objective of using benchmarking was to increase the knowledge about the supply chain management process, to identify best practices in the industry and to enable the industrial project partners to learn from the best practice. Studies such as Prado (2001) focus on the face-to-face interaction and teamwork between participants in the benchmarking process, highlighting the importance of the information sharing or dissemination stage of benchmarking. The benchmarking process usually results in the development of a series of actions within each company involved in the exercise.

Thus, benchmarking involves:
- identification and examination of specific key areas or performance areas in the process under study;
- identification of firms with best practice in the area;
- exchange of information and experiences;
- production of an action plan.

A benchmarking framework is given by Shah and Singh (2001):
- Stage 1: selection of performance measures, depending on the firm's competitive focus, market niche and strategy.
- Stage 2: benchmarking exercise on the firms in the industry, using the selected performance measures. This enables the identification of firms with 'best performance' in terms of the selected measures.
- Stage3: information about specific strategies of the 'best performance' firms to be obtained from sources in the public domain. This information can be related to the specific performance measures of the firms.
- Stage 4: leveraging this knowledge to find what bearing the firms' performance measures have on their specific practices and policies.

For this study, specific performance measures were identified for application across the sectors. By carrying out case studies of exporting firms, the relationship between producers and exporters was examined and a comparison with existing best practice in infrastructure and management practices carried out. In this benchmarking exercise, a qualitative rather than quantitative approach is used to explore each Key Performance Indicators (KPI). This is due to the difficulty of assigning quantitative measures to the supply chain characteristic indicators which are being examined here.

Benchmarking is a tool for improving performance by learning from best practice and understanding the process by which they are achieved. This project in particular focuses on 'process benchmarking' by comparing operations, work practices and business processes in the fresh produce exporting industry in Morocco and Turkey, with those in Spain.

Specific performance measures (KPI) were identified for application across industries. The indicators were decided upon through an examination of the supply chains for each target sector and a study of the areas and levels in which safety and quality systems could be controlled through the supply chain. (Some indicators were also based on EUREPGAP (2001) and Güngor and Güngor (2000)).

Each KPI was explored using questions, which made up a questionnaire for use as a discussion guide during visits and interviews with exporters. A qualitative rather than quantitative approach was initially used to compute each KPI due to the difficulty of assigning quantitative measures to the supply chain characteristic indicators examined in the study. Qualitative data was then classified into three levels. The different elements within each of the three levels in the framework aim to characterise that level, indicating the firm's policies and practices in this aspect, rather than specifying certain criteria, which they must meet. Some points of the framework depended upon a combination of answers in the questionnaire.

The benchmarking project considers several areas for the analysis of food safety and quality management (Table 1). For each area, a number of KPIs were developed comparison in the benchmarking process.

In order to compare the firms involved qualitatively, a framework was developed. This framework followed the structure of the questionnaire, and classified the information gained from each firm into three levels. Some points of the framework depended upon a combination of answers in the questionnaire. For example, 'Production flexibility' was based upon the producer's ability to change crops/varieties grown in response to market demands as well as the exporters' ability to source from different producers with different product bases, in order to meet different market requirements. This, itself, is dependent on the producer-exporter relationship, and the nature of the contract between them.

The different elements within each of the three levels in the framework aim to characterise that level, indicating the firms' policies and practices in this aspect, rather than specifying certain criteria which they must meet. For example, classification of production practices in terms of safety and quality management depends firstly upon the actual production practices that take place and, importantly for this study, the exporter's knowledge about these production practices and ability to control them or gather information about them.

6. Final remarks

Evolution of international trade will rely, beside the traditional factors (comparative advantage, cost, price policies) in other less traditional elements, such as food safety and traceability. Consumers in developed countries, especially in the EU, are very much concerned upon quality control and food scandals. Therefore new strategies should be applied by potential exporters from LDC, in order to get into the EU market.

In the case study we include traceability in the Spanish fresh product export trade as a new strategy in the coming future. Food safety and quality control may allow to get an adequate traceability and thus to get consumer and retailer confidence in the competitive markets. We show in this paper the results of the benchmarking analysis carried on in exporting companies. The identification of 'best practice' in leading enterprises may help a better performance to others and facilitate indirectly the traceability of the products

Table 1. Key performance indicators (García, 2003).

Areas of analysis	KPI for comparison in benchmarking framework
1. Supply base	1.1. Degree of specialisation
	1.2. Export volume
	1.3. Number of producers and fragmentation of supply
	1.4. Varieties
	1.5. Forecasting systems
	1.6. Production flexibility
2. Supply chain management	2.1. Producer-exporter relationship (Type of producer-exporter relationship, e.g. cooperative, private firm)
	2.2. Vertical integration
	2.3. Degree of coordination of operations
	2.4. IT infrastructure and integration for supply chain management
	2.5. Customers: countries exported to
	2.6. Customer contracts
	2.7. Customer visits
3. Traceability and tracking	3.1. Traceability systems
	3.2. Segregation
4. Crop protection	4.1. Producer practices
	4.2. Exporter communication
5. Harvesting	5.1. Harvest hygiene
	5.2. Harvest quality (Product homogeneity, effect of climate, consistency in production)
6. Processing and packaging	6.1. P&P technology
	6.2. P&P quality
	6.3. Labelling
7. Storage and transport	7.1. Exporter storage knowledge
	7.2. Storage capacity
	7.3. Storage quality
	7.4. Transport quality
8. Export quality control (QC) Process	8.1. Quality certification
	8.2. QC staff
	8.3. Worker knowledge
	8.4. Product sampling for QC
	8.5. Laboratory access
9. Pack house worker health, safety and welfare	9.1. Training
	9.2. Worker welfare, health and safety
10. Environmental management	10.1. Environmental management

References

Ait-Oubahou, A. and M. El-Otmani, 2000. Quality assurance for export-oriented citrus and tomato fruit in Morocco in quality assurance in agricultural produce. In: G.I. Johnson, Le Van To, Nguyen Duy Duc and M.C. Webb (Eds.) ACIAR Proceedings 100. Available at: http://www.aciar.gov.au/publications/proceedings/100/H28_Ait-Oubahou.pdf.

Alvarez, D., 2005. El impacto del comercio internacional en el Sur. Boletín Manos Unidas, 161: 15-22.

Andersen, B., T. Fagerhaug, S. Randmoel, J. Schuldmaier and J. Prenninger, 1999. Benchmarking supply chain management: finding best practice. Journal of Business and Industrial Marketing, 14: 378-389.

Briz, J. (Ed.), 2003. Internet, trazabilidad y seguridad alimentaria. Madrid: Mundiprensa.

Briz, J. and I. Trueba, 2006. Comercio Mundial y Seguridad Alimentaria. El fin del hambre en 2025. Madrid: Mundiprensa.

Busch, L., J. Bingen, C., Harris and T. Reardon, 2000. Markets, rights and equity: food and agricultural standards in a shrinking world. Institute for Food and Agricultural Standards, Michigan State University.

Buzby, J.C. and L. Unneveh, 2003. International trade and food safety. Aer-828. ERS. USDA.

Camps Th.W.A. (Ed.), 2004 The emerging world of chains and networks. Reed Business Information.

EAN International, 2001. Fresh Produce Traceability (FPT) Guidelines. Available at: http://www.ean-int.org/agro-food/Opmaak%20tekst%20Fresh%20Produce%20.pdf.

FAO, 1999. The importance of food quality and safety for developing countries. May 31-June 3. Rome: Committee on World Food Security, 25th session. Available at: http://www.fao.org/docrep/meeting/x1845e.htm.

García, M. (Ed.), 2003. Impact international safety standards in Mediterranean fresh products. ICA 3-CT-2000-2001 (2000-2002).

Güngor, H. and G. Güngor, 2000. Managing the quality chain in citrus fruit industry: a case study in Çukurova region of Turkey. Acta Horticulturae, 536 (XIVth International Symposium on Horticultural Economics).

IMF/World Bank, 2002. Market access for developing country exports - selected issues. September 26. Available at: http://econ.worldbank.org/files/18875_market_access.pdf: Staffs of the International Monetary Fund and the World Bank, Washington, DC.

Mitchell, L., 2003. Economic theory and conceptual relationships between food safety and international trade. International Trade and Food Safety, AERERS.

Poole, N.D., F. Marshall and D.S. Bhupal, 2002. Air pollution effects and initiatives to improve food quality assurance in India. Quarterly Journal of International Agriculture, 41: 363-386.

Prado, J.C., 2001. Benchmarking for the development of quality assurance systems. Benchmarking. An International Journal, 8: 62-69.

Shah, J. and N. Singh, 2001. Benchmarking internal supply chain performance: development of a framework. Journal of Supply Chain Management, 37: 37-47.

Unnevehr, L. and H.H. Jensen, 1999. The economic implications of using HACCP as a food safety regulatory standard. Food Policy, 24: 625-635.

Unnevehr, L.J., 2000. Food safety issues and fresh food product exports from LDCs. Agricultural Economics, 23: 231-240.

Zepp, G., F. Kuchler and G. Lucier, 1998. Food safety and fresh fruits and vegetables: is there a difference between imported and domestically produced products. Vegetables and Specialities/VGS-271. April: Economic Research Service, USDA, pp. 23-28.

Competitiveness of Turkey in organic exports to the European Union market

R. Funda Barbaros, Aykut Lenger, Sedef Akgüngör, and Osman Aydoğuş

Abstract

The exports of organically produced agricultural products in Turkey have been rapidly growing, particularly in response to an increasing demand for organic products in European Union countries. The common view and findings of research in organic trade in Turkey confirm that the European market is expanding. There is a missing focus in the prevailing research on this issue. None of the studies have focused on modelling the exports market for organic products in order to estimate price and income elasticities for these goods; this would enable an analysis of the issue for the derivation of policy implications. Another related missing focus in the present research literature is the lack of understanding of the current competitiveness of Turkish exports in the European Union market and the sources of this competitiveness. The paper has three major objectives and, thus, seeks to produce three outputs: The first objective is to explore Turkey's export competitiveness in organic products in the European Union market. In order to fulfill this objective, we will estimate the indices measuring competitiveness of Turkish organic products. The second objective is to uncover the components of export performance. Through constant market share analysis, we seek to determine the key factors underlying the growth or Turkish organic exports. Finally, the third objective is to estimate an export demand model for these products in the European Union market. Thus, it will be possible to estimate price and income elasticities of demand as well as make projections. The study reveals that export demand for Turkish organic products in European Union market is growing, and exports to the European Union market are sensitive to the price and income changes of the target countries. Turkey has a clear comparative advantage versus the competitor EU countries in selected products. Competitiveness, in particularly, relies on relative prices, and thus, does not suggest any sustainability.

Keywords: organic agriculture, Turkish agriculture, export demand, competitiveness, Turkish organic exports

1. Introduction

The objectives of the paper is (1) to understand Turkey's export competitiveness of organic products in the European Union market; (2) to identify the components of export performance; and (3) to calculate price and income elasticities by estimating an export demand model for Turkish organic products in the European Union market. The analysis undertaken in this paper will also enable us to derive some policy recommendations.

Studies on the economics of organic agriculture in Turkey go back to the early 1990s. The studies cover topics such as differences in production costs (Akgüngör, 1996) and domestic market demand (Akgüngör et al., 1999). The common finding of these studies is that organic agriculture in Turkey has been growing rapidly, both in terms of production and in export. The studies also reveal a potential for development of a domestic market for organic products in Turkey. Previous studies point out that Turkey's exports of organic products have a comparative advantage in the European Union market, both with respect to product variety and product quality. However, a sound export policy should take into consideration how price and income changes affect export competitiveness. Existing research emphasises the growing potential of Turkish organic exports yet does not look into estimating the parameters of demand for organic food import of Turkish origin to the EU countries.

Another missing component of the existing research is that none have analysed the level of competitiveness of Turkey's exports against its major competitors and whether and how competitiveness has changed over time. Existing studies also neglect on the analysis of the major determinants of export.

2. Data

Export data is compiled from the organic product export figures collected by the Aegean Region Exporters' Union and published by Turkey's Ministry of Agricultural and Rural Affairs. This is the only available data on exports of organic goods. The data covers the organic products' country of destination, quantity and value and is broken down into product items. Systematic data collection on organic exports in Turkey started in 1998. The dataset on organic food exports used in this study covers the period 1998-2005.

In order to complete the first and second objectives of the study, comparative data on competitor's organic product exports is needed, yet data on world organic food exports as well as organic exports of individual countries (Turkey's competitors of organic exports in the European Union market) do not exist. We, therefore, have used Eurostat external trade dataset without distinguishing across organic and conventional food exports.

The estimation of the export demand model for organic exports from Turkey to the European Union utilises the data on the organic product exports of Turkey. The income and population series of the EU countries were obtained from International Financial Statistics. Consumer price index series which convert dollar values into constant prices were obtained from the OECD online database.

Export competitiveness indices used in this study are calculated for the 1999-2005 period using the European Commission Intra and Extra Trade (COMEXT) database. The indices are based on the incoming and outgoing export flows of Turkey and selected competitor countries in four target countries. Since the Eurostat database does not include organic trade flows, and no other dataset is complete and sufficient enough to do the calculations of the indices, we use conventional products flows as a proxy for organic external trade flows.

3. Methodology

The competitiveness of Turkish Organic Products will be measured with two approaches. The first relies on the calculation of competitiveness indices whereas the second is an econometric investigation. The indices give a rather more general picture on exports competitiveness whilst the econometric estimation provides a more specific picture of the competitiveness. In order to reach an in-depth analysis, we employ both approaches. This section briefly explains these approaches.

Admittedly, the calculation of the competitiveness of all organic product export of Turkey versus all the countries in the world is cumbersome. Therefore, in this study we will focus on selected organic products. The organic products chosen were selected on the basis of the export shares that are higher than 10% of the total organic exports in Turkey. Those products are *raisins, dried figs, dried apricots* and *hazelnuts*. These four products made up 77.7% and 59.7% of total organic product exports in 1999 and 2004, respectively. In order to decide on which target markets, we selected four EU countries with the highest shares in total organic exports to the EU. These countries are selected Germany, the Netherlands, France and the United Kingdom, with approximately 60-70% of total organic exports.

In order to determine the competitors of Turkey in the EU market, we selected those whose export shares are larger than 20% for the identified products in selected target markets during the period 1998-2005. Table 1 summarises the distribution of the value of the identified products imported by the selected countries. As the import value figures in Table 1 shows, the vast majority of exports are made from the extra EU countries.

Turkey has the highest share of imports of the selected four products in selected target markets. However, small as they are, the EU member countries whose export shares in intra-EU trade are higher than 20% are presented in Table2.

Table 1. Distribution of EU imports of selected organic products across intra and extra EU countries.

	1999		2005	
	Intra EU	Extra EU	Intra EU	Extra EU
Hazelnuts				
France	19,421	40,167	27,273	20,261
The Netherlands	3,859	8,183	5,423	15,713
Germany	43,708	210,123	90,769	160,797
United Kingdom	446	9,126	4,343	11,079
Dried Fig				
France	1,157	12,855	1,151	16,506
The Netherlands	139	3,513	338	3,998
Germany	3,409	14,318	3,069	14,687
United Kingdom	1,257	3,769	3,358	2,889
Raisin				
France	8,230	18,907	5,619	23,395
The Netherlands	11,364	38,653	6,651	36,466
Germany	16,624	56,403	11,285	59,833
United Kingdom	33,747	111,164	23,309	105,391
Dried Apricot				
France	755	14,163	650	18,447
The Netherlands	1,288	5,431	334	54,457
Germany	1,524	12,521	3,513	15,235
United Kingdom	3,527	16,773	7,471	18,231

Table 2. Selection of competitor countries.

Target country	Hazelnut		Dried fig		Raisin		Dried apricot	
	Competitor countries (intra EU)[1]	Competitor countries (extra EU)[2]	Competitor countries (intra EU)[1]	Competitor countries (extra EU)[2]	Competitor countries (intra EU)[1]	Competitor countries (extra EU)[2]	Competitor countries (intra EU)[1]	Competitor countries (extra EU)[2]
France	Italy (64%) Spain (20%)	Turkey (98%)	The Netherlands (29%) Italy (22%) Germany (35%)	Turkey (99%)	Greece (38%) Belgium (33%)	Turkey (59%) South Africa	The Netherlands (30%) Germany (42%) Belgium (43%)	Turkey (99%)
The Netherlands	Germany (72%) United Kingdom (26%)	Turkey (88%)	Germany (87%) Italy (26%)	Turkey (99%)	Greece (75%)	Turkey (68%) USA	France (37%) Germany (35%) United Kingdom (49%)	Turkey (90%)
Germany	Italy (76%)	Turkey (81%)	France (22%) The Netherlands (53%) Austria (41%)	Turkey (99%)	The Netherlands (26%) Greece (52%) Belgium (33%)	Turkey (67%)	France (64%) The Netherlands (45%)	Turkey (95%)
United Kingdom	Italy (76%)	Turkey (99%)	France (63%) The Netherlands (28%)	Turkey (96%)	Greece (87%)	Turkey (61%) ABD	France (89%)	Turkey (97%)

[1] Figures in parentheses represent export share of intra EU countries within the total intra EU exports.
[2] Figures in parentheses represent export share of extra EU countries within the total extra EU exports.

In exports of hazelnuts, dried fig, dried apricot, and raisin, Turkey's competitor EU member countries are Germany, Belgium, France, the Netherlands, Spain, Italy, and Greece. The competitiveness indices are obtained for these competitor EU member countries.

3.1 Measuring competitiveness of Turkish organic product: an index approach

We calculated two indices to capture the competitiveness of Turkish organic products in the EU market. These are the *Revealed Comparative Advantage Index (RCA)* and *Constant Market Share Analysis (CMS)*.

Revealed comparative advantage (RCA) index

The revealed comparative advantage index explores whether a country has a comparative advantage over the competitor countries. The formula of the RCA index is presented below:

$$RCA_{ijt} = \ln\{(X_i^B) \mid (X^B) / (X_{ij}^A \mid X_j^A)\} \quad i=1,\dots,4; j=1,\dots,4; t=1,\dots,T \tag{1}$$

Where:
X_i^B = Turkish exports of product i to the EU market;
X^B = total organic exports of Turkey to the EU market;
X_{ij}^A = the exports of product i to the EU market by competitor country;
X^A = total organic exports to the EU market by competitor country;
The positive index values indicate a comparative advantage.

Constant market share analysis

The constant market share analysis developed by Tysznski (1951) and further developed by Leamer and Stern (1970), and Richardson (1971) explores whether the growth in exports is due to export performance or the country's export competitiveness. The CMS equation is specified below:

$$\Delta q = \sum_i \sum_j s_{ij}^0 \Delta Q_{ij} + \sum_i \sum_j Q_{ij}^0 \Delta s_{ij} + \sum_i \sum_j \Delta s_{ij} \Delta Q_{ij} \quad i=1,\dots,4; j=1,\dots,4 \tag{2}$$

Where:
q = target country's organic exports (value);
s_{ij} = Turkey's export market share of product i in country j;
Q_{ij} = total imports of country j;
the subscript 0 on the variables indicates base year value.

The *CMS* analysis identifies three contributing factors which explain why exports by a country grow more quickly than world exports: the first is related to the growth of the export market relative to the world export growth (the structural effect). The second is improvements in competitiveness of the exporting country (the competitiveness effect). The third is the combined effect of the competitiveness and structural effects. The first term of the above equation represents 'the structural effect', the second term represents 'the competitiveness effect', and the third term represents 'secondary effect' (or combination of both effects) (Chen and Duan, 2001).

3.2 Import demand function

We estimated a simple econometric demand model for Turkey's organic exports to the European Union market in order to find out the price and income elasticities of the Turkish organic product exports. An ordinary demand function that economic theory suggests is specified as:

$$X^d = f\ (P, Y^{for}) \tag{3}$$

Where, X^d is export quantity, P is the price of the goods and Y^{for} is the income level of the importing country. Obviously, an econometric estimation of Equation 3 would yield price elasticity and income elasticity of import demand. However, this simple form of the function can be criticised due to lack of cross price effects. In addition, there may be other factors exerting their influence on the import demand. In order to obtain a more accurate estimation of the function, one should extend the above definition. So we consider the following model:

$$X^d_{ijt} = \alpha^d_i + \gamma_1 P_{ijt} + \gamma_2 Y^{AB}_t + \gamma_3 P^w_{it} + \gamma_4 P^{dom}_{it} + \gamma_5 CPI_{jt} + \gamma_6 E_t + \gamma_7 Xinc_t + t + \varepsilon_t \tag{4}$$

Where:
i stands for products and j for selected countries; $i=1,...,4$; $j=1,...,4$; $t=1998,..., 2005$;
X^d_{ijt}= per capita export quantity of product i to country j;
Y^{AB}_t = per capita real income (in ppp) in the selected EU countries;
P_{ij}= real export price of product i for country j;
P^w_{it} = real world price of product I;
P^{dom}_{it} = real domestic market price of product I;
E_t = real exchange rate;
$Xinc_t$ = export incentives.

The estimation of the cross price elasticities is a difficult task since it requires a detailed data collection process for the substitute goods. The organic export goods from Turkey can be substituted by either importing from another competitor, or through domestic purchases. We have attempted to estimate cross price elasticity by utilising real world prices for product i for the first alternative. Since no detailed data of the prices of domestic substitutes is available, we attempt to approximate cross price elasticity of domestic substitutes as all other goods in the importing country that is, by employing the consumer price index inspired by Thursby and Thursby (1984). Obviously, we expect a positive sign for the cross price elasticity for the estimation of the relevant coefficients.

The export of organic products can also be affected by other factors. In order to control such variation, we have also included remaining variables in the Equation 4. For example, if organic food producers feel secure in the domestic market, they will make little effort in exporting activity. Thus, the prices in the domestic market should also be considered in modelling. We expect a negative sign for the coefficient of domestic prices of organic products.

The export prices of organic products are also sensitive to the changes in the exchange rates. So, we have also included the exchange rates in the model with expectation of a positive estimated sign due to the fact that increases in the exchange rates will lower the export price for foreign purchases. Finally, incentives provided to domestic producers may also affect the organic product export. So, we also included a time variable to control any trend in the independent variable.

We employed the Generalised Method of Moments (GMM) estimation procedure, as developed by Arellano and Bond (1991), in order to eliminate any probable endogeneity problem of explanatory variables such as price and income. In order to avoid an arbitrary choice of the model specification, we estimated variants of Equation 3. The simplest form of the restricted equation consists of price and income variables with a time trend. Other variables were included one by one, and the results are compared.

3.3 LA/AIDS model

The Linear Approximation (LA) / Almost Ideal Demand System (AIDS) model is a convenient tool to estimate own price, cross price and income elasticity coefficients of organic products (Deaton and Muellbauer, 1980). The model consists of the consumers' budget shares of the products, product prices and consumers' disposable income:

$$w_t = \alpha_i + \sum_j \gamma_{ij} \log p_j + \beta_i \log(x/P) \qquad i = 1,\ldots,5; j = 1,\ldots,4 \tag{5}$$

In the above model, i stands for products whereas j stands for countries. The demand system under the LA/AIDS model covers five products. In addition to previous organic products, all other products are also included as the fifth product. The assumption is that the consumer expenditure for organic products in selected countries can be modelled independently from all other products and expenditures. In the Equation 5, w_t is the budget share of the ith good, x is the total consumption expenditure, p_j is the price of the jth good and P is the Stone's approximation of the price index specified below:

$$\log P = \alpha_0 + \sum_i \alpha_i \log P_i + \tfrac{1}{2} \sum_j \sum_k \gamma_{ij} \log p_i \log p_j \qquad i = 1,\ldots,5; j = 1,\ldots,4 \tag{6}$$

The model has the following restrictions:

$$\sum_{i=1}^{n} \alpha_i = 1 \tag{7}$$

$$\sum_{i=1}^{n} \gamma_{ij} = 0 \tag{8}$$

$$\sum_{i=1}^{n} \beta_i = 0 \tag{9}$$

$$\sum_j \gamma_{ij} = 0 \tag{10}$$

$$\gamma_{ij} = \gamma_{ji} \tag{12}$$

Under the above restrictions, the AIDS equation is a demand system. The price and income elasticity estimates are defined as below, following Green and Alston (1990, 1991):

The expenditure elasticity is:

$$\varepsilon_i = \left(1 + \frac{\beta_i}{w_i}\right) \tag{12}$$

and the *price elasticity* is:

$$\varepsilon_{ij} = \delta_{ij} + \frac{\gamma_{ji}}{w_i} + \frac{\beta_i w_j}{w_j} \tag{13}$$

Using the LA/AIDS model, we get own price, cross price and income elasticity estimates for Turkey's exports of organic products to the four importing EU countries, under the assumption that consumer expenditure for organic products can be modelled independent of consumers' other expenditures. We employ an OLS method for the estimation procedure for Equation 5.

4. Empirical results

This section is devoted to the discussion of the evidence obtained from the above analyses.

Revealed comparative advantage (RCA) index

Tables 3 and 4 summarise the *RCA* index for selected competitor countries and products. The tables present the variants of the indices. The first variant of the index RCA_1 is the ratio of Turkish exports of a product to the sum of the four selected products to the EU market. The second index number, RCA_2, represents Turkish competitiveness for selected products, when the total exports in the denominator are replaced by the total exports of fruit and vegetable products (not only the selected four products) as defined in the Comext database (Comext Code 08). So, as mentioned above, any calculated positive value for the *RCA* index indicates a comparative advantage for Turkey over the competitor country for which exports figures are used in any product, and vice versa.

Table 3 reveals that regarding RCA_1 the competitiveness of Turkey in organic products is generally lower than her competitors. We calculated a positive value for RCA_1 only for apricots which indicates a comparative advantage over the Netherlands and Belgium, and for hazelnuts over Germany and the United Kingdom. This suggests that the competitiveness of these products is higher compared to those of these countries. However, negative values in the table show that the competitiveness of the countries in these products for which the value is calculated is higher than Turkey.

Regarding RCA_2, Turkey has a noticeable competitive advantage over her competitors. For example, we found a lower competitiveness in terms of RCA_1 in *raisins* export for Turkey versus Greece, the Netherlands and Belgium. However, in terms of RCA_2, Turkey has a clear competitive advantage over these countries. For *dried apricots*, similar results were obtained. In terms of RCA_2, Turkey's competitiveness can be clearly identified. A comparative advantage can also be identified for the *dried figs* export versus the Netherlands, Italy, Germany, France and Austria and for the hazelnuts exports over Italy, Spain, Germany and the United Kingdom.

The different results obtained from RCA_1 and RCA_2 are interesting. However, employing RCA_2 can be misleading. Because competitor countries export many different products to the EU market whereas the majority of fruit and vegetable exports of Turkey are composed of raisins, dried apricots, dried figs and hazelnuts. The unexpectedly high calculated values for RCA_2 are probably due to the specialisation of Turkey's exports on the selected four products. Therefore, the first variant of the index, RCA_1 is of greater value for the evaluation of competitiveness of Turkey which suggests a lower competitiveness for most of the products versus competitor countries in the EU market.

Constant market share analysis (CMS)

Table 5 and 6 summarise the results of the *CMS* analysis which suggests a positive structural effect. However, one can hardly mention the competitiveness of Turkey given the evidence of the competitive effect which is estimated as negative. Thus, the export growth of Turkey in the selected four products is due to the growth of the EU 25 market. In brief, Turkey has no competitiveness in the selected four products in terms of the *CMS* analysis.

Table 3. Export competitiveness of Turkey in raisin versus competitor countries (RCA index) 1998-2005, on average.

	The Netherlands	Germany	Belgium	France	Greece	Italy	UK	Austria	Spain
RCA_1									
Raisin	-0.5999		-0.9589	-1.4763	-1.2397				
Dried apricot	0.0023	-0.0153	0.8199				-0.907		
Dried fig	-0.2553	-0.185		-0.6938		2.4516		-0.341	
Hazelnut		0.0337				-0.5324	1.3403		-0.3976
RCA_2									
Raisin	3.402		3.4483		0.8329				
Dried apricot	4.004	2.8289	5.2267	2.2263			2.0617		
Dried fig	3.7461	2.6593		3.0091		5.5143		2.8177	
Hazelnut		2.8779				2.5299	4.309		4.663

Note: RCA_1 index measures export share of the product in the sum of selected four organic products whereas RCA_2 index measures the export share of the product under consideration in the all fruit and vegetable sector.

Table 4. Comparative export performance of Turkey in raisins versus competitor countries (CEP index), 1998-2005, on average.

	The Netherlands	Germany	Belgium	France	UK	Austria	Italy	Spain	Greece	Turkey
CEP_1										
Raisins	137.2714	119.5429	194.2571	521.6629					255.3429	185.4857
Dried apricot	117.0257	109.6714	56.14571		298.8357		8.495714			118.5571
Dried fig	114.8757			177.2671		171.1757	193.6986	169.3057		88.95571
Hazelnut		111.3757			34.12286					114.7
CEP_2										
Raisins	92.42857	239.3686	88.5						1195.086	
Dried apricot	74.76	219.7171	25.17429	435.6714	535.7229	205.3857	13.90143			
Dried fig	73.91429			153.5629			321.2486	39.48857		
Hazelnut		228.5571			61.55571					

Note: CEP_1 index measures export share of the product in the sum of selected four organic products whereas CEP_2 index measures the export share of the product under consideration in the all fruit and vegetable sector.

Table 5. Turkey's export share of four products in the EU 25 export market (in million $).

Product	Year	Exports of four products by Turkey to the EU 25	Total imports by the four products EU 25	Market share of Turkey in the EU 25	Change in market share of Turkey	Change in the value of imports by EU 25 from Turkey
Raisin	1999	164,096.27	384,245.79	0.43	0.03	1,388.66
	2005	176,014.88	385,634.45	0.46		
Dried apricot	1999	57,863.37	71,644.05	0.81	-0.04	28,209.42
	2005	77,096.07	99,853.47	0.77		
Dried fig	1999	53,873.66	66,875.91	0.81	-0.06	17,693.48
	2005	63,283.43	84,569.39	0.75		
Hazelnuts	1999	375,602.81	493,820.24	0.76	-0.15	309,800.34
	2005	494,011.19	803,620.58	0.61		

Table 6. Sources of export competitiveness of Turkey in the EU 25 market (CMS analysis).

	Value	Share (%)
Change in total exports (total of four products)	158,969.46	100.00
Structural effect	273,265.96	171.90
Competitive effect	-67,128.8	-42.23
Secondary effect	-47,167.7	-29.67

4.1 Econometric analyses

This section discusses the estimation results obtained from import demand function for the selected organic products and the LA/AIDS model.

Import demand functions

We estimated the Equation 4, but were unable to obtain significant coefficient estimates for any of the explanatory variables. Nor, did the z tests suggest an acceptance of the unrestricted model. Thus, we do not include the estimation results of the unrestricted model but rather, included other control variables one by one.

Table 7 displays our estimation results for the restricted and unrestricted forms of Equation 4 by utilising the GMM procedure to control endogeneity. Sargan's χ^2 tests the validity of instrumental variables. The rejection of null indicates the validity of the instruments. z statistics test the existence of the AR (1) and AR (2) process in the residual. AR (1) is acceptable whereas AR (2) is not.

Table 7. Estimation results of import demand function.

Variables	Model 1	Model 2	Model 3	Model 4	Model 5	Model 6
X_{it-1}	0.92 (0.12)[1]	-1.03 (0.36)[1]	0.74 (0.12)[1]	-0.48 (0.33)	0.81 (0.00)[1]	0.85 (0.13)[1]
P_{it}	-0.003 (0.04)	0.26 (0.10)[1]	0.06 (0.05)	0.14 (0.11)	0.05 (0.05)	0.035 (0.05)
Y_t^{EU}	0.11 (0.4)[1]	0.00 (0.04)	0.10 (0.03)[1]	0.02 (0.04)	0.10 (0.00)[1]	0.11 (0.04)[1]
P_{it}^w	-	-0.03 (0.01)[1]	-	-	-	-
CPI_t^{EU}	-	-	0.001 (0.00)[1]	-	-	-
P_{it}^{dom}	-	-	-	-0.01 (0.001)[5]	-	-
E_t	-	-	-	-	0.004 (0.00)[10]	-
$Xinc_t$	-	-	-	-	-	-1.9e-6 (0.19)[10]
T	-0.18 (0.10)[10]	0.00 (0.00)	-0.001 (0.00)	0.00 (0.00)	0.00 (0.00)	-0.00 (0.00)
Obs	96	80	96	80	96	96
Sargan χ^2	76.00 (0.00)	55.84 (0.00)	78.27 (0.00)	69.18 (0.00)	78.65 (0.00)	72.50 (0.00)
Wald	76.24 (4)	26.19 (5)	114.30 (5)	12.61 (5)	96.06 (5)	88.86 (0.00)
z_1[AR(1)]	-5.84 (0.00)	0.54 (0.60)	-4.63 (0.00)	0.20 (0.84)	-5.40 (0.00)	-5.63 (0.00)
z_2[AR(1)]	2.08 (0.04)	-1.74 (0.08)	1.50 (0.14)	-2.13 (0.03)	2.08 (0.03)	2.84 (0.01)

Standard errors in the brackets for independent coefficient estimates; degree of freedom chi-square distribution for Wald test p-value for the Sargan and z tests. Superscript of the brackets indicates the percentage significance level. Sargan χ^2 tests the validity of instrumental variables. The rejection of null indicates the validity of instruments. z statistics is for testing the existence of AR(1) and AR(2) process in the residual. AR(1) is acceptable whereas AR(2) is not.

The coefficient of the lag dependent is significant for each model reported in the table. This supports the suggestion that once an export relationship was established it facilitates exporting activity in the following year. However, the time trend provides controversial results.

The first model gives an ordinary demand function estimate. However, the estimation did not yield a significant price elasticity of import demand, though the sign of the coefficient is negative. The income elasticity, on the other hand, confirms our expectations. It was estimated as 0.11 at the 1% significance level. In this model, the existence of the AR (2) process was not rejected since the null of no autocorrelation is accepted at the 5% significance level.

The other columns of Table 7, exhibits the extension of the ordinary demand function estimates. As mentioned above, in order to approximate cross price elasticity, we used the world price of the selected organic products. Contrary to our expectation; we obtained a negative and significant coefficient for this variable. The inclusion of this variable distorted our estimation of the function, and the income elasticity estimate became insignificant. Again, the existence of the AR (2) process was not rejected. The other approach to the cross price elasticity is to estimate the model by using domestic consumer price index in the importing countries. We obtained a significant positive estimate for this variable at the 1% level confirming our expectations although the magnitude of the coefficient is very limited, that is 0.001. Sargan and z tests also support the accuracy of this model. This might, of course, arise out of the fact that this variable includes the prices of all other variables and is not an exact measure for cross price elasticity. However, this result can be interpreted in the following manner; when the prices of other goods in the domestic market increases, the export of organic products also increases, albeit slightly.

We also estimated the effects on exports of prices of organic products in the domestic market. The estimated coefficient of this variable, -0.01, significant at the 5% level, confirmed our expectations. The effect of exchange rate movements on import demand for organic products was also estimated as 0.004 at the 10 % significance level. This estimation supports the idea that increases in exchange rates enable exporting activity. Finally, we estimated the effects of incentives provided to domestic producers. Interestingly, though the magnitude of the coefficient is negligible, we found a negative coefficient which is significant at the 10% level. However, the existence of the AR (2) process was not rejected for the last three models and these models are unacceptable.

The third model appears to be reasonable given the results of the other models and test statistics. Model 3 suggests that the income level of the selected importing country has a significant effect on the export of the selected organic products but own price has no effect. On the other hand, because the positive cross price elasticity is estimated, the prices of all other products in the domestic market of the importing countries have a positive effect on the export of Turkish organic products.

LA/AIDS model estimates

In this section, we will estimate price, income, and cross price elasticities with an alternative approach for each selected organic product. The LA/AIDS model needs budget shares of the products in the importing countries. The budget shares of the products in selected countries are presented in Table 8.

In estimating the elasticity numbers, we use the coefficients estimated by the LA/AIDS model as explained above. The elasticity estimates are summarised in Table 9.

The main diagonal of Table 9 indicates own price elasticity of the products while the off- diagonal elements are cross price elasticity estimates. The last column can be interpreted as the income elasticity estimate for the selected products. As expected, all coefficients of the price expenditure estimates are lower than 1 and, except for the price elasticity of raisins, all have negative values that are consistent with economic theory. The findings regarding cross price elasticity estimates suggest that the coefficients are negative and that the products are not considered to be substitutes.

As expected by the economic theory per capita expenditure elasticity of the products is positive. This indicates that increases in real per capita income will cause higher demand for organic products. Here, the income elasticity of dried apricots is 1.24 (elastic); income elasticity of raisins is 1.03 (unitary elastic); income elasticity of dried figs and hazelnuts are 0.51 and 0.33 (inelastic). Consequently, a relative increase in the real per capita income of Germany, France, the Netherlands, and the United Kingdom will cause a larger increase in per capita consumption of dried apricots and a smaller increase in per capita consumption of dried figs and hazelnuts.

Cross price elasticity estimates are mostly found to be negative between other ecological products and the selected four products. In other words, other ecological products are not substitutes for these four products. The negative elasticities also emphasise the mentioned specialisation in the organic products.

Table 8. Budget shares of the target countries (%).

	Raisins	Dried figs	Dried apricots	Hazelnuts	Other organic products	Total
Germany						
1998	14.8	16.0	15.3	30.7	23.2	100
1999	13.6	19.3	11.6	23.0	32.5	100
2000	17.7	10.7	15.6	25.1	30.8	100
2001	18.3	9.9	15.3	23.4	33.1	100
2002	18.8	12.5	19.0	12.4	37.3	100
2003	24.4	11.7	17.0	16.4	30.5	100
2004	17.9	16.6	14.0	22.4	29.2	100
France						
1998	20.40	13.00	25.80	9.90	30.90	100
1999	12.80	7.90	14.80	12.90	51.60	100
2000	17.30	13.70	22.80	21.50	24.60	100
2001	21.40	11.00	26.00	22.40	19.30	100
2002	16.50	19.70	36.10	14.20	13.40	100
2003	23.50	15.40	31.50	14.10	15.60	100
2004	19.60	28.60	20.70	16.40	14.80	100
The Netherlands						
1998	36.60	9.60	3.70	21.20	28.90	100
1999	26.40	5.30	3.10	12.70	52.40	100
2000	28.40	3.40	7.00	12.60	48.60	100
2001	26.60	2.80	7.90	22.00	40.70	100
2002	28.20	1.20	4.50	18.50	47.60	100
2003	18.50	4.30	0.20	10.90	66.00	100
2004	14.20	5.60	5.10	13.80	61.40	100
United Kingdom						
1998	23.40	19.10	5.30	11.20	41.10	100
1999	28.00	26.80	5.60	4.70	34.90	100
2000	40.20	17.90	6.10	4.70	31.00	100
2001	31.70	20.30	1.70	3.60	42.70	100
2002	33.00	22.70	0.60	4.80	31.00	100
2003	24.10	35.60	7.40	7.90	25.00	100
2004	27.00	29.80	6.90	5.70	30.70	100

Table 9. Price and expenditure elasticity estimates.

Product	Raisins	Dried apricots	Dried figs	Hazelnuts	Other organic products	Expenditure elasticities
Raisins	0,322376	-0,565286	-0,03903	-0,71081	0,198181	1,030152
Dried apricots	-0,18686	-0,785876	0,492266	0,49808	-0,20693	1,244081
Dried figs	-0,64962	0,5945413	-0,61122	-0,29826	-0,46101	0,513532
Hazelnuts	0,917386	-0,426732	0,108115	-0,88962	-0,61329	0,326822
Other ecological products	-1,1529	0,2344764	-0,22388	0,41458	-0,70931	1,292636

Diagonal elements of the above table show price elasticities, whereas off-diagonal elements indicate cross price elasticities.

5. Final remarks

The findings indicate that Turkey's exports of organic products to the EU market are growing, although it has been limited to traditional export items (dried products) in recent years. Our analysis in this paper, in particular, the *Revealed Comparative Advantage*, seems to suggest a modest competitiveness for Turkey in the selected four organic products which constitute the highest exports share. Turkey is superior to her many competitors in some products although a small number of countries have a better performance. However, the *Constant Market Share* analysis reveals that this relatively modest advantageous position of Turkey over her competitors in selected products in the EU market is mainly due to the growth of market share in the EU. In other words, organic consumption is becoming more popular and this benefits organic products producers. In addition to market expansion, we also consider that specialisation in a few products contributes to competitiveness. Specialisation in these four products may have led to the accumulation of knowledge and expertise which facilitates exporting activities. The growth in organic product exports is also due to the advantages of lower prices. Were prices of Turkish organic products to become higher than competitor countries, it is quite likely that Turkey would lose its market share. Therefore, competitiveness of Turkey is not strong enough to be sustained, for example, in a contraction of the EU market. This situation can be attributed to the lack of added value in the export items.

This conclusion is also supported to a large extent by our econometric investigation which seeks to find the sources of competitiveness of Turkey in the EU market. Demand conditions for the organic products of Turkey seem to be beyond the control of Turkish producers and policymakers. For example, we were unable to obtain a significant estimate of the price effect, but income elasticity was estimated as positive and significant. Similarly, the prices of all other products in the EU market have a small positive effect on the organic export of Turkey. However, incentives for organic products, for instance, have no statistically significant effect on exports to the EU market. Even though we did not obtain a significant result for price elasticity in our first econometric analysis, we estimated a positive and significant coefficient for the exchange rates movement implying that export competitiveness is cost oriented in the EU markets.

Our second econometric approach measured price and income and cross price elasticities for each product. We estimated price elasticities as significant and negative for each product except raisins. The price elasticities are close to unity but still inelastic. This, in fact, provides us with a more optimistic view since should the demand for organic products be elastic, then we could infer that export competitiveness of those product is mainly driven by price. However, more inelastic organic products would be more competitive and not only relies on cost considerations but also quality. Except for hazelnuts and dried figs, the expenditure elasticity of selected products (raisins and dried apricots) are higher than unity thus suggesting that as the income level of countries increases, demand for these organic products also increase. Cross price elasticities between other ecological products and four organic products under study are estimated as negative, implying that other products are not substitutes for the four organic products. This emphasises the specialisation in these four products.

Although the paper tends to offer the evidence of cost competitiveness, it is unable to compare the conditions of Turkey and the competitor EU countries for sustained competitiveness. In order to have a more stable and growing export demand in the EU market, competitiveness must shift from comparative advantage (low cost labour or natural resources) to a competitive advantage due to more productive and distinctive products and processes. Sustainability in competitiveness depends on the political, legal and macroeconomic foundations as well as microeconomic conditions, such as company operating practices and strategies, quality of inputs, infrastructure and institutions. The focus of further research should consider exploring the conditions for sustainable competitiveness with respect to microeconomic foundations of productivity and investigate the sophistication of the environment within which firms in respective countries compete.

Acknowledgements

Financial support of the project by TÜBİTAK under Cost Program (Cost 924) is gratefully acknowledged. The authors are thankful to TÜBİTAK Research Group on Social Science and Humanities for providing the research grant.

References

Akgüngör, S., 1996. Türkiye'de Ekolojik Yöntemlerle Üretilen Çekirdeksiz Kuru Üzümün Verimi, Maliyetleri ve Pazarlanması: Salihli ve Kemalpaşa Örneği. Izmir, Can Matbaası.

Akgüngör, S., B. Miran, C. Akbay, E. Olhan and N.K. Nergis, 1999. İstanbul, Ankara ve İzmir İllerinde Tüketicilerin Çevre Dostu Tarım Ürünlerine Yönelik Potansiyel Talebi, Tarımsal Ekonomi Araştırma Enstitüsü Yayın no: 99-2, Ankara, 1999.

Arellano, M. and S. Bond, 1991. Some test of specification for panel data: Monte Carlo evidence and an application to employment equations. Review of Economic Studies, 58: 277-297.

Chen, K and Y. Duan, 2001. Competitiveness of Canadian agro-food exports against competitors in Asia: 1980-1997. Rural economy project report 01-01. AARI Project.

Deaton, A. and J. Muellbauer, 1980. An almost ideal demand system. American Economic Review, 70: 312-326.

Green, R. and J.M. Alston, 1990. Elasticities in AIDS models. American Journal of Agricultural Economics, 71: 442-445

Green, R. and J.M. Alston, 1991. Elasticities in AIDS models: a clarification and extension. American Journal of Agricultural Economics 73: 874-875.

Leamer, E.E. and R.M. Stern, 1970. Quantitative international economics. Boston: Allyn and Bacon, Inc.

Richardson, J.D., 1971. Some sensitivity tests for a constant market share analysis of export growth. The Review of Economics and Statistics, 53: 300-304.

Thursby, J. and M. Thursby, 1984. How reliable are simple, single equation specifications of import demand? The Review of Economics and Statistics, 66: 120-128.

Tyznski, H., 1951. World trade in manufactured commodities: 1899-1950. Manchester School of Economic and Social Studies, 19: 272-304.

Appendix. Export competitiveness of Turkey for certain commodities.

Table A. Export competitiveness of Turkey in raisins versus competitor countries.

Years	Greece	The Netherlands	Belgium
RCA_1*			
1999	-1.280	-0.841	-1.082
2000	-1.102	-0.731	-1.011
2001	-1.383	-0.725	-1.172
2002	-1.280	-0.518	-0.884
2003	-1.110	-0.481	-0.735
2004	-1.129	-0.555	-0.691
2005	-1.394	-0.348	-1.137
RCA_2**			
1999	0.513	3.272	3.557
2000	0.886	3.376	3.293
2001	0.839	3.230	3.366
2002	0.941	3.265	3.476
2003	0.936	3.443	3.447
2004	0.928	3.540	3.518
2005	0.787	3.688	3.481

Table B. Export competitiveness of Turkey in apricots versus competitor countries.

Years	Competitor countries				
	The Netherlands	Germany	Belgium	France	United Kingdom
RCA_1*					
1999	0.060	0.296	0.999	-1.669	-0.672
2000	-0.058	0.056	0.981	-1.762	-0.947
2001	-0.123	-0.345	1.109	-1.890	-1.505
2002	0.009	-0.148	1.019	-1.348	-1.214
2003	0.057	0.391	0.272	-1.115	-0.956
2004	0.222	0.002	0.571	-1.139	-0.530
2005	-0.151	-0.359	0.788	-1.411	-0.526
RCA_2**					
1999	4.173	2.803	5.638	2.374	2.104
2000	4.048	2.769	5.285	2.172	2.213
2001	3.831	2.769	5.646	2.002	1.482
2002	3.791	3.004	5.378	2.504	1.939
2003	3.981	3.213	4.454	2.409	1.915
2004	4.317	2.752	4.780	2.344	2.342
2005	3.886	2.492	5.406	1.779	2.437

Table C. Export competitiveness of Turkey in dried figs versus competitor countries.

Years	Competitor countries				
	The Netherlands	Italy	Germany	France	Austria
RCA$_1$*					
1999	-0.472	2.740	0.218	-0.884	-0.835
2000	-0.206	2.504	-0.016	-0.706	1.648
2001	-0.586	1.641	-0.183	-1.062	-0.156
2002	-0.345	3.026	-0.314	-0.644	-0.063
2003	-0.251	3.242	-0.234	-0.292	-0.812
2004	0.182	2.224	-0.359	-0.748	-1.044
2005	-0.109	1.784	-0.407	-0.520	-1.125
RCA$_2$**					
1999	3.641	5.853	2.725	3.159	3.045
2000	3.900	5.935	2.697	3.229	3.424
2001	3.369	5.241	2.932	2.830	2.485
2002	3.437	5.963	2.838	3.208	3.617
2003	3.672	5.952	2.588	3.233	2.698
2004	4.277	4.952	2.391	2.735	2.315
2005	3.927	4.704	2.444	2.670	2.140

Table D. Export competitiveness of Turkey in hazelnuts versus competitor countries.

Years	Competitor countries			
	Italy	Spain	Germany	United Kingdom
RCA$_1$*				
1999	-0.515	-0.411	-0.143	0.578
2000	-0.525	-0.473	-0.076	1.292
2001	-0.359	-0.178	0.173	2.378
2002	-0.570	-0.379	0.174	2.091
2003	-0.672	-0.452	-0.067	0.791
2004	-0.634	-0.483	0.113	1.100
2005	-0.452	-0.407	0.062	1.152
RCA$_2$**				
1999	2.597	4.615	2.363	3.354
2000	2.905	4.267	2.637	4.452
2001	3.241	5.182	3.288	5.365
2002	2.367	5.098	3.327	5.243
2003	2.037	4.961	2.754	3.662
2004	2.094	4.625	2.863	3.972
2005	2.468	3.893	2.913	4.115

Impact of export control policy measures in an attempt to mitigate Argentina's inflation

P. Rossi, M. Kagatsume and M. Prosperi

Abstract

The economic and political climates in Argentina are heavily influenced by the agricultural sector. The degree of government intervention in the sector is high and the use of administrative measures to control the general inflation level has been a recurring feature of the authorities of the country. The effectiveness of Argentina's beef export controls in an attempt to mitigate inflation on retail beef prices is examined by means of a dynamic simultaneous equation model which allows for interaction between production, consumption and exports. The most significant result of this study is that, owing to the structure of the beef sector in Argentina, export restrictions are not always an effective long term policy to control domestic beef price inflation.

Keywords: Argentina, beef, inflation, export controls

1. Introduction

With more than 3 million tons cwe (carcass weight equivalent) produced in 2007, Argentina is the fourth largest beef producer in the world, ranking below the United States, Brazil and China (USDA – FAS; www.fas.usda.gov/psdonline/psdHome.aspx).

The better part of the production is consumed domestically (an average of about 85% in the last decade). Beef is a staple food in the country and the Argentines are the world's largest per capita beef consumers (above 60 kg annually in 2006) (Argentine Beef Promotion Institute; www.ipcva. com.ar). Of all foods, beef is arguably the most sensitive when it comes to its impact on the Consumer Price Index, accounting for 4.5% of its composition and, thus, having a significant influence on the country's general inflation level (National Institute of Statistics and Census of Argentina; www.indec.gov.ar).

In 2004, as a result of an increasing demand caused by an improved domestic purchasing power and growing exports, pressure was put on retail beef prices. Authorities mostly attributed the increase in prices to growing international sales and, since early 2006, the government has had a complete control on what can be exported via the approval of export permits and quantitative export restrictions. Through export controls, the government aims to manage the beef supply in the domestic market in order to maintain retail prices as low as possible.

The use of restrictive trade policies as a means of reducing domestic and/or world prices has been extensively analysed. A considerable amount of research have addressed the situation, particularly during the 70s when commodity trade restrictions were imposed in several countries to hold down price increases caused by the strength of export demand. Good examples are Shei and Thompson (1977), who applied a small quadratic programming model of world wheat trade to simulate the effects of abrupt quantity changes on prices in the world wheat market under different degrees of trade restriction; and Wong (1978), who constructed an econometric model to measure the welfare and transfer effects of export premiums in Thailand's rice sector. Later, Carter *et al.* (1980) distinguished between producers and consumers effects of the formation of export cartels in the wheat market; and Sedjo and Wiseman (1983) analysed the effectiveness of log export controls in lowering domestic lumber prices. More recently, Takacs (1994) developed a simple partial equilibrium model of export controls on raw materials to investigate the impact of the restrictions on domestic prices of Mongolian cashmere and Romanian wood products.

The latest structural changes in the beef worldwide markets and global exports shares (Table 1) have made of Argentina a main player in the international beef markets. Therefore, exploring the implications of the recent restrictive trade policy taken by the Argentinean authorities is relevant not only for domestic consumers and the producing industry but also for Argentina's trade partners and competitor countries.

The purpose of this paper is to analyse the effectiveness of the current export control policy in Argentina as a means of mitigating domestic beef price inflation. In the next section the conceptual framework is presented. Section 3 and 4 discuss the data set used and the econometric model formulated to simulate the effects of the beef export restrictions. Our intention is not to evaluate the ability of the model to simulate conditions during the period under consideration, but rather to predict the consequences of the restrictive trade policy. Regression results and the relation of the variables underpinning the model are summarised in section 5. Section 6 analyses the transmission effect of export quantity on domestic prices. Simulations of different policy scenarios are discussed in section 7. Finally, in section 8, some limitations and concluding comments are offered.

2. The conceptual framework

There is a cyclical behaviour of the beef output explained by the fact that beef cattle are both *capital and consumption goods*. Calves are generally weaned at 6 months of age and then fed until they reach an adequate weight to be sent to the market. While it takes approximately 3 years to breed a heavy weight cow (Argentine Beef Promotion Institute, 2006), the shortest period of time producers have to wait from breeding until an offspring can reach the market is between 18 and 20 months (Figure 1). Because of the length of the biological cycle, production cannot react immediately to current market prices and producers base their output decisions on expected

Table 1. Worldwide beef exports. Source: data from USDA-FAS.

Total exports[1]	2000	2001	2002	2003	2004	2005	2006	2007
Argentina	354	168	345	382	616	754	552	532
Australia	1,316	1,376	1,343	1,241	1,369	1,388	1,430	1,400
Brazil	488	741	872	1,162	1,610	1,845	2,084	2,189
Canada	563	619	657	413	603	596	477	457
EU-27	663	610	580	438	363	253	218	139
India	344	365	411	432	492	617	681	735
New Zealand	473	483	475	548	594	577	530	496
Paraguay	58	62	80	78	115	180	232	197
United States	1,120	1,029	1,110	1,142	209	316	519	649
Uruguay	236	145	225	282	354	417	460	385
Others	305	244	321	361	321	348	304	426
World total	5,920	5,842	6,419	6,479	6,646	7,291	7,487	7,605

[1] Units: thousand tons cwe (carcass weight equivalent).

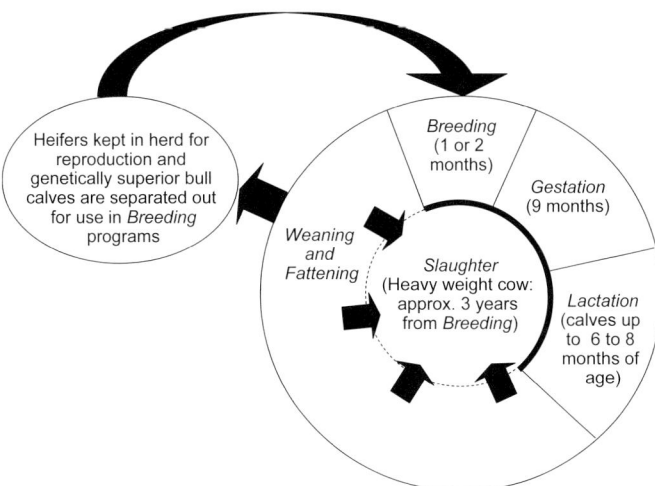

Figure 1. Dynamics of the beef cycle.

prices at the time of sale of the commodity. The general framework used for estimating producers' slaughter decisions was Nerlove's partial adjustment-adaptive expectations framework, which allows the incorporation of price expectations to enter the model.

The impact that exports exert on consumer prices was analysed through the assessment of the impact international sales exert on producer prices and how these in turn affect consumer prices.

Consumers were considered price takers and quantity adjusters. Thus, the domestic demand side was modelled with a regular demand function in which the variables over which the decision maker has no control appear on the right-hand side of the equation.

Imports were not considered in the model since Argentine beef imports have traditionally been negligible (Table 2). The bulk of the domestic consumption (an average of 2,260 thousand tons cwe per year in the last 3 decades) is served exclusively by local production (Argentine Beef Promotion Institute, 2006). There are two main reasons for this. First, cattle ranching and beef consumption play a major part in the traditions of the country and the consumption of domestically produced beef is imprinted in the culture of the Argentines. Second, the duties levied on beef imports from non-members countries of the Mercosur Regional Trade Agreement are high. They range from 10% and 12% for unprocessed beef to 16% in the case of processed beef (Federal Administration of Public Revenue of Argentina; www.afip.gov.ar).

3. The data set

The basic definitions for the variables used in the empirical model are given in Table 3. Monthly data was used. The choice of the time frequency was made in accordance with the research objective. Quarterly or annual data would not have allowed researchers to fully capture the impact of the export controls. Furthermore, richer information about the dynamics of the model can be better obtained from a monthly cycle analysis.

Table 2. Argentina's beef imports.

	2000	2001	2002	2003	2004	2005	2006	2007
Fresh meats								
Uruguay	9.05	7.54	4.22	7.84	2.64	2.52	3.12	2.55
Brazil	0.12	0.66	-	-	-	-	0.02	0.07
Paraguay	0.02	-	-	-	-	-	-	-
Subtotal fresh	9.19	8.20	4.22	7.84	2.64	2.52	3.14	2.62
Processed meats								
Uruguay	1.92	1.16	1.24	0.14	-	0.28	0.16	0.14
Brazil	0.02	-	-	-	-	-	-	-
Subtotal processed	1.94	1.16	1.24	0.14	-	0.28	0.16	0.14
Total beef imports	11.13	9.36	5.46	7.98	2.64	2.80	3.30	2.76

Source: Authors using data from the Ministry of Economy and Production of Argentina.
Units: thousand tons pwe (product weight equivalent).

Table 3. Variables definition.

Endogenous variables	Exogenous variables
S: slaughtered animals, heads.	PPC: producer price relative to the price of corn.
YD: yield, tons per head (cwe[1]).	RF: average rainfall in main productive regions (Pampas and North-East Region), mm.
PP: producer price, pesos/kg.	DSLC: prohibition of slaughtering lightweight cattle, 2005:11 ~ 2007:03=1, and 0 otherwise.
CP: consumer price, pesos/kg.	PFS: proportion of female cattle slaughtered, female cattle slaughtered over total slaughtered heads.
DC: domestic consumption, tons (cwe).	MEEA: monthly estimate of the economic activity of the country.
XQ: export quantity, tons (cwe).	XP: export price, pesos/ton.
Q: production, tons (cwe)	XQUOTA: indicator of the constraining effect of the export restrictions.
	D_i: monthly dummy = 1 if i^{th} month, and 0 otherwise (benchmark: January).

Since stock data are unavailable, data on DC refer to 'apparent consumption' and are derived from DC = Q – XQ.
[1] cwe: carcass weight equivalent.

Due to data limitations, the sample period extends from January 1995 through March 2007. Data were obtained from the Argentine Beef Promotion Institute, the Ministry of Economy and Production of Argentina, and the National Institute of Statistics and Census of Argentina. Data on rainfall was collected from the website Tutiempo.net (www.tutiempo.net). All monetary variables are expressed in Argentine pesos. The Consumer Price Index and the Producer Price Index were used as deflators to account for changes in consumer prices, the former, and producer and exporter prices, the latter. Consumer price refers to the retail price of short ribs (*asado*), one of the most popular cuts of beef consumed in Argentina. Producer price refers to the live price of steers, which is a reference price for all the cattle categories traded in the country.

Regarding the stationarity of the data, the presence of a unit root in the variables was tested by conducting the conventional Augmented Dickey-Fuller (ADF) test. For some of the variables, the null hypothesis of the existence of a unit root was not rejected. We will return to this point in the following section.

4. Model specification

The model was specified according to the conceptual framework. It consists of 6 behavioural equations (S: slaughter, YD: yield, PP: producer price, CP: consumer price, DC: domestic consumption, and XQ: export quantity) and 1 identity (Q: production). All variables except $XQUOTA_t$ are taken in their logarithmic form, so coefficients can be interpreted as elasticities. The specification of the model is as follows:

$$S_t = \alpha_{10} + \beta_{11}S_{t-1} + \beta_{12}S_{t-12} + \beta_{13}PP_{t-1} + \beta_{14}PPC_{t-1} + \mu_{1t} \tag{1}$$

$$YD_t = \alpha_{20} + \beta_{21}YD_{t-1} + \beta_{22}RF_t + \beta_{23}DSLC_t + \sum_{i=2}^{12} \gamma_i D_{it} + \mu_{2t} \ (i = i^{th} \ month) \tag{2}$$

$$PP_t = \alpha_{30} + \beta_{31}PP_{t-1} + \beta_{32}Q_t + \beta_{33}XQ_t + \beta_{34}PFS_{t-24} + \mu_{3t} \tag{3}$$

$$CP_t = \alpha_{40} + \beta_{41}CP_{t-1} + \beta_{42}PP_t + \gamma_{12}D_{12t} + \mu_{4t} \tag{4}$$

$$DC_t = \alpha_{50} + \beta_{51}CP_t + \beta_{52}MEEA_t + \mu_{5t} \tag{5}$$

$$XQ_t = \alpha_{60} + \beta_{61}XQ_{t-1} + \beta_{62}XP_t + \beta_{63}XQUOTA_t + \mu_{6t} \tag{6}$$

$$Q_t = S_t \cdot YD_t \tag{7}$$

The coefficients of the endogenous and predetermined variables are represented by the β s, and the coefficients of the monthly dummies are represented by the γ s. The μ s are stochastic error terms.

Lagged endogenous variables were used for several reasons. First, they were used as a means of capturing the dynamics of the model. Second, using lagged dependent variables allows for the easy calculation of long-run effects. Third, their inclusion allows capturing the trend present in the dependent variables. Fourth, and perhaps most importantly, lagged dependent variables make sense conceptually because of the partial adjustment of production, prices and exports. A lagged dependent variable was not included in the Domestic consumption Equation 5 since it was not found to be significant. This might be explained by the fact that beef is a staple food in Argentina and, therefore, a change in either price or income in period t is fully reflected in consumption in the same period t.

Equation 1 is the Nerlovian supply equation with adaptive expectations. The number of cattle producers are willing to offer for slaughter (S_t) is a function of the price they expect to receive for their live animals (PP_{t-1}). We do not impose any restriction on the coefficient of PP_{t-1} and therefore the sign it may attain in our model can be either positive or negative. In Argentina, beef cattle production is mostly extensive. However, an important change in the sector in the past couple of years has been the utilisation of corn as a feed supplement. To allow for producers' response to alternative feed prices, the expected producer price relative to the price of corn (PPC_{t-1}) was included as one of the explanatory variables. We expect the coefficient of PPC_{t-1} to be negative, suggesting that when the expected relative producer price in terms of a feed input increases (it becomes relatively cheaper to incorporate corn grain into cattle rations) producers keep their animals on farm to make them gain weight. The inclusion of the slaughter variable lagged 12 months (S_{t-12}) allows capturing the observed seasonality in producers' slaughter decisions.

In Equation 2 yield (YD_t) is specified as a function of its value lagged one period (YD_{t-1}), rainfall (RF_t), a policy dummy accounting for the prohibition of slaughtering lightweight cattle ($DSLC_t$), and monthly dummies (D_{it}) for seasonal variations in yields. The regulation which prohibited the slaughter of lightweight cattle was implemented by the Argentine government in November 2005 aiming at increasing supply.

In Equation 3, producer price (PP_t) is a function of its own lagged value (PP_{t-1}), production (Q_t), exported quantity (XQ_t) and the 24-months lagged proportion of female cattle slaughtered (PFS_{t-24}). The inclusion of the variable XQ_t in the formation of producer prices is crucial since it will allow us to assess the impact exports exert on consumer prices via producer prices and, eventually, to evaluate the rationale underlying the use of export controls as a means of lowering retail prices. The observation of the relation between producer prices and the proportion of female cattle slaughtered in the past motivated the inclusion of the variable PFS lagged 24 months (PFS_{t-24}). We expect the coefficient of PFS_{t-24} to be positive, since an increase in the proportion of female cattle slaughtered today will have a negative impact on cattle inventory in the next couple of years, which will in turn lead to higher producer prices.

In Equation 4 consumer price (CP_t) is explained as a function of its own value lagged one period (CP_{t-1}), producer price (PP_t) and the dummy variable D_{12t} since it is customary in most Argentine families to increase their beef consumption during the holiday season (Christmas and New Year). The temporary pressure on prices that the seasonal strong demand may exert motivated the inclusion of D_{12t}.

Equation 5 regards domestic consumption (DC_t) as a function of the current price of a popular cut of beef (CP_t) and a monthly estimate of the economic activity of the country ($MEEA_t$) used as a proxy variable for income. The $MEEA_t$ is a Laspeyres quantity index published by the National Institute of Statistics and Census of Argentina constructed to give an estimate of the monthly economic growth. The price of possible beef substitutes (chicken, pork, fish) were also considered as additional explanatory variables but not found to be significant and therefore dropped from the equation.

Equation 6 specifies export quantity (XQ_t) as a function of its own lagged value (XQ_{t-1}), export price (XP_t) and a policy variable ($XQUOTA_t$). The $XQUOTA_t$ is an indicator of the constraining effect of the export restrictions. Export restrictions were implemented in March 2006. Since then and for certain periods, the government of Argentina authorised exports as a percentage of the volume exported in the corresponding period of the previous year but without setting a monthly quantity. Reflecting this condition, the policy variable $XQUOTA_t$ was used not as a fixed quantity but as a ratio which takes as its base the corresponding period of the preceding year. It is exogenous to the model, since it is imposed by the Argentinean authorities for every period *t*.

Once producers' slaughter decisions are made, beef production (Q_t) is determined by Equation 7.

The equations were estimated jointly. The method of estimation applied was Three-Stage Least Squares (3SLS) to control for the endogenity bias and cross-equation correlation of the residuals.

It has been argued that in regression analysis the presence of a unit root in one or some of the variables will result in spurious regression. To tackle the non-stationarity issue, the variables can be made stationary by differencing. The shortcoming of this procedure is that 'differencing the data removes the information about the long-run relationships among the variables' (Hsiao and Fujiki, 1998: 58). Moreover, if the variables are also cointegrated, an error-correction term must be added in order to appropriately identify long term relationships and causalities (Engle and Granger, 1987).

In a simultaneous equation model the presence or absence of cointegration is pre-assumed from the way the model is specified (Hsiao, 1997a,b). Hsiao and Fujiki (1998) stated that 'a dynamic structure introduces trivial cointegration between the current and lagged variables' (Hsiao and Fujiki, 1998: 77). Therefore, testing for cointegration of the variables using, for instance, a simple Augmented Dickey-Fuller test on the residuals is not relevant.

Moreover, for the case of structural dynamic models of nonstationary and cointegrated variables Hsiao (1997a,b) and Hsiao and Fujiki (1998) have demonstrated that 'conventional structure equation estimators like 2SLS and 3SLS still possess desirable statistical properties' (Hsiao and Fujiki, 1998: 77) under certain conditions.

For these reasons, we applied the 3SLS estimation method to our model without showing the stationarity test on the residuals.

5. Regression results

The 3SLS estimation results are reported in Table 4. The coefficient parameters are provided in each equation. The standard errors and the t-statistics are shown to the right of each corresponding estimate. The results of the $\overline{R}^2$ are provided at the end of each estimated equation. The 3SLS procedure in our software (Eviews 4.0) does not allow for Breusch-Godfrey tests for serial correlation of the residuals. Therefore, for those equations containing lagged endogenous variables, the Durbin-h statistic is reported. The Durbin-Watson statistic is reported for Equation 5.

The coefficients of the lagged dependent variables are statistically significant, positive and less than unity in all cases, suggesting that more than one month is required for the sector to fully adjust to the demand and supply interactions.

We will consider each equation in turn.

In Equation 1 we find that producers respond positively to expectations of higher prices. This could be explained by that while higher expected prices may induce producers to retain female cattle in order to expand the future herd (cattle as *capital goods*), they may also react pushing more animals close to the time of slaughter into the market in order to take advantage of the increased price (cattle as *consumption goods*). The coefficient of the expected producer price relative the price of corn variable (PPC_{t-1}) is negative as anticipated.

In Equation 2 we find that yield (YD_t) is negatively related with rainfall (RF_t). This could be explained by the stress rain causes on animals and how this, in turn, negatively affects cattle performance. The policy dummy for the prohibition of slaughtering lightweight cattle ($DSLC_t$) was found to be somewhat significant and with a low impact on yields, possibly due to the biological constraints to increase the weight of the animals in the short run.

Table 4. 3SLS estimation results.

Explanatory variables	Coefficient	Std. Error	[t-stat]
(1) Dependent variable: LOG(S$_t$)			
CONSTANT	-0.327	0.988	[-0.331]
LOG(S$_{t-1}$)	0.901***	0.076	[11.895]
LOG(S$_{t-12}$)	0.130**	0.060	[2.173]
LOG(PP$_{t-1}$)	0.133**	0.056	[2.372]
LOG(PPC$_{t-1}$)	-0.040*	0.023	[-1.770]
Adjusted R-squared:	0.476		
Durbin-h stat:	-9.298		
(2) Dependent variable: LOG(YD$_t$)			
CONSTANT	-0.261***	0.064	[-4.046]
LOG(YD$_{t-1}$)	0.810***	0.042	[19.246]
LOG(RF$_t$)	-0.004***	0.001	[-2.910]
DSLC$_t$	0.007*	0.004	[1.874]
Adjusted R-squared:	0.803		
Durbin-h stat:	-0.710		
(3) Dependent variable: LOG(PP$_t$)			
CONSTANT	2.364**	0.990	[2.387]
LOG(PP$_{t-1}$)	0.788***	0.044	[17.956]
LOG(Q$_t$)	-0.227**	0.092	[-2.466]
LOG(XQ$_t$)	0.054***	0.017	[3.249]
LOG(PFS$_{t-24}$)	0.180***	0.063	[2.859]
Adjusted R-squared:	0.841		
Durbin-h stat:	0.668		
(4) Dependent variable: LOG(CP$_t$)			
CONSTANT	0.165***	0.037	[4.436]
LOG(CP$_{t-1}$)	0.893***	0.025	[36.292]
LOG(PP$_t$)	0.097***	0.024	[4.118]
D$_{12t}$	0.027***	0.006	[4.436]
Adjusted R-squared:	0.939		
Durbin-h stat:	0.689		
(5) Dependent variable: LOG(DC$_t$)			
CONSTANT	9.567***	0.264	[36.288]
LOG(CP$_t$)	-0.326***	0.078	[-4.194]
LOG(MEEA$_t$)	0.640***	0.072	[8.938]
Adjusted R-squared:	0.339		
Durbin-Watson stat:	1.793		
(6) Dependent variable: LOG(XQ$_t$)			
CONSTANT	-0.980	0.904	[-1.084]
LOG(XQ$_{t-1}$)	0.855***	0.039	[22.054]
LOG(XP$_t$)	0.332***	0.123	[2.694]
XQUOTA$_t$	-0.469***	0.151	[-3.097]
Adjusted R-squared:	0.780		
Durbin-h stat:	-0.199		

Note: due to space limitations the coefficients of the monthly dummies in Equation 2 are not shown.
Significance levels: *10%, **5%, ***1%.

The dynamics of producer prices are presented in Equation 3. An increase in production (Q_t) has the expected effect of decreasing producer prices (PP_t), while an increase in exports (XQ_t) results in a positive producer price increase. The coefficient of the lagged proportion of female cattle slaughtered variable (PFS_{t-24}) is positive and significant, as expected.

Consumer prices in Equation 4 are positively influenced by increases in producer prices (PP_t). The dummy variable D_{12t} is also found to have a positive effect on CP_t as anticipated.

Estimated parameters in Equation 5 show that consumers, as expected, respond negatively to a retail price increase (CP_t) and positively to an income increase ($MEEA_t$). The significant but rather low own price elasticity of demand may be due to Argentina's cultural dependence on beef.

As expected, the export quantity (XQ_t) responds positively to increases in the export price (XP_t). Exports shrunk significantly after the restrictions were imposed, as the negative and significant coefficient of the export quota variable ($XQUOTA_t$) captures in Equation 6.

6. Impact of exports on domestic prices

Having established that the positive effect of the export quantity on producer prices is robust (coefficient β_{33}), we now proceed to assess the indirect effect exports exert on consumer prices via the producer price equation.

The multiplier effect of a percentage change in exports can be easily derived from our estimates of the structural coefficients shown in Table 4, as follows:

$$PP_t = \alpha_{30} + \beta_{31}PP_{t-1} + \beta_{32}Q_t + \beta_{33}XQ_t + \beta_{34}PFS_{t-24} + \mu_{3t} \tag{3}$$

$$CP_t = \alpha_{40} + \beta_{41}CP_{t-1} + \beta_{42}PP_t + \gamma_{12}D_{12t} + \mu_{4t} \tag{4}$$

And substituting Equation 3 into Equation 4,

$$CP_t = \alpha_{40} + \beta_{41}CP_{t-1} + \beta_{42}\alpha_{30} + \beta_{42}\beta_{31}PP_{t-1} + \beta_{42}\beta_{32}Q_t + \beta_{42}\beta_{33}XQ_t + \beta_{42}\beta_{34}PFS_{t-24} + \beta_{42}\mu_{3t} + \gamma_{12}D_{12t} + \mu_{4t}$$

Where, β_{33} (= 0.054) is the impact multiplier of exports on producer prices in the short run, and $\beta_{33} / (1 - \beta_{31})$ (= 0.255) is the long-run multiplier. And where, $\beta_{42}\beta_{33}$ (= 0.005) is the impact multiplier of exports on consumer prices in the short run, and $\beta_{42}\beta_{33} / (1 - \beta_{41})$ (= 0.047) is the long-run multiplier.

The impact multiplier shows that, *ceteris paribus*, a 1% increase (or decrease) in exports would result in a 0.054% increase (or decrease) in producer prices in the same period, and in an increase (or decrease) of about 0.26% in the long run. Similarly, *ceteris paribus*, a 1% increase (or decrease) in exports would be reflected in a rise (or reduction) of 0.005% in the price paid by domestic consumers in the same period, and in a rise (or reduction) of about 0.05% in the long run.

The multipliers reveal that the effect of a reduction in exports on producer prices would be greater than that on consumer prices and that the dampening effect in the long run would be greater than in the short run. The difference mainly arises because of the high costs and margins of the intermediary activities (cutting and cold storage plants, commercialisation and others) that characterise the sector in Argentina.

The key factor that proponents of export quotas pose is that restricting exports has an immediate impact on the domestic supply: the less it is exported, the greater the amount available for the domestic market. And an increase in the amount supplied to the domestic market will, naturally, tend to lower retail prices. However, export restrictions may have a subsequent inflationary effect on the economy. We will return to this in the following section.

7. Policy simulations

The model was solved with a Gauss-Seidel algorithm. We evaluated the predicting ability of the model against our historical data using forecasts from previous periods when assigning values to the lagged endogenous variables (ex post simulation). Satisfied with the performance of the model, we used it to forecast future values of our endogenous variables, deciding on values for our exogenous variables (ex ante simulation). Finally, for the purpose of policy analysis, we examined how the model behaves under alternative assumptions with respect to the exogenous policy variables ($XQUOTA_t$ and $DSLC_t$). We simulated the behaviour of the model under the following scenarios until December 2010:
- *Baseline:* continuity of the current degree of government intervention.
- *No intervention:* absence of government intervention.
- *Liberalisation:* immediate liberalisation of the current restrictions.

Simulation results under these three scenarios are presented in Figure 2. Forecasts assume that the strong domestic demand continues and that no retaliation to Argentina's export controls by other countries occurs. We will examine each scenario in turn.

Baseline

The simulation outcome of the continuity of the current degree of government intervention gives support to the assertion made in the previous section regarding the subsequent inflationary effect of the export restrictions. The diversion of additional quantities of beef to the domestic market as a result of the export controls, in effect, immediately lowers consumer prices and increases domestic consumption. However, a decrease in exports caused by the imposition of an export quota results in an adverse effect on producers' revenue. Producers, whose revenues were curtailed, may not find motivations to expand their offer. Therefore, supply would eventually fall, resulting in rising retail prices.

Domestic consumers would not be the only ones suffering the harmful consequences in the long run. The continuity of the current restrictions results in a significant reduction of exports.

No intervention

This scenario is the antithesis of the *Baseline* simulation. It implies the inexistence of any kind of government intervention with the market forces, neither by imposing minimum slaughter weight restrictions nor by establishing export controls.

In the absence of government restrictions, a strong domestic and foreign demand raises consumer prices and this in turn affects consumption negatively. In our simulations, consumer prices rise continuously since 2005 and maintain their upward trend for almost 2 years. Higher prices encourage producers to expand their herds. During the herd expansion more female cattle are diverted to the breeding herd but biological constraints (Figure 1) impede beef production to adjust immediately. Given enough time for the herd to increase, output expands sufficiently

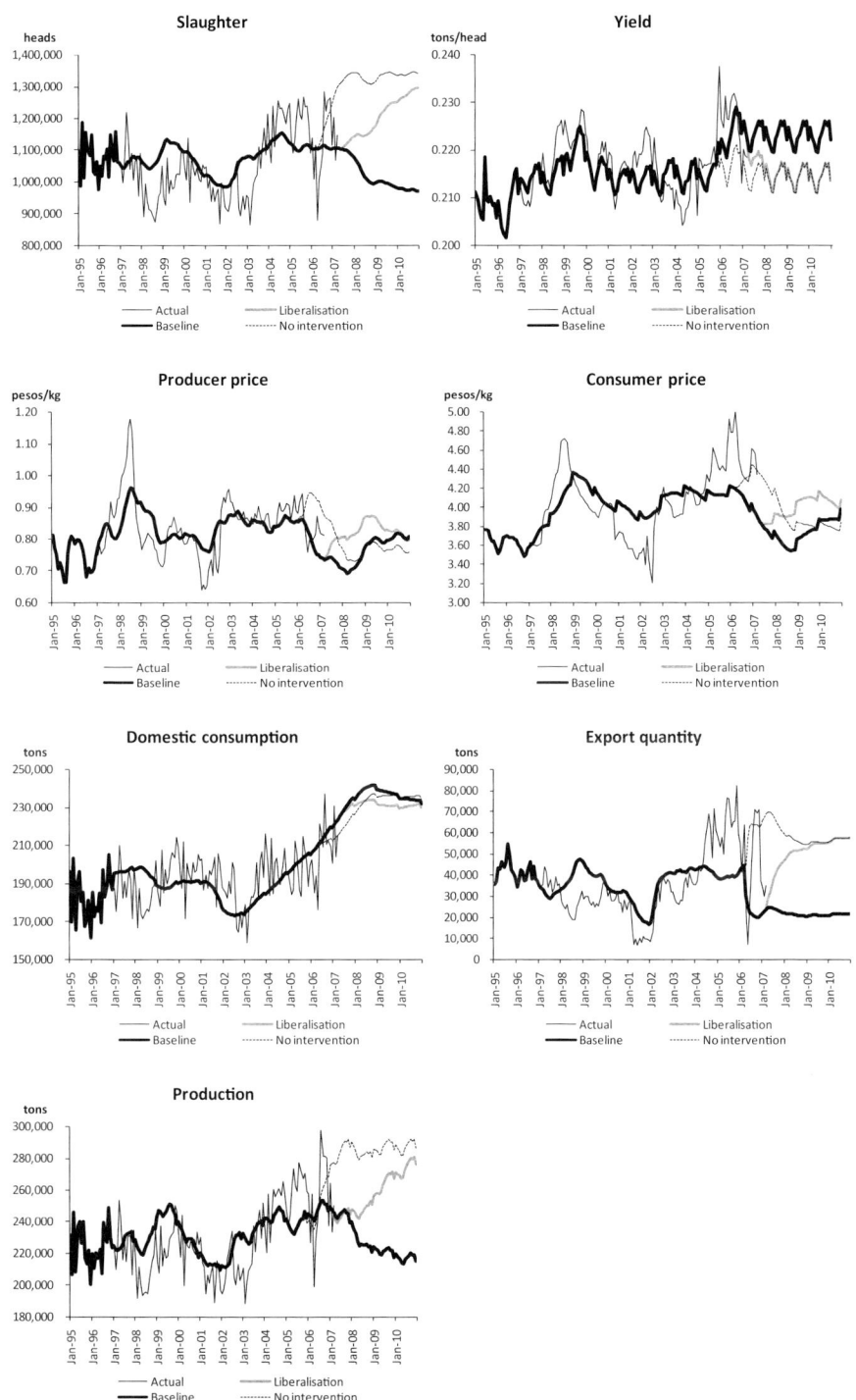

Figure 2. Simulation results.

to push consumer prices downwards. Interestingly, in the end of our simulation period, the retail prices attained under this scenario are the lowest of the three scenarios analysed while the country enjoys high-level exports.

Liberalisation

The liberalisation scenario is a midpoint between the two scenarios formerly presented. It simulates a situation of putting an immediate end to any kind of government intervention in the market, i.e. lifting the export and the minimum slaughter weight restrictions from the last historical data available for the simulations. It is interesting to analyse this option since it is a concrete measure the authorities of Argentina can immediately put into practice.

Simulation results show that despite an immediate domestic price increase would occur upon lifting the current restrictions, the upward trend could eventually be reversed. Simulations confirm the same pattern observed in the previous scenario: the initial rise in domestic prices caused by the exports liberalisation allows a gradual recovery in production which consequently brings down retail prices.

As for the export sector, simulation results show that exports would quickly recover upon lifting the restrictions, reaching the same level than under the *No intervention* scenario.

8. Concluding comments

The most significant result of this study is that, owing to the structure of the beef sector in Argentina, export restrictions are not an effective long term policy to control domestic beef price inflation.

Results obtained from the analysis of the impact of exports on domestic prices suggest that the constraining effect of a reduction in exports on producer prices would be greater than on consumer prices and that the dampening effect in the long run would be greater than in the short run. The structure of the sector in Argentina, characterised by inefficiency and high transaction costs, causes that intermediaries, rather than final consumers, appropriate most of the benefits of the lower cost of the commodity input.

The biological constraints to increase the weight of the animals in the short run explained the low impact of the prohibition of slaughtering lightweight cattle on yields and, therefore, the almost null impact on supply and domestic prices.

The uptrend in consumer prices was immediately reversed after the government announced the export restrictions. This was mainly a consequence of the allocation of additional quantities of beef to the domestic market and the drop in producer prices caused by the export quota. However, export restrictions are only effective for a brief period of time. Results from the simulations performed indicate that the continuity of the restrictive trade policy would eventually cause prices to rise since lower producers' revenues discourage future production. Conversely, simulations confirm that a future scenario of lower retail prices and high-level exports is attainable without government intervention. This would be possible because current higher prices encourage producers to expand their herds. Then, given enough time to complete the biological cycle of the cattle, supply would eventually increase enough to satisfy both domestic and foreign demand, reducing consumer prices.

There are, however, some limitations that should be borne in mind when interpreting the results. The foregoing analysis ignores the longer run effects that may arise from the fact that export controls may stimulate importing countries to diversify their sources of supply in order to counteract the effects of Argentina's restrictive trade policy. The unreliability of Argentina's exports could make importing countries seek for alternative sources of supply, either domestic (by increasing their own production) or foreign (by negotiating with other producing countries for a greater portion of their supply). If such is the case, exports would not be as high as the simulations suggested and the export levels under the *No intervention* scenario would be higher than those under the *Liberalisation* scenario. Another limitation of the analysis is that beef is not a homogeneous commodity as assumed in the model. The cuts of beef consumed domestically are not the same as those preferred by the importing markets. Furthermore, due to data limitations, it was not possible to account for the stock-flow relationships in livestock production or the margins in the commercialisation chain. Had this information been available, researchers would have been able to build a more complete production system and perform a more detailed price transmission analysis. It would be interesting to consider these factors that have not been taken into account explicitly for further research.

The persistence of the export controls would result not only in loss of earnings for the country and a reduction of tax collection (beef exports are levied, in average, with a 15% export tax) but also in loss of ground in the highly competitive foreign markets and, perhaps more seriously, in a significant damage to Argentina's reputation as a reliable supplier. Markets unattended by Argentina quickly find alternative sources of supply even within Argentina's neighbors. Such has been the case of the European Union, where Argentina has positioned itself as a high quality beef supplier. During the first half of 2007, traditional European trade partners for Argentina significantly increased their imports from Brazil (Argentina's main competitor in the EU) and reduced those from Argentina, if compared with the same period of 2006 (Ministry of Economy and Production of Argentina; www.sagpya.mecon.gov.ar/new/0-0/programas/dma/index.php and Association of Brazilian Beef Exporters; www.abiec.com.br/index). Even when the quality of the Argentinean beef is highly valued in the international markets, these shifts in the trade patterns should be read as a sign of warning to the Argentinean authorities when deciding to self-isolate the country's export sector.

References

Argentine Beef Promotion Institute, Catholic University of Argentina (Ed.), 2006. Guidelines for the Formulation of Scenarios for the market of bovine meat in Argentina. Buenos Aires, Argentina.

Carter, C., N. Gallini and A. Schmitz, 1980. Producer-consumer trade-offs in export cartels: the wheat cartel case. American Journal of Agricultural Economics, 62: 812-818.

Engle, R.F. and C.W.J. Granger, 1987. Co-integration and error correction: representation, estimation and testing. Econometrica, 55: 251-276.

Hsiao, C., 1997a. Statistical properties of the two-stages least squares estimator under cointegration, Review of Economic Studies, 64: 385-398.

Hsiao, C., 1997b. Cointegration and dynamic simultaneous equations model. Econometrica, 65: 647-670.

Hsiao, C. and H. Fujiki, 1998. Nonstationary time-series modeling versus structural equation modeling: with an application to Japanese money demand, Bank of Japan's Institute for Monetary and Economic Studies (IMES). Monetary and Economic Studies, 16: 57-79.

Sedjo, R.A. and A.C. Wiseman, 1983. The effectiveness of an export restriction on logs. American Journal of Agricultural Economics, 65: 113-116.

Shei, S.-Y. and R.L. Thopmson, 1977. The impact of trade restrictions on price stability in the world wheat market. American Journal of Agricultural Economics, 59: 628-638.

Takacs, W.E., 1994. The economic impact of export controls: an application to Mongolian cashmere and Romanian wood products. Policy Research Working Paper 1280. World Bank. Washington, DC.

Wong, C.M., 1978. A model for evaluating the effects of Thai government taxation of rice exports on trade and welfare. American Journal of Agricultural Economics, 60: 65-73.

Part 2
Food quality, supply networks and competition

Exploring hybridity in food supply chains

B. Slee and J. Kirwan[30]

Abstract

Much of the theoretical debate about alternative food supply chains or networks has been couched in dualistic conceptions of a mainstream and an alternative food sector. This paper asserts that there is much evidence of hybridity in evolving food chains and networks. Different theoretical perspectives in the social sciences offer spaces in which to explore the idea of hybridity. This evident hybridity in the agro-food sector makes assertions of a new European paradigm of rural development based on a relocalised food sector appear more a normative hope than an empirical fact.

1. Introduction

In recent years, a number of dynamic aspects of food supply chains (FSCs) have attracted great interest among social scientists investigating rural restructuring and change. These include: the expansion of organic agriculture; the development of new value added enterprises at farm level and the revitalisation of traditional and new-old artisanal production practices; the expansion from a low base of the market share of 'alternative' short supply chains, such as farmers' markets; and the so-called quality turn, riding on the heels of another turn in rural social research - the consumption turn (Goodman, 2003, 2004).

All of these changes come together in a vision of alternative agro-food networks (AAFNs) that has been built around empirical and theoretical work from a number of predominantly European social researchers, centred on Wageningen, but conducted in a number of countries in Europe. These and other associated changes in the composition of farm-based economic activity are seen to be constitutive of a new paradigm of rural development comprising an alternative network of producers, consumers and other actors in relation to the mainstream agro-food system (Van der Ploeg *et al.*, 2000; Van der Ploeg and Renting, 2004; Renting *et al.*, 2003).

The theorisation surrounding this work on AAFNs has been sharply criticised by Goodman (2004). He challenges the vision of certain European social scientists of an alternative food sector rising like a phoenix from the ashes of the commodity-based food system to constitute a new paradigm of rural development. In particular, he challenges the binary categorisation into alternative and mainstream and is deeply sceptical as to the existence of a new paradigm while, at the same time, highly cognisant of dynamic changes within the agro-food sector. This dynamism is discussed by Murdoch and Miele (1999: 469) in terms of 'changing worlds of production', and the way in which an apparently bifurcated FSC is in fact becoming increasingly fragmented and diversified as producers respond to changing demand patterns by moving between the 'two main zones of production: standardised, industrialised global food networks on the one hand, localised, specialised production processes on the other' (e.g. through large companies highlighting the 'naturalness' of their production methods, or alternative producers increasing the range of outlets they sell their produce through). Sonnino and Marsden (2006: 181), while acknowledging the potential role of FSCs in rural development, also caution that there is a need for 'a much more nuanced and complex understanding of the relationships between conventional and alternative

---

[30] This paper is informed by the rich debate surrounding the authors' participation in the EU funded Suschain project Contract no. QLK5-CT-2002-01349 and their other work in this field.

food chains' and the 'mutation and evolution' of these networks over time (*ibid.*: 191). These relationships extend into the sphere of regulation which may span both mainstream and AAFN production processes (something that was recognised by Higgins *et al.* (2007: 226) within the context of Environmental Management Systems as 'hybrid regulatory space'). Furthermore, although abstract distinctions can be, and often are made between different FSCs, the reality is that there are frequently 'no clear boundaries between them' (*ibid.*: 184) and no food network (whether conventional or alternative) operates in isolation.

Moreover, in their review of existing research into AAFNs, Sonnino and Marsden (2006) suggest that although the diversity of the phenomenon has been empirically objectified, there has been a distinct lack of theorisation concerning the dynamics of their evolution across the full range of spatial scales within FSCs. This, they suggest, is likely to be because of their highly contextualised nature. In order to partially address this deficiency, they focussed on the notion of 'quality' within FSCs: utilising the concept of embeddedness to specifically examine the relationship between food and territory. While this is certainly very pertinent, the aim of this paper is to take a broader view of existing theoretical perspectives, motivated by a desire to explore the manner and extent to which different theories can help interpret and explain some of the most dynamic areas of agro-food systems that belong neither in mainstream nor alternative FSCs. We argue that it is vital to examine the hybridity evident in food supply chains and networks in that currently there is too often a tendency towards dualistic interpretations, which do not give an accurate reflection of the realities of the modern FSC. A more explicit acknowledgement of this hybridity will allow for the development of conceptual frameworks that can better examine the nature, implications and impacts of the current dynamism within 21st Century FSCs. In the first part of the chapter we provide an overview of the relevance of the notion of hybridity within this context, before briefly reviewing the role and relevance of a number of existing theories in terms of understanding the dynamics associated with FSCs. The chapter then highlights a number of examples of where there is evidence of hybridity with UK FSCs, most notably in terms of the conventionalisation of the organic food sector as well as the growth of major retailers. Finally, the implications of hybridity in UK FSCs for rural development are discussed before concluding with a number of interim conclusions.

2. Hybridity

In this paper, as elsewhere in the social sciences and more widely, hybridity is characterised by 'both, and' categories rather than 'either, or' categories. Thus, rather than exploring opposites, whether expressed as ideal type categories or nature:culture type dualisms, the exploration of hybridity entails the identification of co-constitutive socio-economic and biophysical phenomena. It constitutes a challenge to, and a deconstruction of, previous dualistic thought (Cloke, 2003); or, as Ilbery and Maye (2006: 355) describe it, 'a 'mixing together', rather than separate systems'.

The original use of the term hybridity in social sciences is found in the literature surrounding the study of post-colonialism. Since the rather specific early use of hybridity, the variety of contexts in which the term has been used has multiplied (Whatmore, 2002). In particular, the term is widely used in Actor (or Actant) Network Theory, which draws together the study of the natural and social worlds in a mutually constitutive study of process and practice. This study of co-constitutive relationships is often described as an exploration of hybridity. This recognition of complex hybrid mixes of people, animals, plants and things challenges the previous one dimensional exploration of political economic structures (Cloke, 2003: 6). Much of the discussion of hybridity is framed within heterogeneous interactions of heterogeneous actors (both human and non-human) in networks. Networks, in Murdoch's view are necessarily hybrid (Murdoch, 2003: 269).

Additionally, the term hybrid is used to describe situations where elements of more than one policy perspective manifest themselves, not as separate entities but as interconnected parts of the same policy or governance framework or where theoretical explanation draws on more than one theoretical perspective to explain socio-economic phenomena. In essence, the exploration of hybridity entails the study of relationships between phenomena frequently categorised in terms of opposites and which are often theoretical constructs or ideal types, rather than observable realities. Thus, the exploration of hybridity necessarily entails exploring straddling, crossing and threatening conventional categories of, and approaches to, analysis.

The term hybridity has been used in a rural context by Higgins and Lockie (2002) and Lockie and Higgins (2007). In their work, the term hybridity refers to the emergent forms of governance in natural resource management, where elements of neo-liberal economic policy are juxtaposed with social and environmental resource management practices constituted at local level. This intermixture of policies is seen to underpin the operation of the neo-liberal policy agenda through hybrid 'policies of rule' (Higgins and Lockie, 2002: 420). This same sense of mixing of values is evident in the way both UK and Italian governments in the early 2000s have fostered a neoliberal policy regime while at the same time nurturing localised food supply chains through specific policy means (DuPuis and Goodman, 2005).

Specifically in relation to food, Ilbery and colleagues have suggested the usefulness of the concept of hybridity as a means of better understanding the relationship between different FSCs. For example, they caution against making simplistic distinctions between conventional (or mainstream) and alternative FSCs, in that many producers clearly sell into both types of supply chain and the consumers of 'alternative' foods such as local foods are often not concerned about the methods used to produce those 'local' and 'alternative' foods (Ilbery *et al.*, 2006). Similarly, it is clear that retail outlets associated with the mainstream are increasingly selling produce (e.g. organic and 'local' food) that was previously associated with alternative FSCs. In another paper, Ilbery and Maye (2005: 840) argue for an approach that recognises upstream relations within FSCs, in that even where a product appears to be alternative at the retail end of the FSC, it may in fact have used inputs that are very much sourced from the mainstream FSC (e.g. conventional distribution or input supplies), as 'many alternative producers…'dip into' more conventional nodes'. Similarly, Watts *et al.* (2005: 34), in their examination of alternative systems of food provision, argue that alternative food supply chains can be understood 'as weaker or stronger on the basis of their engagement with, and potential for subordination by, conventional FSCs operating in a globalising, neoliberal polity', resulting in a spectrum of alternativeness dependent on the way in which individual FSCs are constructed. All of this leads Ilbery and Maye (2006: 355), like Goodman (2004) above, to argue that binary opposites between conventional and alternative are difficult to maintain because they 'are linked together in an overall agro-food system [and that] it may be more instructive therefore to think instead about hybrid food spaces'; spaces which have 'blurred edges' in relation to whether they are part of alternative or conventional FSCs.

We agree with this observation and, furthermore assert that the areas of dynamic change in food markets, whether in AAFNs or the mainstream conventional sector, are often better understood through an analysis of hybridity rather than through representation as inflexible dualisms. Indeed, food is a core context in which hybrid theories have been explored whether in relation to technical human natural interactions in Callon's work (1986) or in more recent studies of GM food (Whatmore, 2002: 120ff.). In this context, a recognition of hybridity can help reveal the complex, mutually constitutive and heterogeneous nature of 21st Century FSCs which other theories, such as those reviewed in the following section, may be less able to encapsulate.

3. Understanding food supply chain dynamics: the role of theory

This section briefly reviews some of the competing theoretical perspectives that have been used to explore change in FSCs. Some of these theoretical perspectives are rooted in economics, some in political economy and some in rural sociology, while others span many of the social sciences. In each case, the consideration ends with brief observations on their limitations as comprehensive explanatory models.

3.1 Neoclassical economic theory

Neoclassical theory focuses on resource allocation and price determination in food markets. The principal lessons that can be drawn from neoclassical economic analysis of developed country food markets are: an expectation that food purchases will absorb a reducing share of the consumer's pound (following Engels' Law); an expectation that, of that pound, a greater share will be spent on eating out as part of expanding expenditures on leisure (because of the positive income elasticity of eating out). Additionally, a cost price squeeze is widely evidenced in the primary production sector, largely a result of supply curve shifts in the face of an inelastic demand for most commodity products. However, there may remain scope for niche and speciality products to absorb an increasing share of the affluent consumers' retail pound spent on food (Van der Ploeg, 2006).

Because of the dependence of much food production on biophysical resources and the attendant uncertainties of the natural world, yields can vary and prices can prove very volatile. Further, the movement of resources out of the farm sector is often impeded by factors that induce asset fixity[31], which compounds the free market outcome of low and declining farm incomes and exacerbates the cost-price squeeze. Buffered as they have been by decades of protectionist policies, Western European commodity food production has become relatively high cost compared to Latin America, Australia or New Zealand. However, there is a long tradition of long-distance food imports into the UK from its former colonies, which was challenged by the policy consequences of the UK's entry into the European Union. However, as the GATT and WTO have turned their attention to agricultural protection over the last decade, so the more highly supported commodity regimes have come under intense adjustment pressure. Especially in certain sectors such as poultry meat, sourcing has globalised and significant imports arrive in the UK from Brazil and South East Asia. The inevitable consequence in highly supported markets has been a search for a new rationale for farming, either through niche production or the delivery of environmental services. This search for alternative production and supply chain models can be seen as a defence and survival mechanism against the seemingly inexorable forces of globalisation, exacerbated as they are by the new global policy settlement. Indeed, Marsden *et al.* (1999: 295) suggest that AAFNs can 'create positive 'defences' for rural regions against the prevailing trends of globalisation and further industrialisation of markets'; while Winter (2003) suggests that some of the emergent AAFNs can be described as 'defensive localism' (Winter, 2003).

The nature of contemporary food market structures, with their increasingly concentrated power[32] (Dobson Consulting, 1999), coupled with the inevitable tendencies of a primary industry with a propensity to overproduce, has exerted general downward pressure on food raw material prices

---

[31] For example a dairy farmer's fixed assets in milking machinery and parlour are not much use for other enterprises.

[32] The extreme case of concentration in Europe in food retailing is Finland with the top five firms controlling 97% of the food market; the UK is at 67% and Italy, the lowest at 30% (Dobson Consulting, 1999: 45).

(Padberg *et al.*, 1997). This downward pressure, coupled with a growing interest in speciality and local food, has undoubtedly triggered a push factor into farm-level diversification and value-added projects and a range of initiatives, some collectively organised and often with public sector support assisting the realisation of these new opportunities (Slee, 1989). AAFNs are thus both demand-driven by the emergent markets, a supply response to the cost-price squeeze in contemporary agriculture, a lifestyle choice for some food producers and a policy response in the form of increased public sector support to local and regional food initiatives.

While market analysis through the neoclassical lens can expose disequilibria and enhance understanding of market prospects, it is less able to explain the drivers of demand change and the new institutional structures which have emerged to support AAFNs. Neither can neoclassical economics readily explain the remarkable shifts in market power towards the food retail sector, away from processors and producers (Burt and Sparks, 2003). Instead of stagnation in food markets as a result of the food sector absorbing a declining share of the retail pound, as might reasonably be predicted, the food retailers have been amongst the most dynamic and rapidly growing businesses in Europe, although much of their recent dynamism is not only to do with food, but broader product diversification (Competition Commission, 2000).

3.2 Political economy of agriculture

The political economy perspective, with its roots in Marxian political economy, posits that there are inherent monopolistic tendencies in capitalist markets and that there are likely to be periodic crises in their operation. Inequalities in power and access to resources will lead to adjustments in the structure and organisation of food production and distribution. In addition, a general process of subsumption has been observed in the farm sector which has drawn farming into wider circuits of capital and subordinated farming interests to those of more powerful agents in the agro-food sector. These processes have been described by Goodman *et al.* (1987: 2) as *appropriationism* 'in which elements once integral to the agricultural production process are extracted and transformed into industrial activities and then re-incorporated into agriculture as inputs': and *substitutionism* 'in which agricultural products are first reduced to an industrial input and then replaced by fabricated or synthetic non-agricultural components in food manufacturing'.

Although not exclusively a concept within the political economy perspective, the concept of globalisation can be seen as the outcome of a set of internationalised processes in food production, processing and distribution. Supported by an internationalisation of policy regulation by the WTO, itself underpinned by a broadly neo-liberal economic and political agenda, both commodity and speciality food can be expected to figure prominently in international trade. The evolving food market entails global sourcing and the characteristic time-space compression observed in globalisation, with substantial long-distance movement of food to meet the diverse and increasingly de-seasonalised demands of consumers and retailers. *Inter alia*, the political economy perspective stresses the changing structural and power relations in the food sector, the globalisation of food procurement and the unequal relations between capital and workers.

AAFNs can be seen as a multi-stranded counter-culture which challenges the hegemony of the corporate giants in the food sector. The early origins of many AAFNs were extra-market phenomena, such as the pursuit of self sufficiency through organic farming. Over time, AAFNs have developed substantially as market phenomena, driven by the antagonism of some consumers towards large-scale food production, who are 'voting with their mouths' in preferring alternative production and distribution models. However, the alternative sector is by no means clearly differentiated from the mainstream and is subject to corporate predation, when profitable niches expand.

One way in which the products of AAFNs have sought to gain higher prices within the marketplace is through certification and the effective creation of a restricted supply which has the potential to secure an 'economic rent'. Indeed, Goodman (2004: 8) argues that 'the ability of quality food products to secure premium prices and so generate excess profits [or economic rents] is a central plank of the market-led, value-added model'. Guthman (2004a) explores this potential in relation to the use of organic food standards and certification as a 'tradable descriptor' and means of achieving a price premium for the producers involved. Nevertheless, both she and Goodman (2004) recognise that organic certification, in having the potential to generate excess profits and an economic rent, inevitably attracts the attention of mainstream actors who may then appropriate this economic rent without adhering to the underlying precepts of the certification. In this way, certification schemes become a site of 'social struggle' between different stakeholders (such as state agencies, local social or environmental interest groups) each of whom is seeking to mould them to their own best ends. Moreover, the site and scale of these struggles is now global in many cases (involving such bodies as the FAO and IFOAM) as certification schemes that started out as local and alternative in nature have in many cases become inexorably drawn into the global mainstream.

These issues are comprehensively examined within a special issue of the Journal of Rural Studies: volume 21, 2005 (most notably Mutersbaugh *et al.* (2005), Klooster (2005) and Renard (2005)). These papers highlight the proliferation of such certification schemes since the early 1990s and their growing importance in terms of market penetration. However, they reiterate that this success has attracted the attention of mainstream businesses, resulting in the 'increasing integration of alternative products into traditional product distribution systems' (Mutersbaugh *et al.*, 2005: 382). They note that this tendency has governance implications, as schemes that were set up to help particular groups of relatively disadvantaged producers become taken over by 'private interests'. In the process, certification becomes the 'norm', effectively isolating the producers they were originally designed to help because the latter can no longer afford to produce the higher quality product in the absence of a price premium. The underlying concern of these papers is that 'mainstreaming' pressures effectively dilute, and in some cases eradicate, the benefits and identity of certification schemes that were set up to provide an economic rent for 'alternatives'.

Nevertheless, while these papers provide a useful and comprehensive examination of power and regulation within certification schemes designed to produce an economic rent, there is still a danger that the alternatives and the mainstream are viewed dichotomously. Certainly, there is an examination of the inherent tensions between the two perspectives, but the essential focus of a political economy approach on large-scale structures inevitably restricts its ability to understand the micro-dynamics of AAFNs. For example, whilst political economists might have foreseen the concentration of corporate power within the food supply chain, it is less evident that the dominance of the retail sector was so readily predictable. In addition, the nature of lifestyle businesses and the different ethical drivers of many actors in AAFNs rather undermines the notion of self-interested, profit-seeking behaviour which underpin the political economist's conception of the farm or food business.

3.3 New institutional economics

New institutional economics focuses on transaction costs and the institutions that underpin the operation of markets. Transaction costs are the costs of using the market that include information costs, negotiation costs and enforcement costs. Transaction cost analysis can provide 'an explanation for the structure of forms and for the nature of vertical coordination within a supply chain' (Hobbs, 1996). The contemporary major food retailers have managed to strip out transaction costs by reducing the number of actors in food supply chains. However, emerging new technologies also afford new opportunities for smaller operators to reduce transaction costs and develop new AAFNs, for example through Internet marketing.

Among the drivers of change in the mainstream food sector, it is clearly evident that reducing the costs of using the market is an important factor. The reduction in the number of food chain actors and the field-to-shop control of production processes helps large retail firms reduce transaction costs and has dramatically weakened some of the traditional components of marketing chains such as wholesale markets. However, in their pursuit of homogenous standards and year round availability of commodities by retailers, supply chains have been lengthened in physical distance terms and this has generated much debate, especially in the UK, about food miles.

Many traditional food retailers have suffered from the competitive supply chain efficiencies introduced by the large retailers, and there are likely to have been negative impacts on some AAFNs operating outwith the mainstream sector. However, there are several reasons deducible from the analysis of transaction costs that expose why AAFNs might now constitute preferred marketing channels for some primary producers. First, the corporate muscle of the retailers can drive down prices received by farmers, so that they now receive about 10% of the final product price compared to 50%, 50 years ago (Pretty, 1998). This enormous change might be expected to incentivise the development of alternative short food chain marketing channels or the development of value-adding enterprises by primary producers as survival strategies for deeply pressured farm businesses arising at an individual farm level or through collective action by groups of farmers. Second, supermarkets are regarded by many critics as agencies which threaten local food systems and generate a wide-ranging but similar offer to the consumer (NEF, 2003), undermining traditional food outlets such as bakeries, butchers and green-grocers as well as competing with emergent AAFNs (Lang and Heasman, 2004). The path dependency created by their national level distribution systems (particularly in the UK) may limit the opportunities for exploitable local niches for alternative supply channels. Third, constellations of local agencies have sometimes come together to address these problems and new alliances have emerged, often with public funding, to build constructive partnerships which support the partial relocalisation of food markets. A partial public sector shift from sectoral to spatial policy has enabled new locally based coalitions and new forms of rural governance to shape at least some facets of rural support.

It is possible to explore the possibilities of AAFN development through the lens of transaction costs, both to explain the rejection of the mainstream marketing channels and the emergence of the new networks. However, it seems likely that competitive localism - essentially different regions competing in the regional food market (see Morris and Buller, 2003) - might increase the total transaction costs of AAFNs and that, rather than offering an opportunity of reducing transaction costs the development of AAFNs may constitute an exercise in self-interested rent seeking by powerful or articulate local interests who are able to extract public money from a range of sources.

3.4 Endogenous development

Since the early 1990s, an alternative model of developed country rural development has been actively promoted both at a theoretical and a policy level. Indeed, this model can be seen as the intellectual underpinnings of the new European paradigm of rural development (Van der Ploeg and Renting, 2004). The endogenous development model is articulated as both a survival strategy and development option for farmers and as a redoubt against the modernisation model. It is seen as development from within or from the bottom up, built on locally nuanced farming systems and value-added production and the cultivation of rural distinctiveness.

Over the early 1990s, a research group at Wageningen University led an EU project (Van der Ploeg and Long, 1994) which endeavoured not only to explore the agro-technical manifestations of endogenous development, but also to provide a theoretical rationale for both its existence and its dynamic potential in a wider rural development context. Van der Ploeg and Long (1994) explicitly challenge a unilinear model of development, arguing that at any point in time a farmer faces choices and that certain critical decisions can move the farm to a more developed state either by adopting modern farming practices, essentially buying into the modernisation of farming; or alternatively, developing the market potential of endogenous enterprise. Amongst the diverse observable styles of farming, it is often possible to identify some farmers who retain elements of traditional practice and engage in a process of deconstructing and reconstructing core knowledge and adapting it to their specific circumstance. This is the antithesis of the modernisation approach and offers scope for a range of value-adding and/or differentiated forms of production and marketing.

It is possible to rationalise the development of endogenous enterprise by reference to transaction costs or by reference to the new market opportunities created, *inter alia,* by rural repopulation, rural tourism and the development of local and/or distant niche markets. It is further possible to explain the existence of endogenously rooted enterprise by recognition of different farming styles (Van der Ploeg, 2003). Van der Ploeg's rationalisation is principally rooted in an analysis of the supply side - the farmer's attributes, indigenous technical knowledge and the desire to develop effective survival strategies in the face of market price pressure - rather than in acknowledgement of changing demand, although the changing demands patterns and the decline of trust in the commodity food system are now also articulated as major drivers of change.

The articulation of the endogenous development model as a vehicle for sustaining traditional agricultural and food processing practices is not without some foundation. In fact, in some areas, in particular in areas with residual traditional agricultures that were less fully penetrated by the processes of modernisation, a significant proportion of the food system may revolve around AAFNs. However, over large swathes of Europe, the endogenous mode of production has been marginalised to such an extent that a neo-endogenous model seems more apposite, whereby farmers or small-scale processors and retailers (or indeed any development actors) assert a distinctive regional provenance, whether or not it is rooted in traditional practice (Ray, 2003). However, the scale of endogenous and neo-endogenous enterprise is such that it has not become a major driver of rural change in north-west Europe, though it probably figures more prominently in countries and regions where old and traditional production practices can be effectively melded to new demands from counter-urban growth or tourism. The scope for neo-endogenous development may be enhanced by incursions of urban wealth and purchasing power, whether through tourism or residence in rural areas.

4. Some evidence of hybridity in UK food supply chains

The above theoretical explanations of change in food supply chains offer an economic or political economy context in which these changes can be framed. However, the tendency to polarise the food chain into two components: a mainstream and the AAFN sector tends to obscure the analysis of the interface and dynamics between the two. This section explores two important arenas where hybridity in contemporary food supply chains is strongly evident. First, the growth and change in the organic farming sector is examined in a UK context. Second, the backward-reaching capacity of the highly concentrated food retail sector towards speciality products is explored.

4.1 The conventionalisation of the organic farming sector

In the early 1990s, one of the authors was working on a project to explore the potential for the development of the organic sector in the Highlands and Islands of Scotland (Daw *et al.*, 1991). Part of this study involved looking at another region of the UK with more highly developed organic supply chains. South West Wales had emerged as a leading region in the development of organic food in the UK. As part of the research project, a number of key actors were interviewed and it was evident at that time that there was much disagreement in the organic sector in Wales between the purist organic farmers whose ambitions were to create an alternative food system and whose motivations were more ethical than commercial, and another set of pragmatists who were prepared to develop global sourcing in order to feed the demand from supermarkets. This debate is highlighted by Morgan and Murdoch (2000: 168-169) who argue that organic 'producers face a Faustian bargain: while the supermarkets provide a large and ready market, they seek to tailor organic produce to the conventions of the industrial market…. This problem is especially acute with regard to "quality" conventions: supermarkets set a premium on cosmetic appearance, which in turn leads to waste and packaging. In contrast, the organic community understands "quality" in terms of taste and nutrition, and it accepts blemishes as natural and sees little or no need for packaging'. This dualistic division into purists and pragmatists is clearly a simplification, but over the last decade and a half a debate has continued with the purists still driven by a desire to create an alternative food supply system and the pragmatists eager to sell through supermarket channels.

On that same visit to Wales in the early 1990s, Rachel's Dairy, a dynamic West Wales organic business that had developed through adding value to organic milk on the first Soil Association certified organic dairy farm in the UK, was held up by local academic researchers as an exemplar of what organic agriculture could do for local development (see Lampkin 1990: 482-485). Its website still asserts its local embeddedness and the narrative on the website is a personal history of its founders (http://www.rachelsorganic.co.uk/about/history.html) and their connection with the un-named current owners. Nowhere on that website is it mentioned that in 1999 a large US-based organic milk company, Horizon, had taken over Rachel's Dairy in a multi-million pound deal. Rachel's Dairy now supplies a range of supermarkets, as well as international hotel chains.

The growth and concentration of organic production and retailing has led to the emergence of two major box scheme suppliers in southern England (Abel and Cole; http://www.abel-cole.co.uk/Home.aspx, and Riverford Organic Vegetables; http://www.riverford.co.uk/) who both use large articulated lorries in the relatively long-distance transport of their products, even though the origins of the box schemes were to provide a mixed box of local food, occasionally supplemented by bought-in extras, in order to keep food miles to a minimum and freshness to a maximum. The local franchises of these schemes may still use a significant proportion of local produce but the scale of enterprise and the business models used suggest anything but alternative food networks. This transformation of what were historically highly localised distribution systems may reduce some of the generally accepted environmental benefits and weaken the close ties with consumers which have been identified as two of the key characteristics of AAFNs.

At various times, organic farming has been incentivised by policy support, largely on the basis of widely asserted beneficial effects of organic farming on the environment, as well as a range of other asserted benefits relating to rural employment, and other more controversial assertions about benefits to health. This public support has led to organic farming methods being adopted by new entrants for narrower commercial reasons, rather than embracing the traditional organic ideologies that might resonate more closely with those associated with AAFNs. Indeed, the former minister of agriculture in the UK (David Miliband), a strong advocate of the adoption of radical environmentally friendly policies and of the Worldwide Fund for Nature's One Planet Living, has dismissed organic agriculture (in January 2007) as a 'lifestyle choice' by consumers. In the light of those comments, the debate about the merits of organic farming has been widely aired in the public arena and organic advocates have used both environmental and health reasons for justifying their approach to farming.

Although organic food is only 4% of the UK food market, it has experienced rapid growth and as such has become increasingly contested territory between purists and pragmatists. It has also become a symbolic battlefield amongst the major retailers who are using organic products to jockey for position with food purchasers. Yeo Valley Organic (http://www.yeovalleyorganic. co.uk/) is the first fully organic firm in the top 100 food firms in the UK by turnover, growing its business by 25% in 2005 (Guardian 21st February 2007). At various times organic food has been used as a loss-leader to give particular supermarkets a green identity. The rapid growth in demand has required overseas sourcing of many products (about 70%) which is necessarily underpinned by long-distance food supply chains. The resultant hybridity of food supply chains/ networks in the organic sector is an inevitable consequence of this contestation being played out in the market place. This is not exclusively a UK issue, as work in both the US and Australia has explored what is termed the 'conventionalisation debate' which examines the growing pressures on organic producers and certifiers to morph their values as they are drawn into conventional FSCs (e.g. Guthman, 2000; 2004b; Lockie and Halpin, 2005).

There are many features of the organic sector that display hybridity in the tensions between its original 'purist' form and the current manifestations of organic food supply chains/networks. Our contention is that those areas of the organic sector with the fastest growth and the greatest potential to contribute to rural development can often be found in the boundary area between purists and pragmatists and in the evolving marketing structures associated with this hybridity. Although some authors assert conventionalisation, there is also evidence of a resilient purist organic movement and at the same time the growth of enterprises that retain some elements of the purist organic agenda while using modern retailing techniques to reach a wider range of customers. For example, the Soil Association (2005) found that:

> *'Sales of organic products through box schemes, farm shops and farmers' markets increased by 33% in 2004. Sales through independent shops also rocketed, increasing by 43%. The supermarket share of the market fell from 81% to 75% but still accounts for £913 million in sales.'*

As indicated in the preceding paragraphs, the evidence for large scale supermarket organic retailing must be set alongside the hybrid forms such as Riverford Organics, who deploy similar distributional practices but retain certain core organic ethics and values. The organic sector is evolving and dynamic with many hybrid features, notwithstanding that there are clearly AAFN components to some parts and highly conventionalised components in others (most notably the supermarket sector).

4.2 The growth and adaptive capacity of major retailers

There are widely discussed concerns about the market power of supermarkets, both in relation to the tendency towards monopoly at a local level, their buying power and ability to drive down prices received by suppliers (including farmers), and their control over development land through speculative purchase which might lead to the exclusion of competition. Many supermarkets in the UK have also entered the local convenience store market where it has been argued that they have created even greater pressure on small independent food retailers (Lang and Heasman, 2004). Supermarket power is undoubtedly a concern of regulatory bodies dealing with workable competition in many countries, but supermarkets have also been at the forefront of introducing regulatory practices with respect to food hygiene and safety (Flynn *et al.*, 2003). There have been several inquiries into monopolistic practices by food retailers in the UK and one is currently under way (http://www.competition-commission.org.uk/inquiries/ref2006/grocery/index.htm), but the evidence to date is inconclusive, except in the areas of land banking (accumulating development land possibilities in ways that restrict competitors' access) and in recognition of their ability to drive a hard bargain with farmer suppliers.

In general, until recently, supermarkets in the UK have not exhibited a marked tendency to purchase significant volumes of produce from the immediate locale. In the UK, at least, this is often attributed to their centralised distribution systems and onerous quality control systems, which may require the long-distance movement of supplies from a region of production to a central distribution point and then back to the same area for consumption (Vorley *et al.*, 2006). This contrasts somewhat to other European countries where in France, for example, considerable shelf space is committed to local and regional produce and the organisation of procurement is very different.

Taking the UK as an example, the 'big four' supermarkets now control nearly three quarters of the food retail market, with the largest, Tesco, now accounting for roughly one out of every three pounds spent on food purchases in the UK (Lang and Heasman, 2004). Supermarkets such as Tesco use sophisticated customer profiling techniques to maintain customer loyalty and are acutely aware of their customers' aspirations and interests regarding food. Tesco have moved from the 'pile it high sell it cheap' business approach that it used to break into the food market place in the 1960s, and now offers a highly differentiated range of products and encroaches substantially into the market space captured partially by speciality food producers. They have, like most other supermarkets, adapted their offer to include quality labels and have made efforts to present themselves as a convenient exchange location between the individualised farmer producer and the final consumer, wherein convenience is largely based on a one-stop-shop and a wide-ranging offer. The complex relations between consumers and those from whom they buy their food are beyond the scope of this paper, but it is clear that the supermarkets have tried to personalise their shopping space with images of farmers who produce exclusively for them to promote an image of quality and personal relationships between farmer and consumer. The overall evidence suggests that the so-called quality turn regarding food has not so much led to a decline of supermarkets as their continued expansion (Vorley, 2003).

Given supermarkets' highly motivated profit-seeking behaviour, and capacity to garner market information, it is unsurprising that they have sought to accommodate the growing demand for food with a local provenance. The UK supermarket chain, Waitrose, has pioneered the development of short speciality food supply chains and has recently extended this from speciality to more mainstream produce (http://www.waitrose.com/food/originofourfood/index.aspx), but all supermarkets are now showing signs of trying to connect to local food suppliers, particularly but not exclusively in speciality food in order to broaden their offer to the customer. In some cases, the supermarkets will provide substantial support in product development to the supplier. The capacity of supermarkets to sell significant quantities of speciality product is a strong incentive for the small speciality supplier to engage with them. The disincentive to the producer is their dependence on a limited number of buyers with power to impose exacting demands with the associated risk that failure to comply with these demands could lead to the loss of a major sales outlet (Vorley, 2003; Lang and Heasman, 2004).

Supermarkets have thereby entered into new relationships with speciality food producers and small-scale suppliers. How many food products now retailed by supermarkets constitute genuinely locally grounded (endogenous) products and how many, rather than being genuinely traditional, are the invention of marketing consultants or imaginative farmers is not entirely clear. For example, the highly successful Yarg cheese, which is produced in Cornwall, is not a traditional product rooted in the valleys of Cornwall but a marketing opportunity seized by a outward-looking farmer who developed a clearly differentiated cheese product to enhance his survival prospects in a dairy sector feeling the cost-price squeeze (http://www.lynherdairies. co.uk/lyhner.html; BBC, 2007). Even many of the local food initiatives in a country with a deeply traditional food economy such as Italy can often be seen as quite recent examples of innovation and attempts to develop niche products.

As well as the corporate giants that dominate the UK food retailing sector, there are also some regionally based supermarkets operating regional procurement strategies, which have underpinned their commercial success. Booths, a supermarket chain in North West England, is perhaps the best example of this in the UK and in Germany the Tegut supermarket chain has long been operating a similar regional procurement strategy with its food suppliers (Schaer *et al.*, 2006). It is apparent that current market drivers are forcing supermarkets to reduce the dualism between commodity and speciality food. However, we would suggest that this process has actually been going on for some time, with retailers intent on achieving competitive advantage through strategies of differentiation, often involving place. This is particularly evident in the delicatessen and alcoholic drink sectors where Appelation d'Origine Controllee (AOC) and Protected Destination of Origin (PDO) foods have long been widely stocked.

Several studies have pointed to the capacity of larger actors, normally supermarkets, to expropriate the economic surplus of small scale producers. DuPuis and Goodman (2005: 364) note how AAFNs have become a setting for a struggle for the economic rent created by the new market opportunities and talk earlier in the same article about the vulnerability of small producers to corporate 'co-optation'. Tregear *et al.* (2007) echo the general concerns about who actually controls local product designations and point out how conflictual such attempts to create local food certification can be. Mutersbaugh and Klooster (2005) also explore the development of quality certified products and note the increased dominance by the new and powerful private actors' mainstreaming strategies that seek to increase the quantity of certified products sold through conventional markets.

In the case of organic produce, supermarkets now command a very significant share of the UK market at between 65% and 70% of the market (Firth *et al.*, 2004). Sainsburys is contended to be the market leader in organic sales at c. 30% of the supermarket share of organics, but proportionately Waitrose has an even bigger organic proportion of their total food sales. Both firms now offer organic boxes, which have long been a distinctive feature of the traditional short chain direct selling approach, representing a further morphing of the dualism between alternative and mainstream FSCs. Interestingly, the organic market share of supermarkets has dropped in recent years, reflecting at the margin a preference of some consumers for alternative marketing channels (see Firth *et al.*, 2004).

Certain supermarket procurement practices begin to challenge some of the stereotypical views of globalised food supply chains. Given their scale, in some senses it could be argued that they have a greater capacity to engineer a sustainability-enhancing relocalisation of food markets than AAFNs. It is also possible to detect emergent hybridity in their FSCs, including their partial and ongoing reconnection to local producers and processors; again, notwithstanding issues of downward price pressures and an overdependence on a single outlet for local producers.

The degree of hybridity in most UK major supermarket practices is modest. Most appear to be pursuing aggressive policies of gaining market share and differentiating their offer from competitors, which has often meant accommodating more local food into their offer. However, some intermediate scale food retailers such as Booths in North West England and Tegut in Germany operate from a more regional location and have developed a more explicitly regionally connected offer while French supermarkets in general have accommodated a wide variety of local food and drink in their stores. These latter firms exhibit many of the characteristics of larger food retailers but also some distinct differences, which we characterise as hybrid. The accommodation of a large farmers' market in the car park of a major food retailer in Bath in South West England is a further example of emergent hybridity within the FSC.

5. Hybridity in UK food supply chains: implications for rural development

This section draws together evidence from the various theoretical perspectives discussed and available evidence to make the case for the existence of dynamic hybridity in the UK food sector. We extend the debate about hybridity in its narrower ANT context to embrace the possibility of hybrid forms, hybrid theories and hybrid policies. As early as the mid 1990s, Lowe *et al.* (1995) were arguing that the simplistic endogenous: exogenous dualism had limited explanatory power. This judgement has been endorsed recently by a range of commentators from Goodman (2004) to Lockie and Halpin (2005). We too endorse this assertion.

The brief review conducted in this paper of the principal economic macro-theoretical lenses which have been used to explore change in the food sector reveal a hotly contested debate. On the one hand, the neoclassical paradigm offers a world in which some types of AAFNs might be expected to emerge from the crisis-ridden farm sector, but where the evident market power of the major retailers limits the scope for expansion of AAFNs. A political economy approach reinforces this assertion of a challenging business and economic environment for AAFNs because of their predatory capacity on other food chain actors. On the other hand, a much more positive view of rural renaissance can be found by the adherents of the new rural development paradigm who assert that at the heart of rural development are new agrarian and food production and marketing practices, rooted in locale, both in terms of farming style and market output, which offer an economic keystone of the new rural economy. The political economy perspective posits a danger of expropriation of surplus value by larger food chain operators. Such firms can predate on those small scale producers and processors who have developed successful products. The lifestyle individualism of many small-scale processors offers a free market-testing laboratory for the more market-oriented businesses, which will predate, not always successfully, on the small-scale producer, should a bigger market opportunity present itself.

In spite of an enormous amount of literature, there is no unitary body of social science theory explaining rural development. This is to be expected. Different disciplines have addressed rural development through different lenses. Different lenses may throw different light on different facets of rural change in what are acknowledged to be highly differentiated rural areas. In relation to the interactions between food markets and rural development, it is questionable whether any single meta theory from new rural development in the Wageningen agrarian model, to new rural development in the OECD consumption-driven rural economy model (OECD, 2006), to ecological modernisation can adequately embrace the complex range of adaptive responses of rural social and economic actors and their reflexive engagement with new institutional forms and approaches to governance. If a theory is needed it must accommodate the uncertainty of outcomes and the complex interactions of actors. Around the same time as Lowe *et al.* (1995) were criticising the simplifying dualisms that prevailed at the time, so Marsden and Arce (1995:1277) were proposing Actor-Network Theory (ANT) as a lens through which to explore the interaction of local and globalised food supply chains, again pointing to the restrictions of conventional dualisms. Ten years on the dualistic models and polarities have resurfaced with vigour, but there are at least a few examples of the application of ANT which reveal something of the complexity of hybrid forms and the uncertainty of food network outcomes. This is not to suggest that ANT is somehow capable of providing the missing conceptual lens on the complexity and dynamism of the agro-food system (indeed many authors argue that it is largely descriptive and fails to allow for theoretical explanation: e.g. Marsden, 2000 and Buttel, 2001), simply that it provides an example of how hybridity might be analysed within FSC, methodologically.

Network analysis has been extensively utilised within the social sciences to understand relations between social actors, as well as the take up of new technologies, but ANT can be understood as 'a hybrid of these two more traditional forms' (Murdoch, 1994: 3) which allows network construction to be viewed in action (Law, 1992). ANT, or 'the sociology of translation' (Callon, 1986), was conceived by its originators (most notably Michel Callon, Bruno Latour and John Law) as a means of understanding how scientific, technological, natural and social components can form into an interdependent and coherent network. There is no preconceived frame of reference, simply an exploration of network formation that is recognised as negotiated and contingent, whereby 'if the proponents of a new theory fail to gather a large enough network of allies then, in the long run, it will be unsuccessful' (Comber *et al.*, 2003: 303). Crucially, ANT makes no *a priori* distinctions between the various components of a network, thereby allowing for the breakdown of modernist ontological dualisms, such as those between nature and society, structure and agency, production and consumption, and macro and micro-level perspectives (Lockie and Kitto, 2000). In so doing, it facilitates the scrutiny of networks that may be composed of 'hybrid collectives' of actors and mediations in relation to the development of particular food supply chains (Goodman, 1999).

The hybridity of food supply chains is evident in the complex and dynamic relations between small scale localised and often regionally certificated producers and national or even international food retailers. It is exhibited in the early hybridisation of organic food supply chains where the idealism of the early producers has been increasingly compromised by the market penetration practices of the pragmatists, leading to a blurring of the distinctions between them. The sector is now characterised by a range of forms of marketing from traditional local direct sales, to hybrid box schemes, to mainstream supermarket channels. Organic food is now shipped in large volumes over enormous distances and forms a symbolic engagement with the AAFNs. Whilst some, such as Lockie and Halpin (2005), assert that the evidence for conventionalisation is limited, their study does not consider the European context where substantial subsidy has attracted new entrants for opportunistic reasons, who may even have cynically used organic subsidies as a fallowing strategy.

In other work, Lockie with Higgins (Higgins and Lockie, 2002; Lockie and Higgins, 2007) explore hybridity in governance, where elements of neoliberal farm policy are hybridised with community based agri-environmental policy. We detect similar forces in the UK food sector where substantial support is being given to local food initiatives by regional development agencies in a political climate in which neoliberal values and a widening of international trade opportunities are widely extolled.

Hybrid elements are apparent in some of the most rapidly expanding brands in the UK food sector. The company, 'Innocent', which manufactures real fruit 'smoothies' is now ranked 63rd in UK brands and has grown dramatically in the last few years (Guardian 21st February 2007). Although Innocent presents an image of being alternative with a distinctive ethical/green position, it sells the majority of its product through supermarket channels (http://www.innocentdrinks.co.uk/us/?Page=our _ethics). While at first sight this may appear to be symptomatic of mainstreaming processes, the company does still hold to its alternative credentials. In other words, the company's business model exhibits elements of both the mainstream (in terms of the outlets it uses for its products), and the alternative (in terms of its social, economic and natural relations with other actors and agencies). This observation resonates with Goodman (2004: 9) who observes that there are 'some interesting mutations with regard to supply chains…in terms of the types of relations and organisational features they display'. In such instances, we would argue that in evaluations of such food supply chains it is necessary to more specifically acknowledge the observed hybridity, rather than relying on the singular perspective of either mainstream or alternative.

6. Interim conclusions

The postulation of a new rural development paradigm based on the relocalisation of food supplies seems to be based more on normative constructions than strong empirical evidence. It is not that these relocalised food chains are absent, but that their overall impact is uncertain and the calculations of economic impacts to date are anything but robust. It is undeniable that the competitiveness of many rural areas will be contingent on the valorisation of local assets (OECD, 2006) but likely that these assets may depend on much more than the food producing capabilities of the farm sector. The agri-centrality of the Wageningen school's new rural development paradigm differs substantially from the more multi-sectoral consumption-driven OECD perspective. The bulk of evidence about rural demographic and economic change supports the idea of an increasingly consumption–driven rural economy rather more than the impending triumph of a localist counter hegemony (DuPuis and Goodman, 2005: 361).

AAFNs have attracted enormous research attention in Western Europe and more widely. This interest is evident not least because this model affords possibilities of providing an alternative livelihood strategy for some farmers, but because they may act as a harbinger to a stronger relocalisation of food systems. Further, these AAFNs are often contingent on new institutional forms which are often spatially circumscribed and thus different to the predominantly sectoral development policies which have hitherto prevailed. However, this does not of itself amount to the underpinnings of an alternative or new rural development paradigm. It simply exposes a developing arena of interesting activity in food markets, which is perhaps most highly developed in the European Union than elsewhere because of Europe's policy history, though it is by no means an exclusively European phenomenon. Instead of a new paradigm of rural development, we see important development prospects evident and emerging in the hybrid zone of FSCs, both in relation to policy and practice.

The examination of organic farming shows how in practice many of the core ideologies of the organic movement can be compromised by the scaling up of organic production and the engagement with major food retailers. Lockie and Halpin (2004: 304) have argued with some conviction that: 'we need to unpack the concept of conventionalisation and avoid an uncritical aggregation of multiple dualisms between small and large, artisanal and industrial radical and regulatory local and international, regenerative and substitutionist and so on'. However, they also argue that the values of established and new organic farmers in Australia are not significantly different, suggesting that if some elements of organic farming's supply chains are scaled up and internationalised, this does not necessarily impact on the core values of the organic farmers involved. In other words, the resultant supply chain is neither wholly mainstream nor alternative, but a hybrid between the two.

Although the mainstream food system has been challenged by 'food scares' and deserted by some 'discerning' consumers who have shifted their allegiance to alternative production systems and markets, the mainstream food sector still appears to be resilient and in good financial health. Its success is testament to its adaptive capacity in driving a tough bargain with producers and stripping cost (and other supply chain actors' profit) out of food supply chains and in developing a sophisticated awareness of consumer needs. Its market development has been highly innovative in its response to the recognition of local and regional foods. Supermarkets have developed relationships with many suppliers of speciality regional foods. They have adapted their offer so that the consumer is now confronted by an enormous range of choice. They have pioneered the expansion of the organic food market. They have recognised the public concern about traceability and (with public support and policy requirements in the wake of the BSE/vCJD crisis) now operate rigorous traceability systems (Flynn *et al.*, 2003). With their enormous capacity for market research and product development, they have moved on from simple quality control of commodities to embrace other dimensions of quality, including more local foods and new 'taste the difference' or equivalent brands, driven in part by perceived consumer demand but also by pursuit of greater profit margins. Nevertheless, although their short supply chains may be short in terms of numbers of links, they are still often long in terms of distance, despite some evidence of shorter more localised chain development (Vorley, 2006).

The supermarkets have both the power to predate on the producers who might normally be associated with AAFNs, yet at the same time to provide outlets for their produce. For the supplier of a high quality food currently operating in AAFNs, there are at least two possibilities: firstly, to engage with the supermarkets and to accept the significant loss of independence but counter this with the increased capacity for growth; or secondly, to reject any overtures from supermarkets and to continue to use alternative marketing channels which may result in restricted market opportunities. Further, although there are concerns about the path dependencies established by supermarkets' predominantly national and regional distribution systems which compromise (or at least delay) the development of local trading arrangements (Vorley, 2006), their record of flexibility suggests a continued capacity to extract a high proportion of even the discerning shoppers' retail expenditure on food.

We are left wondering why the advice not to get hooked into binaries and dualisms has been so repeatedly ignored. The real interest in food chain dynamics should be in the existing and emergent hybrid relationships between AAFNs and the mainstream. We believe that in this negotiated territory between mainstream and AAFNs there are profound changes afoot, which will manifest themselves in a variety of ways, places and hybrid forms. In the emergent food system we anticipate a dynamic response to emerging policies that address sustainability generally and climate change specifically. This will probably lead to a degree of regionalisation of food supply, in a retail system which remains dominated by major retailers. The latter will continue to offer a mix of commodities and specialities which will increasingly incorporate local produce in their offer through connecting to local supplies. We anticipate continued buoyancy in AAFNs but argue that there will be pressures for scaling up, during which some of the factors that predicated the development of AAFNs will be altered by their association with the mainstream system, further heightening the tendency towards hybridity. We conclude that it is necessary to reject the dualistic interpretations of contemporary food systems and to better understand the expanding elements of hybridity in both process and form.

References

BBC, 2007, Discover yarg cheese. Available at: http://www.bbc.co.uk/cornwall /attractions/stories/yarg.shtml. Accessed: 14.11.2007.

Burt, S.L. and L. Sparks, 2003. Power and competition in the UK grocery market. British Journal of Management, 14: 237-254.

Buttel, F., 2001. Some reflections on late twentieth century agrarian political economy. Sociologia Ruralis, 41: 165-181.

Callon, M., 1986. Some elements of a sociology of translation: domestication of the scallops and fishermen of St Brieuc bay. In: Law, J. (ed.), Power, action, belief: a new sociology of knowledge? London: RKP.

Cloke, P., 2003. Knowing ruralities. In: Cloke, P. (ed.) Country visions. Pearson: Harlow.

Comber, A., P. Fisher and R. Wadsworth, 2003. Actor–network theory: a suitable framework to understand how land cover mapping projects develop? Land Use Policy, 20: 299-309.

Competition Commission, 2000. Supermarkets: a report on the supply of groceries from multiple stores in the United Kingdom. Volume 2. London: The Stationery Office.

Daw, M., R.W. Slee and E. Wynen, 1991. Organic agriculture: a review of the marketing and economics of production with particular reference to Scotland. SAC Economic Report, 32.

Dobson Consulting, 1999. Buyer power and its impact on competition in the food retail distribution sector of the European Union. Report to the European Commission.

DuPuis, E.M. and D. Goodman, 2005. Should we go 'home' to eat? Toward a reflexive politics of localism. Journal of Rural Studies, 21: 359-371.

Firth, C., N. Green, C. Foster, M. Green, R. Hayward and A. Smithson, 2004. The organic UK vegetable market 2002-3 season. Report to Defra.

Flynn, A., T. Marsden and E. Smith, 2003. Food regulation and retailing in a new institutional context. Political Quarterly, 74: 38-46.

Goodman, D., 1999. Agro-food studies in the 'age of ecology': nature, corporeality, bio-politics. Sociologia Ruralis, 39:17-38.

Goodman, D., 2003. The quality 'turn' and alternative food practices: reflections and agenda. Journal of Rural Studies, 19:1-7.

Goodman, D, 2004. Rural Europe redux? Reflections on alternative agro-food networks and paradigm change. Sociologia Ruralis, 44: 3-16.

Goodman, D., B. Sorj and J. Wilkinson, 1987. From farming to biotechnology: a theory of agro-industrial development. Oxford: Blackwell.

Guardian 21st February 2007. Organic food breaks into top 100.

Guthman, J., 2000. Raising organic: an agro-ecological assessment of grower practices in California. Agriculture and Human Values, 17: 257-266.

Guthman, J., 2004a. Back to the land: the paradox of organic food standards. Environment and Planning A, 36: 511-528.

Guthman, J., 2004b. The trouble with 'organic lite' in California: a rejoinder to the 'conventionalisation' debate. Sociologia Ruralis, 44: 301-316.

Higgins, V. and S. Lockie, 2002. Re-discovering the social: neo-liberalism and hybrid practices of governing in rural natural resource management. Journal of Rural Studies, 18: 419-428.

Higgins, V., J. Dibden and C. Cocklin, 2007. Market-oriented initiatives for agri-environmental governance: environmental management systems in Australia. In: D. Maye, L. Holloway and M. Kneafsey (Eds.), Alternative food geographies: representation and practice. Oxford: Elsevier, pp. 223-238.

Hobbs, J. and J. Hobbs Jr, 1996. A transaction cost approach to supply chain management. Supply Chain Management, 1: 15-27.

Ilbery, B., D. Watts, S. Simpson, A. Gilg and J. Little, 2006. Mapping local foods: evidence from two English regions. British Food Journal, 108: 213-225.

Ilbery, B. and D. Maye, 2005. Alternative (shorter) food supply chains and specialist livestock products in the Scottish - English borders. Environment and Planning A, 37: 823-844.

Ilbery, B. and D. Maye, 2006. Retailing local food in the Scottish-English borders: a supply chain perspective. Geoforum, 37: 352-367.

Klooster, D., 2005. Environmental certification of forests: the evolution of environmental governance in a commodity network. Journal of Rural Studies, 21: 403-417.

Lampkin, N., 1990. Organic farming. Ipswich: Farming Press.

Lang, T. and M. Heasman, 2004. Food wars: the global battle for mouths, minds and markets. London: Earthscan.

Law, J., 1992. Notes on the theory of the actor-network: ordering, strategy and heterogeneity. Systems Practice, 5: 379-393.

Lockie, S. and D. Halpin, 2005. The conventionalisation thesis reconsidered: structural and ideological transformation of Australian organic agriculture. Sociologia Ruralis, 45: 284-307.

Lockie, S. and V. Higgins, 2007. Roll-out neoliberalism and hybrid practices of governing in Australian agri-environmental governance. Journal of Rural Studies, 23: 1-11.

Lowe, P., J. Murdoch and N. Ward, 1995. Beyond models of endogenous and exogenous development. In: Van der Ploeg, J.D. and Van Dijk, G. (eds.), Beyond modernisation. Assen: Van Gorcum.

Marsden, T., J. Murdoch and K. Morgan, 1999. Sustainable agriculture, food supply chains and regional development: editorial introduction. International Planning Studies, 4: 295-301.

Marsden, T., 2000. Food matters and the matter of food: towards a new food governance? Sociologia Ruralis, 40: 20-29.

Marsden, T. and A. Arce, 1995 Constructing quality: emerging food networks a in the rural transition. Environment and Planning A, 27: 1261-1279.

Morgan, K. and J. Murdoch, 2000. Organic vs. conventional agriculture: knowledge, power and innovation in the food chain. Geoforum, 31: 159-173.

Morris, C. and H. Buller, 2003. The local food sector: a preliminary assessment of its form and impact in Gloucestershire. British Food Journal, 105: 559-566.

Murdoch, J., 1994. Weaving the seamless web: a consideration of network analysis and its potential application to the study of the rural economy. Working Paper 3, February. University of Newcastle upon Tyne: Centre for Rural Economy, Department of Agricultural Economics and Food Marketing.

Murdoch, J., 2003. Co-constructing the countryside: hybrid networks and the extensive self. In: Cloke, P. (ed.), Country visions. Pearson: Harlow.

Murdoch, J. and M. Miele, 1999. 'Back to nature': Changing 'worlds of production' in the food sector. Sociologia Ruralis, 39: 465-483.

Mutersbaugh, T., D. Klooster, M.-C. Renard and P. Taylor, 2005. Certifying rural spaces: quality-certified products and rural governance. Journal of Rural Studies, 21: 381-388.

OECD, 2006. The new rural paradigm: policies and governance. Paris: OECD.

Padberg, D.I., C. Ritson and L.M. Albisu, 1997. Agri-food marketing. Wallingford: CABI.

Pretty, J., 1998. The living land: agriculture, food and community regeneration in rural Europe. London: Earthscan.

Ray, C., 2003. Governance and the neo-endogenous approach to rural development. Paper presented to ESRC research seminar on Rural Social Exclusion and Governance.

Renard, M.-C., 2005. Quality certification, regulation and power in fair trade. Journal of Rural Studies, 21: 419-431.

Renting, H., T. Marsden and J. Banks, 2003. Understanding alternative food network: exploring the role of short food supply chains in rural development. Environment and Planning A, 35: 393-411.

Schaer, B., K. Knickel and C. Strauch, 2006. Regional embedding as a marketing strategy: Tegut supermarket and Rhongut meat processing. In: D. Roep and H. Wiskerke (Eds.), Nourishing networks: fourteen lessons about creating sustainable food supply chains. The Netherlands: Rural Sociology Group, Wageningen University; and Reed Business Information, pp. 123-134.

Slee, B., 1989. Alternative farm enterprises, 2nd Ed. Ipswich: Farming Press.

Soil Association, 2005. Organic market report 2005. Press release.

Sonnino, R. and T. Marsden, 2006. Beyond the divide: rethinking relationships between alternative and conventional food networks in Europe. Journal of Economic Geography, 6: 181-199.

Tregear, A., F. Arfini, G. Belletti and A. Marescotti, 2007, Regional foods and rural development: the role of product qualification. Journal of Rural Studies, 23: 12-22.

Van der Ploeg, J.D., 2003. The virtual farmer: past, present and future of the Dutch peasantry. Assen: Van Gorcum.

Van der Ploeg, J.D., 2006. Agricultural production in crisis. In: P. Cloke, T. Marsden and P. Mooney (Eds.), Handbook of rural studies. London: Sage Publications, pp. 258-277.

Van der Ploeg, J.D. and A. Long (Eds.), 1994. Born from within, practices and perspectives of endogenous rural development. Assen: Van Gorcum.

Van der Ploeg, J.D. and H. Renting, 2004. Behind the 'redux': a rejoinder to David Goodman. Sociologia Ruralis, 44: 233-242.

Van der Ploeg, J.D., H. Renting, G. Brunori, K. Knickel, J. Mannion, T. Marsden, K. De Roost, E. Sevilla-Guzman and F. Ventura, 2000. Rural development: from practices and policies towards theories. Sociologia Ruralis, 40: 391-408.

Vorley, B., 2003. Food, Inc.: corporate concentration from farm to consumer. A report prepared for UK Food Group. IIED, 3 Endsleigh Street, London WC1H 0DD.

Vorley, B., A. Fearne M. Pitts and W. Farmer, 2006. A flexible procurement system for local sourcing: supermarket sourcing of local and regional food. In: D. Roep and H. Wiskerke (Eds.), Nourishing networks: fourteen lessons about creating sustainable food supply chains. The Netherlands: Rural Sociology Group, Wageningen University; and Reed Business Information, Agriboek, pp. 103-112.

Watts, D., B. Ilbery and D. Maye, 2005. Making reconnections in agro-food geography: alternative systems of food provision. Progress in Human Geography, 29: 22-40.

Whatmore, S., 2002. Hybrid geographies: natures, cultures spaces London: Sage.

Winter, M., 2003. Embeddedness, the new food economy and defensive localism. Journal of Rural Studies, 19: 23-32.

Competitive positioning and value chain configuration in international markets for traditional food specialties

O.J. Borch and I.H.E. Roaldsen

Abstract

In this paper we discuss the relations between up market quality standards for specialty food products and the structuring of the value chain. We elaborate on the process where complex immaterial quality dimensions are built through bundling resources at different levels. We present results from 11 in-depth interviews with representatives from the fragmented Norwegian value chain for lamb products. The results show that to achieve customer-oriented differentiation focus you have to combine different regimes of coordination and control throughout the value chain. Implications for management of the value chain and contract relations between the actors in the chain are discussed.

Keywords: quality differentiation, value chain, configuration, management, contracts

1. Introduction

Today's up market consumers are increasingly quality conscious in their product choice. Quality aspects like different sensory attributes based on traditional conservation and processing methods, regional food culture and an ethical sound production chain has caught the attention of the postmodern consumers (Bijman *et al.*, 2006). This opens up new opportunities for countries with a marginal, small scale food production, focusing on the potential for increased export of agriculture small-scale products based on natural resources and traditions. Quality aspects such as clean nature and plant and animal health are also stressed as comparative advantages. The marketing of higher-order quality dimensions e.g. immaterial quality dimensions such as country of origin, food traditions, processing methods, the history behind the product, have consequences not only for the processing and end product. It also means that we have to develop a strategy emphasising the new quality standards throughout the value chain. The objective of this paper is to show how the advanced quality standards of traditional food products and the competitive strategy positioning needed have consequences for the structuring and management of the whole value chain. To understand how small- and medium-sized firm adapt to new market challenges we need a broad understanding of strategic tools and how they are generated within the organisation. To improve both the rigor and relevance of the strategy construct, several authors contend that business strategy should be conceptualised according to sub strategies at the level of the business unit. This approach would facilitate the study of strategy from a managerial perspective, and would reduce the risk of creating models that are too simplistic (Chrisman *et al.*, 1988; Hofer and Schendel, 1978; Morrison, 1990).

To understand how small- and medium-sized firms adapt to new market challenges we need a broad understanding of strategic tools and how they are generated within the organisation. Limiting the study to positioning and competitive tools may prove insufficient. We have to look into the organisational and managerial configuration of the firm to see if the firm manages to enter more complex differentiation strategies (Chandler and Hanks, 1994; Brush and Chaganti, 1998). A resource-focused organisational approach include the structural configuration of the value chain, functional parts within the firm, as well as parts of the value chain controlled through cooperative relations with other organisations.

Thus, we discuss how different higher-order quality dimensions are integrated into a combined competitive positioning concept. Secondly, we look at how the implications of such a broad and complex set of competitive tools have for the structuring of the value chain. We elaborate on the capability of the firm management to develop an organisation that supports more complex differentiation strategies (Brush and Chaganti, 1998).

In section two we present relevant theory for the discussion of competitive tools and configuration of the value chain. In section three and four we present the methodology and the results of case studies within two value chains. In the final section, we conclude and discuss both scientific and practical implications.

2. Theory

2.1 Strategic positioning and organisation

The choice of market strategy is a complex task within small firms. While larger firms may have the resources available for a fine-grained positioning adapted to power play in the market, the smaller firms often have to be more creative in applying existing resources through organisational and governance oriented tools (Brush and Chaganti, 1998; Borch *et al.*, 1999). We should therefore look at both the competitive tools and the resource configuration of these firms to decide upon their opportunities for creating sustainable competitive advantage (Rangone, 1999). When it comes to competitive strategy, Porters's (1980, 1985) theory of generic competitive strategy has been among the most influential for the last two decades. According to Porter (1980, 1985) firms failing to choose between the alternative strategies of cost leadership and differentiation, risk being ousted on all fronts. However, there has been a critique of the normative postulate inherent in the dominant Porter-inspired paradigm. This critique is related to the opportunities for combined differentiation and cost leadership strategies (Murray, 1988) and the possible success of following a non-distinct flexibility strategy (Cambell-Hunt, 2000).

In competitive markets with increasing internal rivalry, the producer needs to develop additional tools in order to secure future competitive advantages. Not the least we may find such harsh competition in the mature food markets. The increased rivalry from other firms, import products and substitutes, and the increased negotiation power of the wholesaler-retailer chains imply a high degree of focus on the cost dimension *together* with the efforts towards differentiation (Borch and Forsman, 2003; Borch and Brastad, 2003).

New organisational resources may increase flexibility in choosing among strategic tools. One may expect extra opportunities for enterprises that are flexible on different tools to meet new opportunities and changing trends (Cambell-Hunt, 2000; Rangone, 1999). In particular, when few financial resources are available for buying new resources cooperative strategies are at hand. Through including cooperative relations, the small firm may develop bundles of internal and external resources that may increase the range of competitive tools for the small firms including mixed cost and differentiation strategies and non-distinctive flexibility strategies. The strategic advantages of closer links with other firms in the value chain compared to the traditional arm-length market exchange have been highly emphasised within small business research.

Day and Wensley (1988) and Spender (1993) criticised strategy research for not sufficiently addressing the conversion of organisational skills and resources into positional advantages in the market. Including the resource-based dimensions of competence, routines and working culture may accentuate the intra-organisational premises for achievement and the maintenance of competitive advantage (Barney, 1991; Black and Boal, 1994; Leonard-Barton, 1992). An integrated organisation sub strategy is defined as the immaterial quality of an organisation in terms of competence, routines, personal commitments and working culture inside the organisation and in the interplay with partners outside the firm (Cooper, 1993; Brush and Chaganti, 1998).

2.2 The configuration of the value chain

Implementing a more customer oriented adaptation of products with a strongly differentiated strategy implies higher dependency of a quality approach throughout the value chain. As the firm may not easily achieve internal control over the whole value chain, there is a need to have inter-organisational coordination mechanisms. Also, within larger firms the value chain is split into several companies working more or less independently within the corporation. Thus, superior communication, coordination and control mechanisms are needed both inside the single production unit and between all the companies taking part in the value chain of the quality-differentiated products in question.

In this study we emphasise the need for quality improvements throughout the value chain and how this is achieved through new organisational mechanisms. We are here talking about chains and networks (Bijman *et al.*, 2006; Ghisi *et al.*, 2006). Networks may harmonise competition and enhance cooperation and further to aim for a common solution to similar problems faced by small and medium-sized enterprises (Ghisi *et al.*, 2006).

Stabell and Fjeldstad (1998) expand this perspective by presenting three alternative value configurations as a foundation for a theory of value configuring for competitive advantage. Their work builds on Porter's (1985) original value chain framework and Thompson's (1967) typology of long-linked, intensive and mediating technologies. Stabell and Fjeldstad (1998) propose that the *value chain* models the activities of a long-linked technology, further that the *value shop* models firms where value is created by mobilising resources and activities with the purpose of resolving a particular problem related to the consumer, and finally the *value network* models firms that create value by facilitating a network relationship between their customers using a mediating technology. By introducing these three configurations it's also stated that there will be a need for transforming the value chain analysis into a value configuration analysis (Stabell and Fjeldstad, 1998).

All three configurations focus upon critical value activities, the distinction between primary and support activities, and the analysis of cost and value drivers. Stabell and Fjeldstad, (1998) suggest that the *value chain* requires a machine bureaucracy organisation of primary activities, further that the *value shop* is organised according to either the professional bureaucracy of the operational adhocracy, and moreover the *value network* often is organised according to an administrative adhocracy, particularly when the technology of the infrastructure is complex and requires highly specialised development activities.

Mason *et al.* (2005) suggest three main implications for managers. First, there's the appropriate selection of integration typologies in order to facilitate a demand driven supply chain configuration. Second, there's the recognition of the need for careful identification of supply chain partners in order to facilitate supply chain influence. Finally, there's the way firms define and manage supply chain influence with partnering firms. This study support the theory that the level of market orientation achieved will be significant affected by the relationship focus, channel power, channel leadership, communication, and coordination technology present in quasi-integrated forms.

Imbalanced relationships are stated to be of great importance in business to business relationships (Hingley, 2005). Investigations of the nature and management of power in vertical supply chain context, represented by the UK fresh food channels, show that power is imbalanced in fresh food relationships and that such an imbalance is in favour of buying organisations (Hingley, 2005). However, Hingley (2005) pointed out that this does not mean that such relationships cannot benefit weaker partners in asymmetric exchange, neither that it can be workable, nor be successful and longstanding, or finally that those suppliers do not want to enter into such relationships.

Fjeldstad and Ketels (2006) state that the choice of this value creation has implications for the development of the business strategy. Value creation occurs differently within the value chain and the value network. In the value chain, the value creation derives from products implied that the products match customers needs. On the other hand, value networks create value by enabling exchanges. Competitive advantage occurs when the network matches the needs of its members (Fjeldstad and Ketels, 2006). A value chain product or service properties are at the centre, whilst in a value network the customer is placed in the centre.

Different value configurations are suggested to become the starting point for gaining a more systematic understanding of which of these choices are critical, and how they interact for different types of firms and strategies.

The concept of netchains is introduced among others by Lazzarini *et al.* (2001) to fill the voids of the supply chain analysis literature and the network analysis literature. A netchain is defined as a set of networks comprised of horizontal ties between firms within a particular industry or group, such that these networks are sequentially arranged based on the vertical ties between firms in different layers (Lazzarini *et al.*, 2001). An important aspect of the netchain is the fact that this concept explicitly differentiates between horizontal and vertical ties, to point out how agents are related to each other, either within the same layer or between the different layers. The purpose of the netchain is to integrate both the supply chain analysis and the network analysis. This is to be done through recognising that complex inter-organisational settings includes different kinds of interdependencies associated with sources of values like strategic variables yielding economic rents, and associated with coordination mechanisms involved in an inter-organisational collaboration (Lazzarini *et al.*, 2001).

The netchain perspective suggests that the assessment of interdependencies in a given inter-organisational setting should be the first analytical step in a rent creation system. Further they encourage managers to develop social ties where activities are mutually adjusted instead of planned, at the same time as they're pursuing flexibility to position their firms in valuable networks to benefit from new information and knowledge diversity. Moreover, the netchain perspective insists that the design of interdependencies is the first step in the formation of inter-organisational strategies (Lazzarini *et al.*, 2001).

In this study we take as a starting point the basic value chain and look at the changes in value configuration as the complexity of the production increases due to higher quality ambitions. We emphasise that there may be more configurations present simultaneously to manage the demands of both efficient production, continuous improvements of the present products, and for more explorative activities towards new product platforms.

3. Methodology

To investigate the adaptations necessary to improve the quality standards throughout the value chain, we chose a descriptive case study design. The case study approach allows us to maintain a holistic and meaningful characteristics of real-life events (Yin, 2003), which is of great importance when studying the structuring and management of the value chain. Currently, policy makers in several countries are focusing on the potential of increased export of agriculture small-scale products based on natural resources and tradition.

We chose a longitudinal in-depth research strategy focusing on following the efforts of two specific value chains over several years.

3.1 Data collection and analysis

Secondary data about the industry, companies and the market was acquired through industry reports, news papers, minutes from business meetings, etc. Primary data was collected through observation at business meetings and plant visits during the manufacturing process. Secondly, we conducted in-depth interviews with actors throughout the value chain, representing Norwegian companies exporting lamb meat products to Italy. The interviewees, ranging from farmers to top managers, were recruited through their involvement in exporting lamb meat products to Italy. The value chains of two different locations in Norway were investigated, referred to as value chain A and value chain B in the following. More specific, the persons interviewed represented farmers, managers of slaughtering houses, managers of meat processing companies, and managers of distribution, sales and export.

An interview guide was constructed in advance building upon relevant theory and secondary data providing insight into the processes of the lamb meat production. The respondents answered the same questions together with keywords related to the specific part of the value chain they represented. First, the questions concerned their individual idea of a special quality of lamb meat products. Second, they were asked about basic adaptations in their part of the value chain when dealing with an extraordinary quality. Then questions concerning adaptations inside their organisation were asked, followed by questions on adaptations in other parts of the value chain. Finally the respondents were asked to answer questions regarding the performance of the adaptations, in addition to cost implications.

The interviews were conducted over a two month period of time, between February and March 2007. Most of the interviews were conducted by telephone, but some of them were done in person, depending on the availability of the representatives of the two value chains. The duration of the interviews varied from half an hour until two hours. Every interview was audio recorded and transcribed afterwards.

The results of the in-depth interviews were broadly categorised in the light of the theory emphasising on competitive tools and especially quality dimensions. We linked responses along these dimensions to organisational configuration efforts within the firm, and the interaction between the firms. Two tables were made, one for each of the two value chains (A and B) to structure and interpret the findings of this study.

4. Results and discussion

The results of this study are presented in the two tables below, followed by a discussion of the results in line with the theory section. In the tables, the first category 'quality dimensions' refers to the respondents' opinion on the most important quality dimension for specialised lamb meat products, when specialised refers to products with higher-order quality dimensions. Further, the category 'basic production adaptations' refers to the adaptations which have to be made in the basic production in order to improve and maintain a high quality of the products. Then, 'intra organisational adaptations' are those adaptations which have to be made within the organisational structure of the firm, whilst 'inter organisational adaptations' are those which have to be made between each part of the value chain, e.g. between the farmer and the slaughtering house etc. Finally, the category 'performance implications' refers to the consequences all of the above mentioned adaptations have for the firms' performance, measured as cost implications.

4.1 Quality dimensions

As shown in Table 1 and 2, different quality dimensions both material (physical) and immaterial (non-physical) are listed by the respondents according to where in the value chain they are positioned. Further, there is some degree of overlap between the different phases of the value chain.

In every phase of the two value chains, there are a focus upon the physical quality aspect of the product, e.g. meat body, fat content, tenderness and so forth. The terminology differs however, in the different parts of the value chain referring to it as either eating quality, meat quality or physical quality. This creates communication challenges throughout the value chain. Then there's the aspect of focus upon immaterial quality dimensions that varies between the different parts of the value chain. Aspects like origin, history or storytelling, and tradition are to be considered as central findings, appearing in our data material in different forms. However, these quality dimensions may dominate in the primary (farm) part of the value chain and downstream towards the marketing of the end product.

In the farming-, slaughtering- and processing phase the findings show that in value chain A there is much emphasis on the grazing conditions and if the animals come directly from outfield to the slaughters. This is seen as a competitive advantage within the above mentioned phases of the value chain. The manager of one of the processing companies explained this as:

> 'You become what you eat. If the lambs eat garbage before they are slaughtered it reflects on the quality. If they eat herbs, etc. that grows in the mountain areas, then it shows in the meat.'

When looking into distribution, sales and export, the health aspect becomes important, in relations with demands of absence of illness or medicine use. These parts of the value chain also mention origin and history as important quality dimensions. One representative of the slaughtering houses put this in the following way:

> 'The origin factor from the environment we are in, the arctic environment, is very important. And the unique part of the country we live in. Then to use this as an advantage in building a trademark. Grazing areas here should also be seen as an advantage.'

Farmers in both value chains believe that willingness to pay for their products are linked to the immaterial quality dimensions. In addition, they are concerned with the flavour of the meat being in accordance with the wilderness and looked upon as a pure organic product. The slaughtering houses are, not unexpectedly concerned with the physical quality of the meat in a more material way than the others. The processors in both of the value chains mention raw material as an important aspect. In one way the raw material is important for the processing process because of it being done like handicraft work, and in another way the access of the raw material is seen as a quality aspect.

Another finding related to distribution and sales, the representatives are very concerned with the ability to deliver the products at the right time according to the demands in the market. This is not a concern shared by representatives from other parts of the value chains.

To sum up, there exists a preference for a combination of material- and immaterial quality dimensions throughout the whole value chain. Moreover, this aspect of immaterial quality dimensions can occur in different ways according to the part of the value chain that is under scrutiny. When comparing the differences between value chain A and B, there is more focus towards immaterial quality dimensions in value chain A in comparison with value chain B. This may be due to the fact that these two value chains represent two distinct parts of the country.

4.2 Configuration at primary production level- the sequential value chain with some modifications

In every part of the value chain there seems to be a common agreement concerning labelling as a central part of the adaptations in the primary value chain. This is accounted for in both value chain A and B. It is to a large extent possible to standardise some of the steps towards increased quality in a sequential value chain. However, there are questions of bringing in higher quality standards than normal for the average production. This may call for involvement at higher levels, and may cause protests within the primary value chain.

An interesting finding is that several of the respondents initially claim not to adapt in a special manner when dealing with specialised meat products. Instead they claim to be following ordinary production routines and regimes. This is mostly the case for those representing the slaughters as shown below with quotes from two of the managers:

> *'Most of the time we run the same arrangements, at least in the season.'*
> (Manager Value chain B)

> *'As a starting point, there are no significant adaptations that we need to do.'*
> (Manager Value chain A)

When investigating this further, there is no doubt that they actually do make adaptations. A result that also represents several of the respondents is the fact that they sort out the production of specialised meat produce, one way or another. For instance, in the slaughtering part of the value chain, the animals are slaughtered at a specific time during the day, sometimes also at separate weekdays than what is the reality of ordinary produce. In the processors part of the value chain, a separate department is handling these special products. The same occurs for the distribution and sales part of the value chain, where the production of specialities is placed at one regional division only.

Table 1. Value chain A.

Part of the value chain	Quality dimensions	Basic production adaptations
Farmer	• Willingness to pay • Meat body • Fat content • Directly from outfield • Organic • Pure product	• Follow standard production • Keep lambs away from public roads • Grazing geography in line with trademark
Slaughtering	• Origin: arctic environment • Grazing area • Lambs appearance • Meat body • Fat content • Physical quality	• Follow ordinary routines • Classification • Labelling • Documentation • Region of origin specifications • Cutting guidelines • Time of slaughtering during the day • Extra sorting • Extra labelling
Processing	• Feed • Mountain grazing • History • Handcraft • Raw material • Fat content • Tradition	• Control over animals • Process animals from one farm at time • Labelling • Communicate differentiation • Food safety • Hand made prod. • Special spices • Special packaging design • Trademark • Traceability
Distribution & sales[1]		
Export[1]		

[1] Firms operating in both value chain A and value chain B. Results presented in Table 2.

Intra organisational adaptations	Inter organisational adaptations	Performance implications
• 1 day of extra work • Hired extra personals • Special registration and labelling, lambs sent to a special processor	• Deliver to special processor • Lambs to that processor must be separated from others at the slaughters • Extra labelling • Extra documentation	• Follow specific standards • Increased cost (traceability) • Special agreement with processor • Need increased competence
• Extra quality consciousness among staff • Extra control routines • Updating /management of cutting department • Enough human resources • Extra skilled workers • Extra work separating & keeping special from ordinary produce	• Extra work for shops and chains to promote special products • Interested and enthusiastic producers • With farmer: cooperation & documentation • With processor: • take the whole animal not only the 'best parts'	• More demanding specifications • No increased costs • More work • Increased salaries • Shortage of raw material for own produce • Extra costs buying more raw material
• Long term strategy • Influence framework conditions • Adapt our systems • Trained/skilled personals • Handling small streams of goods • Hired product & marketing coordination • Started new business from an old one	• With farmer: lambs directly from outfield. • With slaughter: slaughter animals within limited time frame • Coordination • Transportation • Reporting • Buy services from farmer, slaughter, etc. • Special classification • Coop. with cutter • Every part of value chain acts with our standard	• Special agreement with the farmers: increased cost • Increased production cost • More time consuming work • More following up work • Special agreement with sales firm

Table 2. Value chain B.

Part of the value chain	Quality dimensions	Basic production adaptations
Farmer	• Mountainous taste • Wild game flavour • Willingness to pay	• Sorting the lambs • Weighing the lambs • Extra infield pasture
Slaughtering	• Slaughtering quality • Meat quality • Eating quality: meat body, fat content, tenderness	• Follow ordinary routines • Keep the stream of animals at a steady motion
Processing	• Access to raw material of good quality	• Separate dep. for specialties • Weather conditions • No use of machines (handcraft) • New packaging and labelling
Distribution & sales[1]	• Physical quality: meat body, fat content, bone • Deliverance at the right moment • Origin / storytelling • Health aspect • Outfield • Grazing • No illness or medicine use	• With farmer: slaughter outside season and avoid freezing • With slaughter: avoid stress, make anatomic cutting • Hygiene • Sort female/male • With processing: durability and temp. • With sales: one location, presentation, stable deliverance
Export[1]	• Origin • Health • Outfield grazing • No illness or medicine use	• Positioning in up market segment • Choose the best animals • Careful cutting • Transport • Distribution system

[1] Firms operating in both value chain A and value chain B.

Intra organisational adaptations	Inter organisational adaptations	Performance implications
• Hire extra personals	• Sorting the lambs • More work in processing part	• Increased costs • Crop reduction • Special agreements • More documentation
• Extra planning as to transport • Extra sorting and handling into a small stream of goods • Extra pH- and temp. measure • Competence • Create understanding for small streams of goods	• With producer: gain satisfying animal growth to gain tender meat	• Marginal increase in cost because of small streams of goods
• Employees have specialties as their special field of competence • Small, integrated administration	• With farmer: specification of weight and fat content, no infield grazing, no illness or medicine use • With slaughter: fulfil extra demand	• Special agreement with suppliers • Extra cost of personals • Networking with slaughters
• Special group of products • Separate focus area • Extra personals in sales, managers, product manager, retail chain negotiator		• Slaughtering capacity • Distinction in the streams of goods • Advanced computer systems
• Established export dep. • Economy systems • Language barriers • Be professional in up market	• Special agreement with farmer and slaughter • With farmer: pasture, feed • With slaughter: measure pH, classification, selection	• Increased costs • Differentiation • Selling the world most expensive product • Everyone has to take responsibility

In value chain A there is a strong emphasis on looking after the aspect of origin, through adaptations for the farmer to let the animals graze in a certain geographically area. Then for the slaughtering houses there are adaptations as to extra classification, labelling, and documentation. In both value chains the processors make adaptations to manufacture their products in a handcraft manner based on old, traditional recipes and avoid using machinery as much as possible. In addition, the processors mention adaptations as to packaging and labelling, with special emphasis on the element of design. The representatives of both value chains focus upon anatomic or correctly done cutting of the slaughtered animals.

To sum up, every part of the value chain has made adaptations, even if they might claim not to, initially. Most of the adaptations in the basic value chain have to do with sorting the animals in the farming- and slaughtering phase, followed by extra labelling and a more thoroughly made packaging in the parts of sales and distribution. One other thing to take notice of is the fact that some activities in some parts of the value chain A and B are placed at special locations. This makes the production process more specialised and more easy to introduce new handling procedures and efforts towards improvement of standards.

4.3 Configuration at administrative level-the value shop

The studies of the primary production show that in almost every part of the value chain adaptations had to be made regarding timing, procedures and labour efforts. In the farming phase there was a need for hiring extra personals. In the slaughtering phase there was a need of more skilled and trained personals, with special focus on production of specialised meat products. This created a high strain at the administrative level in communicating the new standards and motivating for improved efforts. This can be described with the following statement from one representative of the processing part of the value chain:

> '(…) this was done by informing and talking to those who do the practical work and for them to become aware of their work and how it affects the raw material in the next part of the chain. In this kind of production line, the product is never better than our weakest link!'

Competence among the workers is important independently of the part of the value chain the workers take part in. Further, adaptations like extra control routines; extra communications concerning updating and additional management capacity for coordination are also mentioned as important aspects of adaptations.

In both value chain A and B, in the farming phase, there are focus upon the need of extra labour. In the slaughtering- and processing part of the value chain there are focus upon the handling and understanding of dealing with small streams of goods. One of the managers of the slaughters put it the following way:

> 'The biggest need for adaptation is within competence. To create an understanding that there's a need for the small streams of goods. During season, they're regarded upon as dirt in the machinery. It creates irritation and then you need to create an understanding.'

We can see how the middle level management had to work in a different way. They had to work closely with the production leaders to achieve improved results and find ways of being more related to quality improvement and more handicraft production without to high increase in costs. This meant a high degree of creativity. There was also a need for close coordination both up stream and down stream in the value chain with the other mid-level managers to find practical solutions. These processes are similar to what we find described as a value shop, with continuous improvements and tailor-made practical solutions. In this process, the R&D team of the joint strategy efforts had to be in close contact with the middle management.

4. Configuration at the strategic level – the value chain

The broader set of quality standards had large consequences at the strategic level. It increased inter-dependencies with every firm in the value chain. Most of these extra efforts resulted in increased costs. This is the case of every part of the value chain except for the slaughtering part, which claims to have only marginal increased costs when dealing with production with special adaptations. The main reason why there is an increased cost combined with production of specialised products, is the handling of small streams of goods.

The following statement from a farmer in value chain A explains why there are additional costs related to this type of production:

> 'I must follow specific standards. Increased costs are related to traceability. Special agreement with processor is needed. There is a need of increased competence and coordination.'

The next statement shows how the increased costs are related to another farmer's production in value chain B:

> 'There is a crop reduction, and a need for special contracts. You also need more documentation.'

In the slaughtering part of the value chain they faced more demanding specifications which lead to more work. Despite this they don't claim to have increased costs, but admit to some extent that handling of these special products is more troublesome work than with other more standardised products.

For the processor's part, there is the aspect of time. A production based on handicraft principles demands more time consuming work. Further, both in value chain A and B focus is put upon the need for extra work following up on the other parts of the value chain. To be able to do this, special agreements and contracts are made between the involved parts of the value chain, together with limited networking.

The challenges at the strategic level were related to payment for extra costs back in the value chain. It was a job translating the quality demands registered by the export organisation to standards within the up stream production as one of the managing directors in the processing plant expressed:

> 'I need to have close interaction with the marketing director. We have to regard him as our man'.

There was a need for close coordination and information, as well as clear decisions to be made. The R&D department had to be in close contact with the project and management group to make things happen, and to allow for extra time costs to appear. At this level there would be several small conflicts as to openness, agreement about the best procedures. One of the managing directors even questioned the methodology of the top researchers involved expressed in the following way.

> *'...The research group is not choosing the correct method for controlling the cooling phase. This has to be changed or we have to consider whether we shall take part in this study. I will make complaints about this'.*

At the strategic level, there was a need for frequent coordination, for high degree of trust as to giving the correct answers, and for reducing the perceived risk of being exploited by the other actors. This meant a lot of new efforts with linking up and reducing conflicts, negotiation of new solutions, and frequent meetings to improve trust. The efforts at strategic level very much took the form of a value network were the managing directors and the research group worked in close cooperation.

5. Conclusions

In this paper we have shown that the market positioning tools have to intertwine quality dimensions at different levels, with increase in complexity due to an increasing degree of immateriality towards ethical, cultural and emotional dimensions (Cambell-Hunt, 2000; Rangone, 1999). This comes in addition to improving the core product to satisfy the highest physical quality standards. The intertwined quality dimensions including a high degree of immateriality dimensions increase the technology level, and put strain on efficiency within the value chain. A very important strategic decision is the acceptance of increased costs, especially in the development process. Earlier studies (Borch and Forsman, 2004) have shown indications of negative relations between product development capability and perceived financial performance. The added quality dimensions will increase product costs strongly especially in the earlier phase of strategic reorientation. Small firms in particular will not have much slack to costly R&D and new product development processes; hence the products introduced have to be in line with customer needs from the very beginning. This implies strong customer relations form the beginning of the product development process to reduce failure rate. This means also, that there has to be a focus both on the differentiation aspects and the cost advantage aspects, with possible experiments along the different factors (Murray, 1988; Cambell-Hunt, 2000).

Such efforts towards an increase in the range of competitive tools have significant consequences for the configuration of the value chain. New resources both at operational and administrative level have to be included to manage the increased complexity following expanded quality marketing efforts (Cooper, 1993; Brush and Chaganti, 1998). Within each phase of production special adaptation measures has to be taken. The measures that make it possible to direct actions towards more targeted competitive strategies have administrative implications. These are related to specific technical and operational adaptations in each part of the value chain, and administrative efforts to coordinate between the different parts. This also included increased uncertainty and needs for reciprocal communication to solve new problems that emerges, and feedback loops from the following stages in the production process as to how the previous steps have performed. We are here talking about the characteristics of a value shop to find the best practical solutions (Stabell and Fjellstad, 1998). From an organisational perspective there may be a need for changes in the whole organisation at its links to cooperation firms. There has to be frequent changes and improvements in both primary production procedures, the environment were production take place, value added processing and marketing. The processing part of the value chain plays a special role in coordinating both upstream and downstream. The value chain may take the form of a value shop in the fulfillment of combined differentiation focus strategies.

The advanced layers of quality also have consequences as to the configuration at top levels. There are risk of conflicts and lack of harmonisation between the partners. Thus, there has to be close interaction and cooperation efforts as well as conflict solving procedures at strategic management level and frequent contact with R&D to find the best solutions and regard the benefits and costs of each alternative discussed. The value chain may take the form of value network in the start up and product development phase.

One important conclusion is that the three types of value configurations may be active at the same time at different levels. The combined high quality differentiation demands high organisational flexibility. Different value chain configurations may be working in parallel. There will be increased complexity towards both value shops and value networks to be built into the organisation.

The results reveal that increased customer satisfaction through quality differentiation has to be a task for the whole value chain. What is of utmost importance is that the configuration of the firm gives increased efforts and strains on the whole organisation. There may be a need for new types of coordinative competence at both primary level and at middle management to deal with the increased communication and control efforts between the different parts of the value chain. Also the top management has to be more dynamic in their strategic decision-making process with frequent considerations of the match between customer satisfaction, competitive tools and organisational configuration. Each level of differentiation has to be followed up by analysis of the costs of adaptation versus the opportunities for increased income within the new niches chosen.

This study was emphasising a specific type of production within a mature industry. Studies should be conducted in a broader range of industries. We still lack studies including both competitive strategy and organisational configuration including the whole value chain. We conducted this study in the development phase. This may cause special challenges as to organisational configuration. Other solutions may be present later in the process of joint quality efforts. We may expect that the need for more complex structures will not be so imminent especially at strategic level.

References

Barney, J., 1991. Firms resources and sustained competitive advantage. Journal of Management, 17: 99-120.

Bijman, J., S.W.F. Omta, J.H. Trienekens, J.H.M. Wijnands and E.M.F. Wubben, 2006. Management and organization in international agri-food chains and networks. In: J. Bijman, S.W.F. Omta, J.H. Trienekens and E.M.F. Wubben (Eds.) International agri-food chains and networks. Wageningen, the Netherlands: Wagening Academic Publishers.

Black, J.A. and K.B. Boal, 1994. Strategic resources: Traits, configurations and paths to sustainable competitive advantage. Strategic Management Journal, 1: 131-149.

Borch, O.J., M. Huse and K. Senneseth, 1999. Resource configuration, competitive strategies and corporate entrepreneurship: an empirical examination of small firms. Entrepreneurship, Theory, Practice, 24: 49-70.

Borch, O.J. and S. Forsman, 2003. The competitive tools and capabilities of micro firms in the Nordic food sector - a comparative study. In: S.O. Borgen (Ed.). The food sector in transition - Nordic Research. Proceedings of NJF-seminar No. 313, June 2000. NILF-Report 2001-2. Oslo: Norwegian Agricultural Economics Research Institute, pp. 33-50.

Borch, O.J. and B. Brastad, 2003. Strategic turnaround in a fragmented industry. Journal of Small Business and Enterprise Development, 10: 393-407.

Brush, C.G. and R. Chaganti, 1998. Business without glamour? An analysis of resources on performance by size and age in small service and retail firms. Journal of Business Venturing, 14: 233-257.

Campbell-Hunt, C., 2000. What have we learned about generic competitive strategy? A meta-analysis. Strategic Management Journal, 21: 127-154.

Chandler, G. and S.H. Hanks, 1994. Market attractiveness, resource-based capabilities, venture strategies and venture performance. Journal of Business, 9: 331-349.

Chrisman, J., W. Boulton and C. Hofer, 1988. Toward a system for classifying business strategies. Academy of Management Review, 13: 413-428.

Cooper, R.G., 1993. Winning new products: accelerating the process from idea to launch. Cambridge, MA: Perseus.

Day, G.S. and R. Wensley, 1988. Assessing advantage: a framework for diagnosing competitive superiority. Journal of Marketing, 52: 17.

Fjeldstad, Ø.D. and C.H.M. Ketels, 2006. Competitive advantage and the value network configuration. Making decisions at a Swedish life insurance company. Long range planning, 39: 109-131.

Ghisi, F.A., D.P. Martinelli and T. Kristensen, 2006. Horizontal cooperation among small and medium-sized supermarkets as a tool for strengthening competitiveness. In: J. Bijman, S.W.F. Omta, J.H. Trienekens and E.M.F. Wubben (Eds.) International agri-food chains and networks. Wageningen, the Netherlands: Wagening Academic Publishers.

Hingley, M., 2005. Power to all our friends? Learning to live with imbalance in UK supplier-retailer relationships. Industrial Marketing Management, 34: 848-858.

Hofer, C.W. and D. Schendel, 1978. Strategy formulation: analytical concepts. St. Paul: West Publishing Company. 219 p.

Lazzarini, S.G., F.R. Chaddad and M.L. Cook, 2001. Integrating supply chain and network analyses: The study of netchains. Journal on Chain and Network Science, 1: 7-22.

Leonard-Barton, D., 1992. Core capabilities and core rigidities: a paradox in managing new product development. Strategic Management Journal, 13: 111-125.

Mason, K., P. Doyle and V. Wong, 2005. Market orientation and quasiintegration: adding value through relationships. Industrial Marketing Management, 35: 140-155.

Morrison, A.J., 1990. Strategies in global industries. How US businesses compete. Westport, CT.: Quorum Books.

Murray, A.I., 1988. A contingency view of Porter's generic strategies. The Academy of Managagement Review, 13: 390-400.

Porter, M.E., 1980. Competitive strategy. New York: The Free Press.

Porter, M.E., 1985. Competitive advantage. New York: The Free Press.

Rangone, A., 1999. A resource-based approach to analysis in small-medium sized enterprises. Small Business Economics. 12: 233-248.

Spender, J.C., 1993. Business policy and strategy: an occasion for despair, a retreat to disciplinary specialization, or new excitement? Academy of Management Best Paper Proceedings: 42-46.

Stabell, C.B. and Ø.D. Fjellstad, 1998. Configuring the value chain for competitive advantage: on chains, shops, and networks. Strategic Management Journal, 19: 413-437.

Thompson, J.D., 1967. Organizations in action. New York: McGraw-Hill.

Yin, R.K., 2003. Case study research. Design and methods. Applied Social Research Methods Series. Vol. 5. Sage Publications.

Marketing elements for launching a new functional food product: a case of Indonesian market

J. Puspa and R. Kühl

Abstract

In the area of functional food (FF) many studies have been performed to focus on explaining the psychological set such as attitude and acceptance, and also to understand the socio-demographic profiles of the functional food consumers. Due to the fact that the FF market is a converging market between food and pharmaceutical markets, in view of a firm, a concern is supposed to set a specific strategic marketing so that it can represent both industries. However, because of the absence of a study that focuses on the practical implication of these consumers' behaviour especially concerning with strategic marketing implication for FF, the aim of this study is to investigate and to understand a communication and segmentation relevant for marketing a FF-product based on the consumers' characteristics such as socio-economic-demographic elements and consumers' psychological set. The results of the factor analysis of the FFs attributes shows that in general there are at least two major communication themes relevant for marketing of a FF i.e. affective and cognitive components. The cluster analysis suggests that it is necessary to segment the market of FFs i.e. based on consumers' perspectives and not based on consumers' characteristics.

Keywords: communication, segmentation, key success factors for market entry, functional food in Indonesia

1. Introduction

The concept of gaining physical health through the consumption of special food items is actually a part of the Indonesian ancient traditional eating pattern. Knowledge about preventing or even curing a certain disease by the consumption of special food items has been passed through many generations. Nowadays, in a modern industry, this concept advances the development of the Indonesian traditional herb based medicines. Therefore, it can be assumed that in principal the concept underlying the development of functional food- the food that provides a real health benefit- is not new to the consumers. However, the modern manufacturing approach (technological innovation) that integrates ingredients performing health benefits to the conventional food items used to develop a functional food product may create a different consumers' acceptance and attitude.

In the advanced countries, in the last two decades, product innovation projects have been directed internationally to redesign conventional food and beverage products in such a way that they can provide additional health benefits beyond their basic nutritive function. An example of this is that researchers have tried to influence the systemic immune reaction by specific nutrients and food ingredients directly (e.g. by triggering immune cell activation or altering immune cell interactions) or indirectly (e.g. by changing substrate for DNA synthesis, altering energy metabolism, changing physiologic integrity of the cells or altering signals or hormones) (Deibert and Berg, 2002). However, there has been a long-standing debate about the definition and limitations of health promoting effects of FFs and about how to prove those effects. Apparently, these issues relate to legal aspects, especially, concerned with the commercial uses of health claims for marketing of functional food products. In many Western countries there is a strict dividing line between food and medicine. Health effects such as prevention and reduced risk of chronic diseases are, more or less, considered as the domain of medical and not of nutritional intervention. Principally, through establishing a legal regulation the government or food authority intends to protect the consumers from the abuse of a health claim. But from the view of the industry, this legal regulation is often being seen as a rigid construct limiting their marketing creativity.

The development of the FF market in Indonesia achieves also such a significant progress as in the Western countries. The FF market in Indonesia is actually an industry driven market. Moreover, the significant trend of the FF market in Indonesia follows the fast market development of supplement products and it is strongly influenced by the international FF market developments. Increasing people's awareness of following a healthy lifestyle through nutrition and consumption upholds the positive movement of the FF market. Since the year 2000 many - so called - FFs were launched in the market with a variety of health claims. Most of them are produced and marketed by locally or even regionally based companies. However, due to the absence of a consistent regulation, which is supposed to be issued by the Indonesian food authority, we can find that many so-called functional food products have been improperly marketed. The usage of a health claim in many advertising campaigns was misleading and elusive. These health claims were not supported with scientific arguments or relevant scientific findings. Therefore, it is not surprising that one can find a product, which is claimed to be beneficial for treating a broad range of diseases such as from skin infection to cancer. This development is aggravated by a strong trajectory competition in this market. In the year 2005 the Indonesian National Agency of Drug and Food Control ('Badan POM') has launched a new regulation (HK 00.05.52.0685) limiting the usage of health claims. With this regulation, a food firm compulsorily requires an approval from the Badan POM for the marketing usage of a health claim. The permission is based on individual peer reviewers' evaluation. In this case, the Badan POM applies a very general definition of a FF i.e. a food that has a certain physiologic function for the body. Therefore, currently, there are in total 2889 individual products classified as FF group with a variety of claims including nutrition claims, nutrition content claims, structural-function claims and health claims.

Many studies have been performed in the Western countries to explain the psychological set such as attitude, acceptance, learning, and also to understand the socio-demographic profiles of the FF consumers (Malla *et al.*, 2007; Labrecque at al., 2006; Verbeke, 2005; Van Kleef *et al.*, 2005; Niva and Mäkelä, 2007; Gilbert, 2000; Bech-Larsen and Grunert, 2003, Poulsen, 1999). But there is no study that directly concerns with strategic marketing of FF such as possible communication and segmentation. Therefore, this study was done with main purposes to understand (1) possible communication messages and (2) consumers clustering for segmentation purpose. These issues were selected setting up a marketing strategy for FF different from the one used for the marketing of conventional food products. This is one of the salient success factors affecting the success of the commercialisation of a new functional food product (Puspa *et al.*, 2007). Moreover, it is interesting for a firm's point of view to set up a consumers-based marketing strategy, especially when it deals with experienced consumers such as the Indonesian FF consumers.

2. Theoretical background

Due to the fact that FF is a mixture product of food and pharmaceutical, marketing a FF logically requires knowledge from both disciplines. Therefore, understanding the marketing management of both industries possibly will enable a firm to better plan its marketing actions. Food and pharmaceutical markets have different consumers' characteristics and a different market structure (see Table 1). We can clearly observe that consumers of conventional food are usually low involved, familiar with products, more price sensitive, and less loyal rather

Table 1. Different marketing characteristics between food and pharmaceutical markets (Puspa et al., 2007).

	Conventional food	OTC and prescription drugs
Consumer characteristics	Familiar with products	Familiarity with and knowledge of products vary among consumers.
	Non loyal, variety seeking or inertia buying is common.	Loyal to certain product classes.
	Low involvement consumers.	Medium to high involvement consumers.
	Price sensitive.	Less sensitive to price.
Distribution	Product is distributed widely	Limited distribution channels only via pharmacies and OTC drug stores.
	Mass market distribution via retailer-chains.	
Segmentation strategy	Wide; product can usually be consumed by a variety of consumers.	Targeted for certain segments according to medical indications.
Communication strategy	Mostly concerned with affective issues such as: taste, product newness, originality, natural product.	Mostly concerned with cognitive issues, or directly concerned with health/medical benefits of the products. Scientific clinical trial results are basis for communication platform.
	End- consumer oriented	Two ways of approaching the end-users: • via medical community, medical community a trustworthy mediator for patients. • via sales persons in stores.
	Mass communication campaign through TV or radio spots, advertising in magazines.	Mostly through personal communication or direct promotion done by sales forces (medical representatives).
Product development strategy	Consumers need-oriented combined with competitor's innovation trend.	Disease management-orientation.
	Short process of product development (It takes only several months).	Long product development process (many years of planning and development)
	Product differentiation strategy based on core of available products (developing new taste, new packaging, new assortment).	Innovative product development strategy.
	Low investment product development.	High investment product development.
Branding strategy	Umbrella branding for new product (company name or established core product as umbrella brand).	Development of new brand name, especially for a new market.
Pricing strategy	Competitor- based pricing policy.	Internal cost and investment based pricing policy.

than consumers of pharmaceutical products (Bröring, 2007; Zaichkowsky, 1985, Assael, 1995; Hupfer & Gardner, 1971). Marketing strategies, especially related to distribution, segmentation, communication and product development are also significantly different. The communication strategy for conventional food is usually focused on emotional issues such as taste, originality, newness and it is communicated through end-consumer advertising media such as TV/radio spots and advertisements in magazines. In contrast to this, promotional campaigns for drugs usually cover cognitive issues which emphasise the scientific benefits of the product. Promotion for pharmaceuticals is not directly targeted to end users but to the mediating medical community such as doctors, nurses, and pharmacies. The information transfer process for pharmaceutical products is mostly effectively accomplished through personal communications (face to face promotions) done by a firm's medical representatives. In order to market food products the respective firms mostly enter the mass market. Food products are usually targeted on a very wide segment. In contrast to that, marketing of pharmaceuticals requires a clear segmentation usually depending on the product indication.

The other most distinctive issue for these two industries is they way they develop a new product. In order to develop a new drug a high investment is needed. It also takes several years until the firm has found a new potent substance. The product assortment strategy also differs in the two industries. Food companies usually apply a product differentiation strategy, which emphasises the continuous modification of an existing product, such as by offering a new taste, a new packaging, a new formula, or a new campaign, etc. The pharmaceutical industry has to present a radical innovation. Therefore, the branding strategy usually used by food firms is umbrella branding with the company's name or top existing product's name as the core brand name, while a totally new brand name has to be used for a new pharmaceutical product.

Although from the strategic marketing point of view there are many salient issues determining the success of launching a product, for FF cases this study argues that firstly the issues of communication and segmentation should be focused on. Communication is important because it puts something in the consumers' minds or changes the consumers' attitudes or induces the consumers to a buying action. Communication messages can be established through emphasising what the targeted consumers want to reach and what response they want (Kotler, 1994). Communication effects are greatest where the message is in line with the existing opinions, beliefs, and dispositions of the consumers (receivers) (Kotler, 1994). In the food and beverage market, basically, we can observe that the communication platform for food and beverage products follows the one usually used for commodity products. In this case consumers are assumed to have low involvement towards all products. Most of the consumers buy a product without making a brand or an alternative comparison and without passing a long process of decision making. Principally, they just pick a familiar brand or make a decision based on trial and error. Therefore, communication of these products is usually simpler, and more animated, which is suitable for passive learning consumers. It has little opportunity for reflection or making connection, and it often deals with unimportant matter (Assael, 1995). Furthermore, apparently, for currently marketed foods and beverage products the main communication contents are related to emotional or affective issues such as taste, freshness, great pleasure or enjoyment, and newness. Due to the fact that FF has clearly distinctive differences to the conventional food, consequently we can argue that communication as prevalent marketing tool should be implemented differently. Urala and Lähteenmäki (2004) reported that consumers willingness to consume FF could be predicted by rewarding feeling and necessity for FF, that describes how essential consumers think that FFs are for oneself or people in general. These two elements can be assumed as important communication messages for FFs. Moreover, since attitudes and beliefs are the major motivations for consuming or adopting a FF (Bech-Larsen and Grunert, 2003; Childs, 1997; Hilliam, 1996; Niva, 2000; Childs and Poryzees, 1997), a communication concept developed based on consumers' characteristics such as attitude and belief may contribute to the success of launching a product. With regards to this issue the first aim of this study is to find out ideas concerned consumer-oriented communication messages suitable for a FF.

The second important strategic marketing is concerned with segmentation. The consumers' needs, preferences, or desired value dimensions are heterogeneous (Blocker and Flint, 2007). Therefore, segmentation is important for a firm. Through segmentation a firm can identify and capture a homogenous consumer group that is important as a target consumer for a firm's marketing activities. By segmentation the representing manager can estimate the marketplace (Wedel and Kamakura, 2002). Urala and Lähteenmäki (2003, 2004) proposed that consumers do not perceive functional food as one homogenous group. This finding may suggest that each functional food product has individual segment. With regards to the segmentation method, the traditional marketing management has suggested that consumers' differences in terms of socio-cultural-demographical issues are considered as one of the prime methods for clustering the consumers (Kotler, 1994; McCarthy and Perreault, 1985). In the area of FFs some studies have been performed to observe the role of these variables towards people's acceptance. They reported that the citizens' views about food, health as well as their eating pattern are related to gender, age, socio-economic status and phase of life (Niva and Mäkelä, 2007). Poulsen (1999) and Bech-Larsen *et al.* (2001) reported evidence of considerable socio-cultural differences between US and European consumers regarding FF consumption. According to Poulsen (1999) people with age older than 55 and women are the main prospective consumers of FF products. Many studies confirmed that female consumers are the most likely buyers or users of FFs (Childs and Poryzees, 1997, Gilbert, 2000). Concerns about healthy eating increase with age (IEFS, 1996). Education plays also an important role in determining eating habit. In Europe, people with better knowledge and higher awareness tend to have a higher willingness to purchase a FF (Hilliam, 1996). Highly educated people maintain more healthful eating habits than other (Roos, 1998). Further, Urala *et al.* (2003) suggested that women perceived the health-related claims to be more advantageous than men.

However, opposing arguments that suggest segmenting the market based on other aspects besides socio-cultural-demographical have been recently well reported. Yankelovich and Meer (2006) argued that for the new criteria of market segmentation, the traditional demographic traits are no longer sufficient to serve as basis for a marketing strategy. Segmenting consumers based on needs, value, behaviour such as price sensitiveness brand loyalty seems to be more successful in the market (Feldman, 2006). Urala and Lähteenmäki (2003) found that in a Finnish study there were some differences among socio-economic groups in the attitude to FF, but the differences were small. Verbeke (2005) explained that socio-demographic characteristics like age, gender, education and the presence of children in the household are not confirmed as significant determinants of FF acceptance. Realising that these studies have been done in different countries (but most of them were done in Europe), and due to different used measurements (such as measurements for identifying acceptance, attitude, motivation, knowledge etc), different results are expected. Based on these backgrounds, the second aim of this study is (2a) to investigate variables important for determining buying decision and (2b) to understand the role of these variables for setting up a segmentation model.

3. Data and methodology

In order to accomplish the research objectives, this study's framework was designed consisting of a combination of an exploratory and a descriptive research. We conducted a consumers' study using an in-depth personal interview method. This field study was being conducted in Indonesia (Jakarta and its surroundings) 2003/2004. Jakarta was selected since it represents a rapidly developing city where the market for such functional products is widely opened up.

Respondents were recruited based on a judgmental sampling method, which selects certain respondents according to the presumable representation of the population of interest (target segment) (Dillon *et al.*, 1994). In general, the inclusive criteria for selecting respondents were as follows:
- persons aged 16 years and older;
- persons, who are living in the selected region;
- persons, who have selected diseases such as CHD, hypertriglyceridemia, obesity, and a healthy group.

In recruiting respondents we used an *a priori* segmentation method, which targets respondents prior to the study by establishing the following groups: (A) patients with coronary heart disease (confirmed by angiography, PTCA, bypass operation or myocardial infarction); (B) patients with high serum triglyceride level (all hypertriglyceridemic patients with plasma triglyceride levels of >200 mg/dl - or according to the NCEP [National Cholesterol Education Program ATP II guideline]), with or without the metabolic syndrome; (C) patients with obesity (according to the ITFO [International Task Force on Obesity] guideline, overweight and obesity is defined when the patients have a BMI [Body Mass Index] of more than 30 kg/m^2), with or without other metabolic diseases; and (D) healthy persons as controls. Due to the fact that awareness of prevention and illness are hypothesised as motivators for FF acceptance (Wrick, 1995; Milner, 2000; Verbeke, 2005) the groups of patients were selected as respondents in order to represent a direct target group for FFs and to represent motivated consumers who are dealing directly with the diseases, who are aware of the food-diseases relation and who have direct access to the scientific information. Patients were recruited in RS. Jantung Harapan Kita (National Cardiovascular Centre Harapan Kita). A healthy person group as control represents unmotivated consumers, it is our initial assumption, who are presumed to be less aware of disease prevention and healthy diet patterns. The respondents were interviewed by using a prepared questionnaire. The total number of recruited respondents was 183. It consists in total of 47, 55, 45 and 36 respondents of group A, B, C, and D respectively.

In order to model the possible communication arguments that can be used for a FF product, all possible positive and negative attributes of FF were collected (see Table 2). These attributes represented the advance benefits received by the consumers of having consumed food items with health claims as compared to the consumption of drug items (prescriptions or OTC-products, and supplements). Some product profiles were included because those are one of the important elements in the marketing of a food product. In general, a product profile includes both intrinsic and extrinsic features. Examples of extrinsic FF profiles important to be mentioned are labeling, brand name, packaging size and design, and price. A factor analysis was used in order to find out some important new factors, which could be extracted from those collected multi-attributes of FFs. The new factors found by factor analysis would then be useful to further determine the general communication concept of a FF. Using a regression analysis the new factors were analysed in order to obtain information regarding the importance of each factor for influencing the consumers' buying behaviour.

To analyse the different responses of consumers' characteristics relevant for segmentation a non parametric test (Jonckhreere-Terpstra-Test) and Univariate General Linier Model were employed. In this case some economic-demographical characteristics (such as age, income, spending income for foods, and education, job specifications) and some consumers' psychological set (such as attitude, motivation, involvement level, and knowledge) were included in the analysis. Attitude towards FF was measured by using Fishbein's and Likert's models. Fishbein's model is a combined evaluation of the consumers' preference to the attributes of FF and their belief in

Table 2. Attributes of functional food and the result of factor analysis. Source: SPSS-Factor Analysis-model result.

Factor	Variables	Factor loading	% variance eigenvalue
1 Functional imaginary			22.1
	FF should contain only natural ingredients (no artificial substance).	0.653	
	A FF is a healthy food.	0.741	
	I will consume a FF that provides enough energy for my activities.	0.725	
	In my opinion FF can make me healthier and fitter.	0.525	
	FF compositions should not exceed the total guided nutrition values.	0.821	
	It is important to consume a FF that contains enough vitamins, minerals and fiber.	0.783	
2 Short term and long term health benefits (health benefit sought)			12.3
	A FF product that can prevent a disease is very important.	0.765	
	In my opinion a FF that can normalise body's physiological system such as normalise my plasma lipid profile is important.	0.798	
	A FF product that can reduce risk of disease is interesting to be consumed.	0.801	
	In the long term of usage a FF product should be able to reduce the usage of drug (cost spent for drug).	0.807	
	A FF product that can protect long term health is beneficial.	0.814	
	In my opinion a FF should be suitable to be consumed by all ages of people (children, young and adult).	0.638	
	FF that can shape my body figure is interesting.	0.341	
3 Potency and efficacy			7.3
	To my opinion a FF should perform the health effect fast (at least as fast as the effect of a drug).	0.710	
	The health effect of a FF should be measurable.	0.710	
	The quantity to be consumed in order to achieve the health effect should be minimal.	0.675	
	The performed health benefits are more important than taste of the product.	0.658	
4 Intrinsic/extrinsic Associations: confidence of outlook			6.9
	I always consider selecting a FF marketed by a well known company.	0.597	
	Brand name is important for my buying decision.	0.803	
	Packaging design can improve my confidence to the product.	0.785	
	It is important for me that the product is packed in appropriate quantity.	0.709	
5 Freedom of choice: price sensitivity			5.4
	In my opinion functional food should not be sold with the high price (30% higher that conventional similar product).	0.640	
	Before I buy, I always look at the availability of bonus program at supermarket.	0.870	
	I like FF rather than a drug because of it has a flexible pricing policy (price reduction through discount program).	0.868	

Table 2. Continued.

Factor	Variables	Factor loading	% variance eigenvalue
6 Intrinsic/extrinsic associations: confidence of usage			4.7
	A FF product with safety and efficacy guarantee from the manufacture will improve my confidence.	0.756	
	My belief in a FF product will improve when the offered health effect has been well proved by many clinical studies.	0.582	
	Before I try a FF, I will look at packaging whether there is a label confirming the quality and safety or country of origin.	0.621	
	A FF that can be consumed for a long time without any cumulated side effect will improve my usage confidence.	0.644	
7 Advance benefits: fast onset of action and interaction effect			3.7
	A certain FF should not have a side effect when it is consumed together with other FF product.	0.813	
	A FF should show its health effect fast (few days after the consumption).	0.647	
	I trust better a FF product which has been developed by an independent research centre.	0.742	
8 Intrinsic/extrinsic Associations: convenience of usage			3.3
	A FF product should be sold with format that makes them easily to be handled (practicability of packaging, easy to be consumption during traveling, etc.).	0.592	
	The presence of variety of FF products with the same health effect will provide a higher usage convenience.	0.585	

those attributes such as health belief, belief in health claim, belief in disease prevention measure through nutrition, belief in potency and efficacy of FF, in safety and naturalness etc. Motivation to take a prevention measure was measured using some indicators (such as I am always concerned with my health status, I do medical check routinely, I follow the healthy diet (avoid foods that contain high fat, high sugar and salt), I do routine sport (with minimal training time of 2×2 hours in a week), I do not concern with total calorie per day should be consumed, I am a vegetarian, I consume healthy foods such as vegetables and fruits (5 times a days), or bio products, functional foods, fortified foods, for prevention purposes, I consume routinely health supplements, I think any action for prevention disease is exaggerated). Knowledge level was measured by combining objective and subjective knowledge assessments (examples of subjective knowledge questions are: I am familiar with functional food concept, I know exactly the brand names of functional food products available in the market, I know where I can buy functional food, my inherent knowledge about health effect offered by functional food products is adequate). While, objective knowledge level comprised of (1) A 6-items multiple-choice test concerning nutrition and health aspects such as which food element is important for reducing cholesterol? (answer: fibre), which fat component is bad for your health? (saturated fatty acids), how many calories per day should be consumed by healthy adult? (answer: 1,500-2,500 Kcal), what is the effect of EPA & DHA? (answer reduce plasma TG level, prevent atherosclerosis, improve the functions of brain, as anti-inflammatory etc.), which element is important for preventing osteoporosis? (answer: calcium), can you mention some examples of natural-conventional foods that are rich in Omega-9?

(answer: olive oil, palm oil, etc) and (2) A brands recalled. Involvement level towards FFs was measured using the theoretical based construct of involvement level (modified construct validity statements) (Assael, 1995; Zaichkowsky, 1985). A short report of individual results of these variables is presented in Table 3. Segmenting the consumer a quick cluster analysis (K-means cluster) was done. In this cluster analysis the three main clusters i.e. group with no intention to consume FF, group with 'may be' consume FF and group with absolutely will consume FF were selected as main parameters in grouping the respondents' profiles.

Table 3. Summary of respondents' profiles.

	Variables	Results
1	Gender	F= 41% M= 59%
2	Age	(18-30)=15.3%; (31-40)=13.1%; (41-50)= 24%; (51-60)= 23.5%; (61-70)= 18.6%; (>71)= 4.9%
3	Income	(<1 Million Rp)= 14.8%; (1.1-3 Million)= 53%; (3.1-6 Million)= 18.6%; (>6.1 Million Rp)= 13.7%
4	Education profile	Low= 39.3%; medium 1= 21.9%; medium 2 (uni)= 27.3%; high (master)= 11.4%
5	Household size	1 person= 3.3%; 2 persons= 7.7%; 3p= 37%; 4p=26.8%; 5p>=47%
6	Usage rate	No experience= 9.3%; once tried but buy no more=24.6%; with enough experience= 66%
7	Attitude	
	• Fishbein	Means= 33.6 (quite positive) (max=100; min -15 with scale -108 up to +108)
	• Likert	Median= +1 (scale -2 to +2)
8	Motivation	Low=12%; medium=41%; high= 47%
9	Involvement level	Low= 4%; medium= 67%; high 30%
10	Knowledge level	Bad=19%; average= 22%: good=30%; very good= 29%
11	Intention to buy	will not buy= 19%; probably will buy=15%; will buy= 51%; absolutely will buy= 15%

4. Results

4.1 Communication platform for launching a functional food product

As the involvement of respondents towards FF achieves a medium to high level (Puspa, 2007) we further assume that consumers will pay a lot of attention to product information. This is different from the involvement level toward conventional food products, which has been suggested to be low. According to the marketing theories the high involvement consumers go through a medium - or high intensity buying decision process. They need more product information in order to perform an intensive attribute evaluation and for establishing a positive motivation, belief and attitude toward the object. These findings confirm the importance of the communication strategy within the whole marketing process of functional food.

Factor analysis of several attributes of functional food has identified 8 new factors of the communication messages appropriate for FF according to the respondents' opinions (see Table 2). Interpretation of these findings results in two major communication themes suitable for marketing of a functional food product in general, such as: (1) affective components, which cover all communicative messages emphasising the emotional aspects of consumers, such as feeling, passion, fear, happiness, etc. (2) cognitive components, which include all logical reasons for buying or consuming a product (see Figure 1). The affective components cover two major issues: (a) *functional imaginary*, which consists of all emotional factors, resulting from the functional consequence of having consumed a functional food product. Examples of this are the feeling of satiety (feeling full), feeling good, and looking good (cosmetic feeling) and (b) *extrinsic/intrinsic associations*, which concern other emotional factors resulting from the intrinsic and extrinsic features of functional food, such as 'confidence of outlook' (resulting from good product appearances), 'confidence of efficacy' (due to product effectiveness in performing the health claim), 'confidence of safety' (because of the presence of product assurance/labeling proved by clinical studies in humans, such as issues concerned with the absence of side effects in long term consumption), 'confidence of usage' (resulting from the presence of a guarantee of quality and appropriate applications), and 'convenience of usage' including features, which offer ease of handling of the product (easy pouring or opening, easy for keeping purposes and for disposal).

The cognitive components are defined by two main issues: (a) *health benefits sought* (relating to all possible health benefits that can be offered by the functional food product). The health benefit sought can be divided into two parts based on the length of the product reaction time to achieve the expected health claim, i.e. (a1) short term effects, such as attaining the expected nutritive value, improving health and fitness, improving the body's defense mechanisms, etc., and (a2) long term effects, such as lowering the risk of disease, prevention of disease, or - may be - the treatment of a disease, and a profile showing that a functional food may reduce drug cost. This

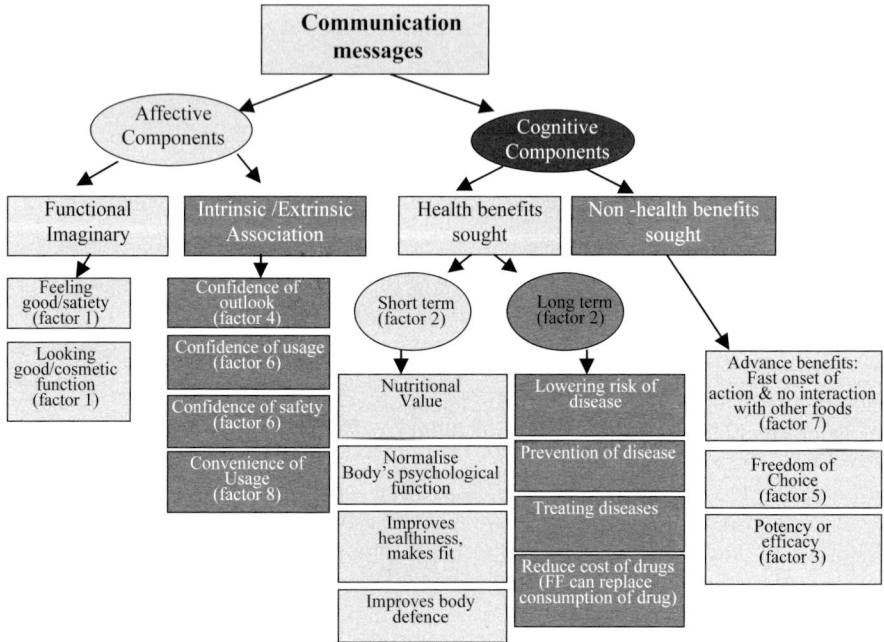

Figure 1. Possible communication messages for functional foods: an interpretation from the result of factor analysis.

issue is important for the Indonesian citizen, because there is no community health insurance program in Indonesia and because of the high price of drug items there. The other cognitive element is (b) *non-health benefits sought*. This relates to all other important features or benefits beyond the offered health aspects/claims. The non-health benefits include many other cognitive components such as (b1) advance benefits concerned with features such as fast onset of action and no interaction with other food substances. The consumers expect to receive a fast performance of health benefits (whenever it is possible, the effect should be reached as fast as the effect of a drug (or supplements) and a better interaction profile. In this case - according to the consumers - a functional food product may be more interesting than a drug item (or supplement), because it has no or lesser interactions with other food substances. Freedom of choice (b2) means that consumers have a variety of ways for consuming foods with a certain expected health claim. In the food industry firms usually provide the consumers with a large variety of similar products. An example of this is the sitosterols/sitostanols containing functional food products. To obtain their lipid lowering effect the consumers can consume a variety of product groups, such as sitosterol/sitostanol containing margarine, milk, yogurt, or a combination of these products. This is one of the apparent advantages of a functional food with a health claim as compared to a drug item. The same example is also valid for freedom of choice due to variety of prices (daily treatment cost) of functional food products. The other non-health benefits sought is (b3) potency and efficacy, a feature that offers a strong effectiveness regarding the health claim (as potent as a drug).

Furthermore, this study shows that for a new product launching the cognitive components seem to be more important to be emphasised in the context of a communication message rather than the affective components. According to the result of a multiple regression analysis the affective components did not significantly influence the consumers' intentional buying behaviour in most cases (see Table 4). In the first launching period of a functional food product, apparently, the affective components would have a lesser effect in building the consumers' belief in the product. Because these messages could not clearly explain and convince the consumers to what extent the product can solve what the consumer tries to get done, the cognitive components seem to be more appropriate for convincing the consumer to intentionally make a choice decision. This finding confirms the existing theories, i.e. consumer-response stage models or response-hierarchy-models such as 'Hierarchy-of-Effects', 'Innovation-Adoption', 'Communication' and 'AIDA model'. All of

Table 4. Regression analysis of new factors resulted from factor analysis.

Model		Regression coefficients		Significance
		B	SD	
1	(constant)	3.906	0.051	0.000
	Factor 1 (cognitive)	0.108	0.051	0.036
	Factor 2 (cognitive)	0.157	0.051	0.002
	Factor 3 (cognitive)	0.111	0.051	0.050
	Factor 4 (affective)	0.020	0.051	0.694
	Factor 5 (cognitive)	0.041	0.051	0.423
	Factor 6 (affective)	-0.087	0.051	0.090
	Factor 7 (cognitive)	-0.033	0.051	0.522
	Factor 8 (affective)	-0.029	0.051	0.576

Dependent variable: intention to buy. Independent variables: factor 1-8.

these models assume that consumers pass through a cognitive, affective and then behavioural stage, in that order (Kotler, 1994). In these models a cognitive stage leads to the establishment of attention, awareness, reception and knowledge level towards an object.

4.2 Segmentation of the consumers

Using 'consumers' intention to buy' of a FF pattern that was mentioned in the concept testing as a variable group ('absolutely will not buy' was grouped in 1 and 'absolutely will buy' was grouped in 5) this study has tested the ordered differences among classes of some independent variables. The Jonckhreere-Terpstra test statistic (see Table 5) showed that all respondents' psychological set variables besides knowledge level showed a significant difference among classes. Furthermore, consumers' motivation level towards disease prevention measures, consumers' attitude level (both from Fishbein's and Likert's models), and involvement level towards FF were significantly different across the variety of consumers' commitment to buy/consume a new FF. According to this result people who will not buy have a different psychological set as compared to people who intend to buy a functional food product. A correlation analysis confirmed this finding. Respondents' psychological sets mostly have a positive correlation with intention to buy (see Table 6). Respondents with a positive attitude towards a FF concept, with a high involvement towards FF, and with a more positive motivation level towards a disease prevention measure have a higher intention to consume or buy functional food than respondents with a negative attitude, low motivation and low involvement. Respondents with health awareness and disease prevention consciousness show a higher intention of consuming a new functional food product.

This study found that there is no difference in terms of knowledge across different 'intention to buy' behaviours. The correlation coefficient between these variables was also not significant. But the other finding confirmed that respondents' consumption experiences of FF (see usage rate

Table 5. Non-parametric test of respondents' characteristics (Jonckhreere-Terpstra-test) with intention to buy scale as dependent variable.

	Observed J-T statistic	Means of J-T statistic	Sd of J-T statistic	N	Significance (2-tails)
Psychological set					
Motivation level	5087.5	4304.0	343.2	182	0.022*
Attitude level					
• Fishbein's scale	5726.0	4303.5	344.8	182	0.000*
• Likert's scale	5244.0	4381.5	321.7	183	0.007*
Knowledge level	4013.0	4381.5	306.5	183	0.229
Involvement level	5248.0	4381.5	287.2	183	0.003*
Usage rate	5622.0	4381.5	296.6	183	0.000*
Demographic-economic variables					
Education level	4108.5	4381.5	332.4	183	0.412
Gender	4600.0	4304.0	293.5	182	0.313
Age	4919.5	4304.0	338.0	182	0.064
Household expense	4209.5	4381.5	293.6	183	0.558
Income level	3699.5	4381.5	320.0	183	0.033*

* Significant different (0.05%).

Table 6. Correlation between intentions to buy variable with respondents' characteristics using Sprearman-Rho.

	N	Spearman-Rho correlation coefficient
Psychological set		
Motivation level	182	0.166**
Attitude level		
• Fishbein's scale	182	0.301**
• Likert's scale	183	0.193**
Knowledge level	183	-0.095
Involvement level	183	0.218**
Usage rate	183	0.306**
Demographic-economic variables		
Education level	183	-0.061
Gender	182	0.075
Age	182	0.138
Household expense	183	-0.046
Income level	183	-0.59*

* Significant correlation ($P<0.05$), ** significant correlation ($P<0.01$).

variable) showed a positive correlation with the intention to buy. The more frequently a respondent consumes FF the more he will tend to try the newly marketed functional food product.

The results of the knowledge test were also contradicting the results of the involvement level measurement. Usually, people used to have a high involvement level will tend to be active in searching, learning, and absorbing product information. Therefore, as a consequence of these activities, they will possibly be more knowledgeable and more willing to consume the product (partly in IFIC; http://ificinfo.health.org). We argue that in order to understand this relationship we need to distinguish between objective (cognitive) and subjective (affective) knowledge because of the different roles of these variables (Puspa and Kühl, 2006). In this case we would suggest that a specific investigation using an appropriate knowledge assessment method is necessary for the confirmation of this finding.

Considering the fact that people's intention to buy is an important indicator for the future success of the marketing of a new functional food product, the result of this measurement can be used as a marketing platform for FF in general. With regards to this analysis our study showed that socio-economic-demographic elements tend to have less influence on determining people's intention to buy than psychological set variables. Furthermore, they would be less important than the psychological set variables. Moreover, the results of our Univariate GLM analysis (see Table 7) showed that psychological factors have a greater influence on the people's intention to buy of FF (especially motivation, and attitude) than the consumer's demographic-economic variables. Furthermore, these results demonstrate the importance of the psychological variables for determining the buying decision. As a consequence of that the consumers' psychological set will be more suitable for determining a marketing platform such as segmentation. This finding argues for the above mentioned statement of Yankelovic and Meer (2006). In summary, our statistical tests confirmed that the prospected customers of FF (represented by the group of respondents willing to consume functional food) have a distinctive bundle of profiles as compared to the

Table 7. Univariate analysis of respondents' characteristics using intention to buy scale as dependent variable.

	Sum squared of Typ III	df	Mean squared	F	Significance
Psychological set					
Motivation level	4.136	4	1.034	3.423	0.012*
Attitude level	36.209	64	0.566	1.873	0.003*
Knowledge level	1.618	2	0.809	2.679	0.074
Involvement level	0.409	2	0.204	0.677	0.511
Usage rate	2.191	3	0.730	2.418	0.071
Demographic-economic variables					
Education level	0.762	2	0.381	1.262	0.288
Gender	1.924	1	1.924	6.371	0.013*
Age	1.556	5	0.311	1.030	0.405
Household expense	0.002	2	0.001	0.003	0.997
Income level	1.532	6	0.255	0.845	0.538

* Significant characteristics (0.05%).

non-prospective customer group. As a consequence of this interpretation it can be argued that a firm still requires segmenting the market when it wants to bring a functional food product on the market. Efforts should be concentrated on capturing the right segment, such as a segment that consists of people who are ready to adopt/consume the newly marketed product.

Cluster analysis also showed that segmentation is still relevant for FF, because consumers are heterogeneous in terms of their psychological set. Each cluster consists of different consumer characteristics (see Table 8). Furthermore, segmenting consumers based on their psychological set including motivation, perception, attitude, and knowledge is possibly more relevant than classifying them based on economic-demographic variables. In the cluster analysis these variables were distributed consistently across all clusters. As a consequence of that a suggestion can be made with regard to segmenting the market, i.e. based on consumers' perspectives and not based on consumers' characteristics. The proposed segmentation for functional food marketing consists of a combination of two elements, i.e. 'Instrument' and 'Value'. Therefore, this process is called 'I-V Segmentation'. 'Instrument' means that a product is being seen as a tool or a stuff that helps the consumer's job to be done (Christensen and Raynor, 2003), while 'Value' refers to the relative importance of a product perceived by the consumers. Therefore, the 'Instrument-Value Segmentation Model' is based on the product as an instrument and on how the consumers value the product. This 'I-V segmentation 'emphasises the necessity to focus on the consumers' perceived value of a product or service. It concentrates more on what consumers really do with their product. This covers all cognitive and affective demands. A certain segment may consist of some different consumer characteristics such as differences in age (young or old), a different education level, family status etc. All groups of consumers will have 'the same job to be done' such as to overcome the high risk of cardiovascular disease through a convenient and safe food consumption.

The practical applicability of this 'I-V segmentation model' for the perspective of a firm is explained in Figure 2. Entering a small and limited market segment can be considered as an appropriate

Table 8.Cluster analysis of respondents' characteristics.

Characteristics	Cluster 1: non perspective buyers (will not buy answers) n=31	Cluster 2: medium perspective buyers (will possible buy answers) n=25	Cluster 3: high perspective buyers (will and absolutely will buy answers) n=118
Motivation level	low	medium	medium to high
Attitude (Fishbein)	quite negative	quite positive	neutral to very positive
Knowledge level	bad	average	bad
Involvement level	high	high	high
Household size	2 persons	2 persons	2 persons
User rate	having no personal experience	with personal experience	with personal experience
Education level	vocational school	with master degree	vocational school-with master degree
Gender	male	female	male & female
Age	61-70 years old	31-40 years old	all groups

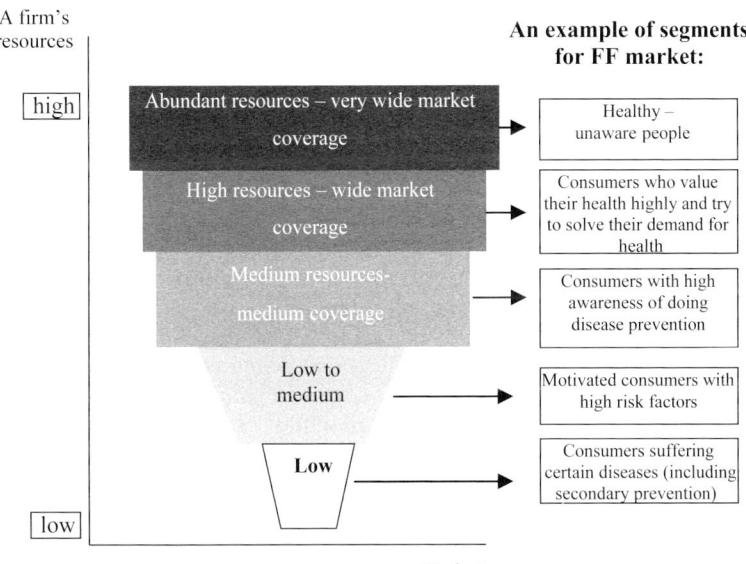

Figure 2. A proposed segmentation model for functional foods.

strategy, especially when the firm has only limited financial and non-financial resources, because -in our opinion- entering a large market requires large (tangible and intangible) resources. Without enough support it will be inefficient. As far as functional food products are concerned, segmenting a product to patients, who can value the product' health benefit, is an example of a narrow but reasonable segmentation strategy. This segmentation strategy is adopted from the strategy usually applied in the pharmaceutical industry. In other cases, e.g. when a firm has higher (moderate) financial resource, covering a larger target market will make much more sense. Based on our 'I-V segmentation model', a firm can target its functional food to consumers, who a have high level of risk factors for certain related diseases. This segmentation can be further developed based on the firm's capability. The final stage will be reached when plenty of financial capabilities and resources are available. In that situation covering a very huge segment like all healthy people – including those who are unaware of a healthy lifestyle – may be possible. In this last segmentation model the firm is assumed to be willing and ready to invest into changing people's awareness, perception, attitude and acceptance towards a healthy lifestyle and disease prevention measures as a prerequisite for their acceptance of functional food products.

5. Conclusions

Considering that the functional food market is a converging market combining the boundaries of the food and pharmaceutical markets, marketing a functional food logically requires knowledge from both disciplines. Therefore, mastering the marketing management of both industries will possibly provide a firm with extra internal capability in pursuing the strategic marketing for its innovative functional food product. Although from strategic marketing point of view there are many salient issues determining the success of product launching, for functional food cases this study argues for a focus being given firstly to the issues of communication and segmentation. The issues related to communication and segmentation strategies (besides networking) in practical senses of marketing of FF are two key important strategic concepts that distinguish the issues of marketing management for FFs and the one for conventional foods (Puspa *et al.*, 2007). Besides that, interest in focusing on these issues was emerging, especially because the abundantly available published literature in the area of FF is rather more emphasising the issues of consumers' behaviour and their acceptance towards the functional food concept. Therefore, our study aiming at modelling a communication and segmentation strategy relevant for marketing a functional food product based on consumers' characteristics such as socio-economic-demographic elements and consumers' psychological set.

Since the consumers tend to be highly involved in purchasing a functional food product firms may require a specific communication measure that ascertains the improvement of people's belief in and attitude towards their product. The cognitive and affective types of campaigns serve as an equal function but provide a different effect depending on the product life cycle. Our study suggests that during the launching period cognitive components especially concerning health related benefits will be more influential in building consumers' belief than the affective components. With regard to segmentation this study shows that socio-economic-demographic elements tend to have less influence in determining people intention to buy than psychological set variables. This was confirmed by both individual statistical tests such as inference test using a non-parametric means different test and correlation analysis and also a multivariate analysis. Furthermore, the cluster analysis shows that segmenting consumers based on their psychological set including motivation, perception, attitude, and knowledge is possibly more relevant than classifying them based on economic-demographic variables. As a consequence of this a suggestion can be made with regard to segmenting the market, i.e. based on consumers' perspectives and not based on consumers' characteristics. An example of this is the suggested I-V segmentation model for functional food.

References

Assael, H., 1995. Consumer behavior and marketing action. Ohio: The International Thompson Publishing, pp. 267-285.

Bech-Larsen, T., K.G. Grunert and J.B. Poulsen, 2001. The acceptance of functional foods in Denmark, Finland and the United Sates. MAPP Working Paper 73. Aarhus: The Aarhus School of Business.

Bech-Larsen, T and K.G. Grunert, 2003. The perceived healthiness of functional foods a conjoint of Danish, Finnish, and American consumers' perception of functional foods. Appetite, 40: 9-14.

Blocker, C.P. and D.J. Flint, 2006. Customer segments as moving targets: integrating customer value dynamism into segment instability logic. Industrial Marketing Management, 36: 810-822.

Bröring, S., 2007. Innovation strategy in the emerging nutraceutical and functional food industry. Proceeding of the International Food and Agribusiness Management Association, Italy.

Childs. N.M. and G.H. Poryzees, 1997. Foods that help prevent disease: consumer attitude and public policy implications. Journal of Consumer Marketing, 14: 433-447.

Christensen, C.M. and M.E. Raynor, 2003. The innovator's solution; creating and sustaining successful growth. Boston: Harvard Business School Publishing Corporation.

Deibert, P. and A. Berg, 2002. Functional food. Der Lipid Report. 200; 1.

Dillon, W.R., T.J. Madden and N.H. Firtle, 1987. Marketing research in a marketingenvironment. 3th Edition. Illinois: Irwin.

IEFS, 1996. A pan-European survey of consumers attitude to food, nutrition, and health. Report Number Four. Dietary Changes. Dublin: Institute of European Food Studies.

Feldman, D., 2006. Segmentation building blocks. Marketing Research, 18: 23-29.

Gilbert, L., 2000. Marketing functional foods: how to reach your target audience. AgBioForum, 3: 20-38.

Hilliam, M., 1996. Functional foods: the Western consumer view point. Nutrition Review, 54(S): 189-194.

Hupfer, N. and D.M. Gardner, 1971. Differential involvement with products and issues: an exploratory study. Proceeding of the second Annual Conference of the Association for Consumer Research, pp. 262-270.

Kotler, P., 1994. Marketing management – analysis, planning, implementation and control, 8th edition. New Jersey: Prentice Hall International Editions.

Labrecque, J., M. Doyon, F. Bellavance and J. Kolondinsky, 2006. Acceptance of functional foods: a comparison of French, and French Canadian consumers. Canadian Journal of Agricultural Economics, 54: 647-661.

McCharthy, E.J. and W.D. Perreault, 1985. Essentials of marketing. Homewood: Irwin.

Malla, S., J.E. Hobbs and O. Perger, 2007. Valuing the health benefits of a novel functional food. Canadian Journal of Agricultural Economics, 55: 115-136.

Millner, J.A., 2000. Functional foods: the US perspective. American Journal of Clinical Nutrition, 71S: 1664S-1669S.

Niva, M., 2000. Consumers, functional food and everyday knowledge. Paper presented at Conference on nutritionists meet food scientists and technologists, Portugal.

Niva, M. and J. Mäkelä, 2007. Finns and functional foods: socio-demographics, health efforts, notions of technology and the acceptability of health-promoting foods. International Journal of Consumer Studies, 31: 34-45.

Poulsen, J.B., 1999. Danish consumers' attitude towards functional foods. MAPP Working Paper 62. Aarhus: MAPP - Center for market Surveillance, Research and Strategy for Food Sector.

Puspa, J. and R. Kühl, 2006. Building consumer's trust though persuasive interpersonal communication in a saturated market: the role of market mavens. In: M. Fritz, G. Schiefer and U. Rickert (Eds.), Trust and risk in business network. Bonn.

Puspa, J., T. Voigt and R. Kühl, 2007. Marketing competencies in the era of globalization: The case of functional food. Proceedings MIC Conference, Slowenia 2007.

Roos, E., 1998. Social patterning of food behavior among finish men and women. Publication A6/1998. Helsinki: National Public Health Institute.

Urala, N. and L. Lähteenmäki, 2003. Reasons behind consumers' functional food choices. Nutrition & Food Science, 33: 148-158.

Urala, N. and L. Lähteenmäki, 2004. Attitude behind consumers' wiliness to use functional foods. Food Quality and Preference, 15: 793-803.

Urala, N., A. Arvola and L. Lähteenmäki, 2003. Strength of health-related claims and their perceived advantage. International Journal of Food Science and Technology, 38: 815-826.

Van Kleef, E., H.C.M. Van Trijp and P. Luning, 2005. Functional foods: health claim-food product compatibility and the impact of health claim framing on consumer evaluation. Appetite, 44: 299-308.

Verbeke. W., 2005. Consumer acceptance of functional foods: socio-demographic, cognitive and attitudinal determinants. Food Quality and Preference, 16: 45-57.

Wedel, M and W. Kamakura, 2002. Market segmentation: conceptual and methodological foundations. Noewell, MA: Kluwer Academic Publishing.

Wrick, K.L., 1995. Consumer issues and expectations for functional foods. Critical Reviews in Food Science and Nutrition, 35: 167-173.

Yankelovich, D. and D. Meer, 2006. Rediscovering market segmentation. Harvard Business Review, 2: 122-131.

Zaichkowsky, J.L., 1985. Measuring the involvement construct. Journal of Consumer Research, 12: 341-352.

Quality management in Polish dairy cooperatives

J.H. Hanf and A. Pieniadz

Abstract

This paper investigates the relationship between the chosen quality strategy and the vertical coordination mechanism of a focal company by using new institutional economics, as well as strategic management approaches. The theoretical findings are tested using evidence from 19 of the largest Polish dairy cooperatives, interrogated in spring 2006. Because the research topic is of sensitive nature and requires detail knowledge on the firm's quality management processes, qualitative and inductive case interviews across different hierarchical levels in each cooperative have been conducted. The results show that all co-ops recognise the changing market requirements and are treating food quality as more than plain food safety and the ability to continuously reproduce an *ex ante* defined set of attributes. However, compared to investor-owned dairies, co-ops are disadvantaged in quality-based competition due to their lower flexibility and access to financial and qualified human resources. To overcome this intense competition, co-ops modify their production profile, which leads to market segmentation. Moreover, the choice of quality strategy is an economic activity, guided by the co-op's profit expectations within the selected market. The chosen quality strategy determines the design of the vertical coordination mechanism. Thus, the higher the requirements for the final product, the further quality management systems go beyond a firm's boundaries, and the higher is the intensity of the relationships between the intermediary stages in the dairy chain.

Keywords: network theory, relationship management, quality management, cooperatives, Poland

1. Introduction

In countries where food is no longer scarce, questions of food security are becoming less important. Instead, issues addressing food safety and quality are gaining in importance. Thus, in most developed countries, food quality has been used as a means of differentiating food products (branded products versus non-branded products) whereas food safety has become a competitive necessity. However, due to food scares such as the BSE- and FMD- crises, or more recently, the 'rotten meat' scandal in Germany, food safety issues connected with firm boundaries that overlap vertical interactions hold differentiating potential. Hanf and Hanf (2005) considered the most striking consequence of these dramatic food scares to be the fact that politicians, consumers, producers and suppliers all assess food quality as no longer the matter of a single firm. Instead, the whole food chain has to work together in order to deliver the 'new quality'. Since food-borne hazards know no geographical boundaries, food safety standards have become a ubiquitous phenomenon that nationally and globally influences agri-food markets. Additionally, as products become more differentiated, commodity requirements are becoming more demanding, which leads to higher and more specific quality demands. Thus, in order to meet the demanded new quality, food processors and retailers have to re-design their food chains in such a way that these standards are adhered to every step of the way; thus, the coordination mechanism of the existing food chain must be altered. Spot market transactions, which are unable to properly coordinate the exchange of trust attributes, are substituted by transactions in vertically coordinated chain organisations. Such higher coordinated chain organisations are either hybrids or vertically integrated firms. For the agri-food business, there is evidence that the majority of these chain systems is organised as vertical networks, i.e. supply chain networks (SCN).

In transition countries or new member states, quality management concepts might still be an emerging field and might be used as a differentiating instrument: through EU-accession, the structure of those markets has shifted towards more globalisation and competition based on quality and price differences, rather than just price. On the one hand, the minimum quality standards of the EU set a bottom line that forces low-quality producers to raise their quality or drop out of the market. On the other hand, private standards such as the 'International Food Standard' (IFS), and Standards of the 'British Retail Consortium' (BRC), as well as industry-wide standardisation systems like the family of ISO standards are diffusing to those markets from Western countries. Concurrently, the new EU member states are seeing changing consumer demands – in terms of incomes and concerns over product standards. The changing environment in those markets, including both mandatory and voluntary standards, and ongoing restructuring processes at all stages in the food chain, may cause unique developments as far as quality management is concerned.

The aim of this paper is to identify the quality perception of the Polish operators in the dairy market and to find out which influence the chosen quality strategy exerts on the vertical coordination mechanism. In the first part of the paper, we present a brief review of the relevant theories. Since the Polish dairy market is dominated by cooperatives, we additionally review the general cooperative literature. The literature suggests that due to their complex governance structures co-ops may face significant hold-ups affecting quality control and management. Following the theoretical discussion, the second portion of the paper details the relevance of quality management thoughts for the Polish dairy cooperatives.

2. Theoretical considerations

2.1 The 'new quality' demand

There were several severe food crises in the years prior to the BSE- and FMD- crises in the winter of 2000/01, e.g. the Coke-scandal in Belgium, the BSE-crisis in the UK, and the wine-scandal in Austria and Germany. However, the crisis in the winter of 2000/01 can be regarded as the straw that broke the camel's back (Hanf and Hanf, 2005). The growing concerns of consumers, producers and governments worldwide have influenced the political debate on food safety. In the European Union (EU) a variety of new standards have been set in order to ensure the demanded minimum level of food quality. The result of these developments is that legal quality requirements are becoming more stringent and comprehensive (i.e. covering more safety attributes), and food policy is becoming increasingly integrated across various sectors (Ugland and Veggeland, 2006).

With increasing knowledge and perception of risk, consumer demand for safety and a willingness to pay for it increases (Antle, 2001). At the same time, as incomes rise, consumers demand even more quality, including, besides safety, such attributes as nutritional value, product diversity and tightness of product specification. Providing credence attributes is becoming an integral and ubiquitous issue for business operators. Indeed, trust-based attributes are expanding and include, besides food safety and nutritional properties, different contextual product properties related to certain public goods or values, such as environmental justice or cultural (traditional) values, etc. (Allaire, 2004). Consumer are, however, not able or willing to intensively and fully ascertain the credence characteristics of food products. Thus, they look for signals to facilitate their buying decisions, e.g. a well-known brand or a certificate of quality, thereby motivating the participants of the food chain to take the appropriate measures and to meet the 'new quality' demand (Hanf and Pieniadz, 2006).

Through the expansion and deepening integration of the EU, the quality-based competition among business operators has intensified. On the one hand, the minimum quality standards of the EU force low-quality producers to raise their quality or drop out of the market (Hockmann and Pieniadz, 2006). On the other hand, the increasing demand for quality signals especially allows supermarkets and manufacturers of branded products to benefit from imposing voluntary, private quality and safety standards, some of which are even more stringent than similar governmental regulations. Hence, the use of private voluntary standards across food categories has been increasing in both long-standing EU members, as well as in transition countries (Swinnen, 2006; Spencer and Reardon, 2005). Fulponi (2006) argues that private standards will become even more prominent in upcoming years as we observe increased market concentration and buying power in the retail sector, as well as its integration with financial markets. Unnevehr *et al.* (1999) assert that since food safety and quality can be successfully managed using private standards, their diffusion will henceforth even reduce the need for direct legal regulations. Thus, in order to meet the demanded new quality, food processors and retailers will have to enact additional mechanisms and re-design their food chains to induce the incentive-compatible behaviour of upstream business operators. Hanf and Hanf (2005) concluded that these demands on quality lead to the conceptualisation of chain quality management concepts by combining these 'new quality' demands with general chain management concepts.

2.2 Verticalisation and chain quality management

Food supply chains can be characterised as pyramidal-hierarchical networks. Such networks have a strategic character, with the focal company being the core element. The focal company is the centralised decision-making unit and may be either the manufacturer or retailer (Jarillo, 1988). Thus, the focal company determines the decisions of all network members, including the choice of measures to ensure the achievement of the super-ordinate network aims (Wildemann, 1997). Efficiency gains, higher profits, and cost reductions are important reasons for building such networks – which can be called supply chain networks – with food quality being regarded as one of the most important. Allaire (2004) mentioned the 'quality turn' as a main reason for the tendencies towards verticalisation in food chains worldwide. The consultancy KPMG (2000) characterises verticalisation as the building of vertically coordinated systems resulting in changing markets for 'fast moving consumer goods' (FMCG). Thus, vertically coordinated systems are understood as the exchange of goods not primarily conducted by market transactions. In other words, verticalisation means intensifying vertical relationships, which can take different forms of bilateral commitment between partnering firms based on implicit and explicit contracts. Generally, we can distinguish between two partnering types: strategic and operational partnering:

Strategic partnering is defined as an 'on-going, long-term, inter-firm relationship for achieving strategic goals, which deliver value to customers and profitability to partners' (Mentzer *et al.*, 2000: 550). The aim of strategic partnering is to improve or entirely alter a company's competitive position through developing new products and technologies and by creating new markets (Webster, 1992). Additionally, strategic partnering should also include exclusivity and non-imitability (Mentzer *et al.*, 2000). Operational partnering is defined as a 'needed, short-term relationship for obtaining parity with competitors' (*ibid.*: 550). Thus, an operational partnering strategy seeks to improve operational efficiency and effectiveness, especially by reducing transaction costs. Such orientation involves shorter time spans and less organisational resources. Therefore, operational partnership is much easier to implement (and also to reverse) than strategic partnership. In addition to such aspects of aligning interests, chain management has to consider aspects of coordination (Gulati *et al.*, 2005). In their framework on chain management Hanf and Dautzenberg (2006) combined these considerations with the thought that networks consist of different levels, namely firm, dyadic, and network levels. They point out that these three aspects have to be mirrored in the collective strategy[33] of a supply chain network.

Thus, if quality is the leading idea or strategy to be coordinated along the SCN, all members must share a homogeneous understanding of quality management, which provides the preconditions for the emergence of a collective strategy, and thus collective actions that address the chosen strategy. In this case, we expect a correlation between the chosen quality strategy and the design of the partnership. Therefore, the following assumption can be made in order to test it empirically in the second part of the study:

- If a firm chooses a pure cost leadership strategy, we expect that this firm will produce products that solely meet the minimum quality requirements (EU/governmental regulations). In this case, we expect that vertical exchange will take place by arm's-length transactions, meaning that vertical coordination is more or less done via the (spot) market. Thus, it will be sufficient for a cost-optimising firm to develop operational partnerships in both upstream and downstream stages.
- If a firm chooses the opposite strategy of product differentiation and quality attributes (especially trust elements) are chosen as the means of differentiation, we expect the firm to develop more sophisticated relationships. Yet we expect that the differentiated firms are more likely to develop strategic partnerships. In this case, vertical coordination can be regarded as highly cooperative or even vertically integrated.

2.3 Quality problems in cooperatives

In the previous section, we argued that food quality is no longer the matter of a single firm, but instead the whole food chain has to work together in order to deliver the 'new quality'. However, Hanf and Schweickert (2003) as well as Hanf and Kühl (2005) mention that due to their organisational form, cooperatives face problems integrating themselves into supply chain networks. A major reason for this are the co-op's internal institutions governing the behaviour of the co-op's members and affecting the co-op's ability to manage the quality of its products. Arguments for this are the following: in the context of increasing vertically coordinated agri-food systems, Sykuta and Cook (2001) showed that at the producer level, the most practical coordination mechanism is contracting. Because of their very own property rights structure, producer co-ops have some advantage compared to investor-owned firms. However, in addition to these benefits, they also face some problems. By using a property rights approach, Cook (1995) pointed out five general sets of problems: Free Riding Problems, Horizon Problems, Portfolio Problems, Control Problems and Influence Cost Problems. As Cook (1995) showed, these sets of problems constrict the various types of cooperatives (Sapiro I-Nourse II) differently. Combining a principal-agent approach with the concepts of opportunistic behaviour, conflicts of interest, asymmetric information and stochastic conditions, Eilers and Hanf (1999) show that it is not clear who is the principal and who is the agent, i.c. both the cooperatives and the members can be principals and agents. For this reason, neither leadership mechanisms nor selective terms of delivery can be enforced by the cooperatives, i.e., the members can deliver all the commodities which alternative dealers do not accept. Cooperatives that are to accept these commodities face the problem of adverse selection. Additionally, Fulton and Giannakas (2001) show that the cross-subsidisation and member heterogeneity in large centralised, multipurpose co-ops may lead to substantial financial pressures for the cooperative because members of such cooperatives do not

---

[33] In general, collective strategies are defined as systematic approaches by collaborating organisations that are jointly developed and implemented (Astley and Fombrun, 1983; Astley, 1984; Bresser, 1988; Bresser and Harl, 1986; Carney, 1987; Edström *et al.*, 1984; Sjurts, 2000).

see a strong connection between the success of the co-op and their own business. Furthermore, Karantininis and Zago (2001) showed, by applying a game theory model, that instead of selling their commodities to open co-ops, farmers would rather sell them to investor-owned firms if they had the choice. Fulton (1995) concludes that if markets disappear as a result of an increased vertical coordination, cooperatives may also begin to disappear. Hendrikse and Bijman (2002) share this assessment if investment on the side of the processor or retailer becomes more important for the total chain value than the investments by the farmers. In an empirical survey, Schramm *et al.* (2006) evaluated German dairy co-ops' brands. Using institutional economic and behaviour approaches, they showed the strengths and weaknesses of co-ops' branding strategies. Even though they were able to locate different factors exerting influence on branding strategies, quality issues were of major importance – negatively as well as positively. Besides these disadvantages, Briscoe and Ward (2006) name some managerial advantages of co-ops, as far as small and medium-sized co-ops are concerned; these include better communications with farmers, staff flexibility, easier (more efficient) control, hands-on management, greater motivation, and identification.

3. Quality management in Polish co-ops

Even though unbranded and branded products co-exist in the Polish dairy product market, an increase in market share of branded (higher quality) products is becoming evident. However, the majority of the branded products are manufactured by large companies. Particularly in the retailer sector, large (foreign-owned) retail chains are gaining market share. For these chains, it is typical to proliferate the food assortments, meaning that their suppliers are forced to produce more differentiated products.

For the producing sector in Poland, it can be said that a consolidation is taking place. However, in 2006 over 300 milk-processing firms with nine or more employees still existed. The majority of these dairies were producer cooperatives with milk processing being their prime economic activity.[34] Additionally, there is a number of small food processing firms manufacturing milk products as by-products. The consolidation process of milk processing is ongoing: whereas before the EU accession it occurred as a result for closures of small, inefficient and/or low quality producers, currently it occurs as a result of merging two or more cooperatives or by taking over financially weaker cooperatives, whereas mergers of prospering firms are hardly observed. In 2006 ten-firm concentration ratio was 55% with regard to the industry turnover. Five of the dairies considered in this ratio were cooperatives. Furthermore, the largest dairy company in Poland was a cooperative and had a market share of 11% (IERiGZ-PIB, 2007).

Because of their relevance for the Polish dairy market, we have chosen co-ops as the unit of empirical investigation. In February and March 2006 we interrogated 19 of the 22 largest of them. In 2005 the investigated cooperatives collected 41% of the raw milk (3.5 billion liters) and sold 31% of the final milk products to the market (1.2 billion Euro). Moreover, they accounted for around 25% of employment in the dairy industry (9.9 thousand) and around 24% of all milk suppliers (71.3 thousand).

The primary emphasis of our study was to gain an in-depth understanding of each individual case in the sample with regard to the research question given in the aim of this paper, that is quality perception and the design of vertical coordination mechanism. Because these questions are of

---

[34] The role of private companies has been increasing on a regular basis, whereas at the beginning of transition (1990) the milk market was almost completely covered by cooperatives. These private companies have been created through overtaking or acquisition of dairy cooperatives or through new direct investments.

sensitive nature and require detail knowledge on the firm's quality management processes, we decided to draw on qualitative and inductive case interviews to collect the data (Patton, 2002). Therefore, semi-structured interviews were conducted across the various hierarchical levels in the co-ops, including chief executive officers, quality managers, and supervisors in the marketing and supply departments. The sequence of the representatives questioned was the same for each co-op. The interviews were conducted by telephone and lasted between 20 and 40 minutes per respondent.[35]

This qualitative approach made particular sense in view of the above-mentioned research questions: on the one hand, chain quality management as well as networks concern activities and processes that are challenging to quantify and may even be ambiguous or misunderstood. On the other hand, the topics are particularly sensitive in emerging markets. Moreover, in those markets there might be some unique and relevant developments, which have to be first recognised, while giving the respondents some freedom to explore our general views. In the following, we elaborate on the relevance of the previously considered quality management thoughts based on the cooperatives questioned.

3.1 General comments on dairy co-ops

Despite the fact that organisational capabilities in Polish agriculture remain relatively low, producers' cooperatives continue to be a significant part of Polish dairy processing. To some degree all cooperatives draw on the long history of cooperative thinking. Most of them were founded in the 1920s and 1960s. According to the statements of the interviewed persons, cooperative values are coming increasingly under pressure. The challenges of maintaining a coherent socio-economic environment have been amplified by ongoing liberalisation, globalisation and standardisation, all of which change trade patterns for agricultural and food commodities and influence production costs and commodity prices. Similarly, the continuing expansion and deepening integration of the European Union, as well as the current reforms of the common market organisation for milk and milk products are redefining the challenges for operators in the European dairy market. Thus, for milk processors that decide to stay in the market, the issue is whether or not to adapt the current business strategy to the changing operating environment. The success of an enterprise not only depends on its ability to reconfigure the production system (technology, management) within the firm and improve the quality of inputs, but also to redesign its food chains, so as to efficiently produce the demanded quality and variety of milk products. In this context, co-ops face additional organisational problems that hamper their flexibility to make the adjustments needed. The complexity increases since the co-ops must meet the interest of their members while also satisfying the consumer. The member-driven orientation makes co-ops fundamentally different from investor-owned corporations in that they are compelled to look for quite stable markets, since they are not able to compete with more flexible and strictly profit-oriented private enterprises.

The interviews showed that on the one hand, all co-ops recognise the changing market requirements (demanded new quality) and understand quality to be an important action parameter for reaching the needs and wants of the consumers. This indicates that even for the Polish co-ops, food quality is more than plain food safety and the ability to continuously reproduce an *ex ante* defined set of attributes. On the other hand, the co-ops are also aware of the strong competition on the product

---

[35] Additionally, some major investor-owned dairies were interviewed as well. In this case, only the quality managers were asked for their analytic expertise, allowing relative statements regarding various quality management issues in co-ops and investor-owned firms.

(consumer) market and of being confronted by multiple problems with regard to their 'inherent charactcristics'. One of the largest constrains seems to be the conflict between the co-ops' status (cooperative principles) and economic goals: for most of the investigated co-ops, 'success' means the degree to which the enterprise has achieved the targeted goals. Since co-ops target different social and economic goals and the decisions are made mostly on a consensus-driven basis, there are plenty of potential conflicts of interest and hold-ups in the decision-making process. For example, with regard to the quality issue, there are significant inherent frictions when selecting small dairy producers–members that deliver low quality raw materials. The co-ops feel, generally, to be disadvantaged by the organisational and management structure, as well as by the limited financial and qualified human resources that would significantly improve both the process and product quality. Some co-ops also mentioned restricted access to foreign capital and know-how as being their main competitive disadvantage in quality improvement. Indeed, investor-owned firms with foreign investments benefit from having better access to approved business concepts and quality assurance systems, as well as capital from the main company. In interviews, the representatives of the two firms with FDI mentioned that they had not noticed any additional costs regarding implementation of higher quality standards in the plant. The implementation of QMS was monitored by representatives of the main company, and the staff in the domestic sub-company was well advised and supported by special training with regard to quality issues. One of the co-op leaders mentioned that 'the domestic dairies with FDI have just to copy the approved business concept and educate their staff on the costs of the mother company, whereas the co-ops have to be very 'innovative' while meeting the current market challenges and dealing with co-op specific constrains'. The 'innovative' thinking refers, however, to finding a creative solution under the given circumstances, while imitating the marketing strategies of private and prospering companies.

The above-mentioned considerations reveal that the lack of investment is one of the crucial hurdles for those co-ops investigated that wish to adopt additional quality improvement instruments. Questioned co-op representatives reported being sceptical regarding the benefits of the quality assurance systems prior to their implementation. In some cases, these doubts had postponed the decision to adopt. Once introduced (i.e. HACCP prior to EU accession) the co-ops acknowledged many advantages, e.g. less variation in quality outputs, better harmonisation of operational sequences, and less variability of staff skills while managing the quality.

Further, co-ops recognise some advantages as far as the relationship with their suppliers are considered: producers tend to trust a cooperative more than (foreign) investor-owned companies. The interrogated representatives pointed out that a farmer is typically risk-averse and seeks stable, trust-based relationships and social acceptance, both of which he can enjoy as a member of a co-op. In most cases, these utilities outweigh pecuniary disadvantages, since most of the co-ops bid lower prices for raw milk. Additionally, their support as 'service providers' enables them to supply some services to the farmers independent of the government or other private services. Besides information transfers between the co-op and the farmers (consulting, choice of production techniques), co-ops offer their members credits or access to credits for investments in the growth and specialisation of the farms. These instruments increase producer loyalty and assure, at least, continuous access to raw materials. However, co-ops still face multiple conflicts when selecting quality suppliers (supplier=member). The organisational 'stickiness' in the selection process of quality producers impedes the manufacturing process and quality output and compels the co-ops to target markets for lower quality. Nevertheless, the co-ops strive to adjust to the market requirements and utilise various instruments to induce the incentive-compatible behaviour of upstream business operators. For example, co-ops use quality-dependent payment schemes to remunerate better raw milk quality. Additional provisions exist as well, including a price premium for extraordinary quality (super extra) and direct delivery for farms either approved by the veterinary bureau or which possess certain breeds of milk cows. All cooperatives pay a price premium upon membership. Thus, payment schemes differ greatly between dairies. However, in all pricing mechanisms, the price increases as compliance with quality requirements set by the purchaser increases. Co-op representatives mentioned that the EU quality regulations have an immense 'educative' influence on the farmers with regard to quality improvements. On the other hand, mandatory regulations take away a co-op's ability to select (passive selection). The co-ops expect some competitive advantages at the procurement stage due to the better 'access' to their local communities, in the medium-term.

Proposition: cooperatives are disadvantaged in quality-based competition due to their lower flexibility and limited access to financial and qualified human resources. Thus, they are often imitators or choose generally stable markets for their proliferation.

Proposition: cooperatives have some advantages over private firms at the procurement stage in the medium-term, owing to their local communities' attachment, and their potential of being a 'service provider' that enables them to supply services independent of the government or other private services.

3.2 The co-ops solution: how to be competitive

Market segmentation

Economies of scale have become a factor of considerable importance in the milk sector and have affected all stages and legal forms of enterprises in Europe. The (largest) Polish co-ops recognise the challenge and strive to expand in the milk market by applying various growth strategies. The most common strategy is internal growth via entering new (export) markets and market penetration with regard to FMCG such as UHT-milk. Moreover, well performing co-ops expand through mergers and acquisitions which, besides rapidly increasing revenue, allow them to utilise economies of scope, e.g. the transfer of capital, technology and know-how within the company, as well as synergies of using common brand names. We observe that all investigated co-ops modify their production profile, which leads to a kind of market segmentation and mitigates direct rivalry among firms. Basically, they move toward specialisation on either the white or yellow production line, or they extend their production, offering highly diversified goods of both lines. The interviews indicate that firms use both cost-leadership, and to different degrees, product differentiation strategies. Product differentiation is important to all investigated co-ops, as they recognise the need to make products more attractive to the target market. However, differentiation takes on various forms, from a simple modification of an existing product (a new flavour of yoghurt) to creating a new branded product, in which factors other than price are taken into account by consumers (market segmentation).

Proposition: to overcome the intensive competition, co-ops modify their production profile, which leads to market segmentation.

Choice of quality strategy

The heterogeneity of the co-ops is even greater when comparing the chosen quality strategies. Co-ops, which take on the role of the focal firm in a dairy chain especially act to escape from price competition by setting themselves apart and bringing quality to a differentiating parameter. Investments in brand, reputation and reduction of information asymmetry about product quality (social marketing, TV adverts, food exhibitions, etc.) are becoming a priority for this group. All of those co-ops use intensive ISO quality standards. Some of them also implemented voluntary ISO standards for environmental management and possess an adequate certificate integrating both systems, whereas the remaining manufacturers of branded products intend to implement them in the near future. The respondents of those co-ops stressed that the main incentive for implementing the voluntary environmental standards was to demonstrate their environmental concerns, and hence to increase their reputation and brand loyalty. Several dairies in this group additionally address region-specific credence attributes, such as cultural and traditional values of the area where the co-op is located, and social justice while stressing the importance of product purchase for employment in rural areas. In most cases, this strategy leads to a kind of 'local patriotism' among consumers, as far as the purchase of the regional milk products is concerned. To stabilise their market shares and to protect their independence, the co-ops with a strong brand reject producing and selling their products under a private retailer's label. This premium-quality strategy, however, usually concerns the largest of the investigated co-ops, and thus seems to be a minority when all Polish co-ops are concerned.

On the other 'end' of the investigated firms are co-ops that utilise a strong cost-orientation for their competitive advantage. Cost leadership is achieved by economies of scale, thus producing basic products and improving the efficiency of all business operations is a priority for this group. In those groups there are usually no dominant standard-setting purchaser, thus the dairies have some freedom in their choice of quality strategies and measures to guarantee the effectiveness of the chosen strategy. Accordingly, those co-ops offer their products at the cheapest price (price leadership) while meeting just the minimum quality as demanded by the obligatory regulations. The representatives of those co-ops argued that there is so far no need to change this strategy, since there is still a profound group of low-income consumers who demand their products, and hence enable attractive profits. Because the firms do not posses a strong brand, they use voluntary public quality certifications and labels to signal quality, such as 'Q' (quality) and 'Eco' (ecological), developed and assigned by the Polish Centre for Testing and Certification (PCBC). Some standards promote national food products of high and reliable quality, such as the 'Try Fine Food' standards (PDZ) designed by the Polish Ministry of Agriculture and Rural Development. Representatives of the co-ops mentioned however, that they recognised that their products are currently threatened by the plurality of signs, which can sometimes even increase the uncertainty among consumers.

Between those two above-mentioned groups, there are co-ops that are strongly dependent on direct purchasers. Usually these co-ops have no brand (or not a strong one) and regard the dominant purchaser as the standard-setting entities; they then adjust their quality strategy and management to the respective requirements.

If the focal company is a manufacturer requiring tightly-specified industrial products, the co-op has to adjust quality assurance systems to the specific requirements (e.g., unique chemical or physical parameters). Quality signals and voluntary quality systems seem to be irrelevant to those co-ops. Some FDI use the possibility of intra-industry trade based on the co-ops' supply, since the co-ops have better access to the local milk suppliers. On the other hand, the co-ops benefit from the financial support of the focal firm, while carrying out relation-specific investments. Joint investments first concerned quality improvements at the procurement stage, and then the adoption of new processing technologies. The adherence to specific requirements is ensured by close business-to-business (B2B) relations, including some knowledge-sharing routines and enhanced monitoring. Additionally, in such direct relationships, the threat of direct and strong sanctions (losing the focal purchaser) limits opportunistic behaviour and facilitates cooperative adaptation by the co-op. At the same time, the high intensity of unexpected controls and enhanced monitoring suggests that the focal firm either does not trust the partner or must steadily improve the knowledge about its capability, as well as the correctness of the process.

If a dairy sells its products to a retail chain and the retailer then sells them as proprietary private label products, the implementation of retailer-specific schemes will be required. Thus, the processors are voluntarily obligated to implement standards for auditing retailer-branded food products, such as IFS and BRC. Interestingly, the retailers are satisfied if those concepts are running but they do not need to be certified, which seems to be specific for an emerging market. In this case, the quality standards are used to coordinate pooled interdependencies. We found that focal firms prefer control-based relationships rather than trust-based ones to govern partnership behaviours and the maintenance of their specific requirements. In particular, retailers with strong bargaining power apply restrictive control mechanisms, even if the running quality concepts are certified. Adjustment to the retailer-specific requirements involves investment in specialised resources, which increases the co-ops' dependence on retailers. However, because IFS and BRC are widely used standards, the co-ops have formal access to alternative institutional customers on the national or international markets.

Proposition: co-ops follow different quality strategies within the chosen production profile. Adoption of higher quality standards is an economic activity, guided by the co-op's profit expectations.

Verticalisation

The chosen quality strategy influences the vertical coordination mechanism along the dairy chain. In the next step, we investigate the linkages between quality performance and the design and intensity of vertical relationships with the upstream and downstream stages. In order to simplify the discussion of the empirical results we generalised the findings through building of four groups of firms that showed similarities with regard to their strategic orientation and the structure of the SCN. The first group (1) represents cooperatives, which take the position of a focal company in the SCN, and hence are standard-setting units exercising chain quality management. In the second (2) and third (3) groups the network is led by a focal firm, being the dominant purchaser. Thus, even if co-ops in those groups deliver their products to different purchasers, the dominant one determines the co-op's quality performance, and hence the design of the relationships with the upstream stages (suppliers). The fourth group (4) includes cooperatives, which had not built its own network as of the time of investigation. In following, we elaborate on the features of each of the four groups mentioned.

If a co-op is a manufacturer of branded products (group 1), it takes on the position of a focal company itself. Producing and delivering quality products requires implementation of superior (or at least higher than average) quality management systems. Those manufacturers have recognised that they must actively create their own distribution opportunities. For all channels – retail, wholesale, and export – they use medium- and long-term contracts, which contain all sorts of details that address product quality matters. Thus, the co-ops control, to some extent, quality measurements that are external to the firm. However, despite reciprocal information exchange and ongoing negotiations, these relationships still have an operational character. However, the co-ops increasingly use partnering mechanisms that are more strategic in nature, so marketing information such as point-of-sale data is exchanged. The co-marketing is particularly intensive in partnerships with retail chains, because it is based on ongoing negotiations and adjustments addressing sales strategies, promotions, and pricing behaviour. Typically, this leads to complex reciprocal interdependencies, which demand well-defined organisational principles and a certain level of management skills to govern the relationships. Such relation-specific systems seem to be unique for an individual chain of branded products manufacturer.

Interaction at the procurement stage can also be described as intensive, especially with the larger and specialised farmers. Using incentives to upgrade the quality of raw milk, the co-ops exert a firm boundary for the overlapping quality scheme. Some of the actions result from the implementation of ISO quality standards, which require quality objectives to be included in the quality policy and to be leveraged to upstream stages. Additionally, the co-ops provide intensive consulting assistance and herd management for their members. One co-op even provided business angles as an alternative know-how source (technology transfer) as early as at the beginning of the 1990s. The parties utilise usually indefinite collaboration contracts covering strategic issues and addressing principles set in the respective co-op act. These activities outperform the industry average and indicate sustainable, long-term collaborations at the procurement stage. Overall, we think that in this case, we can speak not only of a chain quality concept; but rather of a strategic one.

When the focal company is either a manufacturer (group 2) or branded retailer (group 3), we found that purchasers prefer control-based relationships rather than trust-based ones to govern partnership behaviours and the maintenance of their specific requirements. Contracts and managerial discretion are used to meet sequential interdependencies, with the contracts containing specifics on quality and payment. As long as these specifics are met, the duration is prolonged. Additionally, we found some reciprocal interdependencies among the partners in B2B relationships between the co-ops and the industrial purchaser. Overall, the relationships between the focal companies and the dairies are very intense, and not easy to imitate. Therefore, this type of partnering is more strategic than operational.

Regarding the relationship between co-ops and their members, we found that co-ops encourage growth strategies through intensive consulting assistance, which aims to select larger farms. Overall, we conclude that supply chain networks are established and chain quality management is exercised. However, even though the partnering can be described as more strategic in nature, there is a lack of a collective quality strategy. Thus, we would classify the paradigm as an operational chain quality management. Because more and more retailers are bringing their proprietary private label products on the market, there is increasing price competition among the products. For the co-ops concerned, this means that they face strong costs pressure, which precludes resource allocation to more sophisticated quality management systems.

The fourth co-ops group (4) identified in our data concerns dairies not involved in any SCN. In this group, there are no standard-setting purchasers. Furthermore, these co-ops utilise strong cost orientation to deliver basic product meeting solely mandatory standards and schemes. The majority of the produce is distributed via wholesalers and is based on relatively unstable relationships. Both parties apply loose, long-term contracts, in which the partnership duration, payment schemes, and general quality requirements are fixed, whereas the amount, composition and price of delivery are flexible. Since the agreements do not include any partnership-specific agreements, they are easy exchangeable. Because of their cost orientation, it is not surprising that those co-ops restrict their relationships with suppliers to the basic commitments and principals as regulated in the cooperatives' statute. Nevertheless, the co-ops' relationships seem to be better developed at the procurement stage than at the distribution stage. We could identify operational partnerships between the co-ops and their milk suppliers and some dyadic actions addressing the chosen quality strategy at this stage, but there is still a missing recognition of similar interests and initiatives to explore operational advantages in relationships with their institutional customers. Further development of retailers and wholesalers with strong bargaining power will force the dairies either to join their SCN or take the role of a focal company and strengthen their brand. Independent of that, the dairy must first create its supply chain network and develop a chain quality management.

Proposition: the challenge of the focal firm is to choose the quality approach that best fits the overall network's aims as well as its performance.

Proposition: the chosen quality strategy determines the design of the vertical coordination mechanism. The higher the requirements of the final product, the further quality management systems go beyond a firm's boundaries and the higher is the intensity of the relationships between the intermediary stages in the dairy chain.

4. Final remarks

Food today is perceived as a complex bundle of characteristics, with an increasing level of importance placed on credence attributes relating to product and methods of production (e.g. environmental friendliness). Food processors and retailers must re-design their food chains in such a way that all stages of the food chain are involved in meeting the demanded 'new quality'. Therefore, the coordination mechanism of the existing food chain has to be altered, because spot market transactions are not able to properly coordinate the exchange of credence attributes; they must be substituted by transactions in vertically coordinated chain organisations. Such chain organisations are either hybrids or vertically integrated firms. For the agro-food business, there is evidence that the majority of these chain systems are organised as vertical networks, i.e. supply chain networks. Chain management must incorporate the relationships and interdependencies of the member firms, as well as problems arising at the firm level, the dyadic level, and the network level. Applying these thoughts on quality issues, it becomes evident that quality management has to be understood as a firm-boundaries overlapping approach. In this context, we have shown that chain quality management has to be divided into operative chain quality management and strategic chain quality management.

The example of Polish dairy cooperatives provides new insights into quality management issues faced by cooperatives. First, our findings indicate that activities related to quality improvements are generally aligned with current market opportunities for optimal enterprise performance. On the one hand, co-ops recognise that they must deliver safe and reliable food and differentiate their products, at least in a part, to make them more attractive to the consumer. This indicates that even for the co-ops, food quality is more than plain food safety and the ability to continuously reproduce an *ex ante* defined set of attributes. On the other hand, co-ops face various problems, the largest of them being the conflict between the co-ops' principles and economic goals and limited financial and qualified human resources that would significantly improve both process and product quality. The co-ops' specific problems compel them to modify their production profile and usually to tap markets for basic products, since they are hardly able to compete with more flexible and strictly profit-oriented private enterprises on markets for high-value added products. However, our study reveals that there are some exceptions to this general observation, especially when examining the co-ops' chosen quality strategy and the design of the quality management systems.

Overall, we conclude that in most cases, supply chain networks are established and chain quality management is exercised. However, this is only the case if there is a focal actor that influences its network structure. The results show that retail chains and industrial purchasers with foreign investment and strong bargaining power usually take the position of the focal firm in the SCN. In those cases, strategic partnering between the individual chain stages dominates. However, because there is a lack of a collective quality strategy overlapping all actors, quality management initiatives are still operational in this case. There are still some Polish cooperative dairies, which are not embedded in any SCN. These include processors of non-branded goods or those with weak brands that sell their products to purchasers without a focal position. Because there is no powerful focal firm in the chain, no managerial discretion can be exerted and no chain quality management concepts can be installed. Thus, we could only identify operational partnerships between the co-ops and their milk suppliers and some dyadic actions addressing the chosen quality strategy at the procurement stage. In contrast, at the distribution stage, we observed that the partners do not share homogenous interests regarding quality issues; there is even a lack of dyadic initiatives aimed at exploring the operational advantages of the cooperation.

Our empirical results show profound diversity regarding quality management approaches in the Polish milk supply chains. However, one thing is clear: the chosen quality strategy determines the design of the vertical coordination mechanism. Thus, the higher the product requirements, the further quality management systems go beyond a firm's boundaries and the stronger the shift is from operational towards strategic quality management.

Based on the findings, it is suggested that firms can choose between alternative quality strategies and SCN designs. It is argued that it may be important to support the cooperatives in advancing their quality management concepts. It would be especially recommended in regions being in economic transition, especially if customer satisfaction is low, or if the consumer requirements are not sufficiently being leveraged upstream to suppliers.

References

Allaire, G., 2004. Quality in economics: a cognitive perspective. In: M. Harvey, A. McMeekin and A. Warde (Eds), Qualities of food. Manchester: Manchester University Press, pp. 61-93.

Antle, J.M., 2001. Economic analysis of food safety. In: B.L. Gardner and G.C. Rausser (Eds.), Handbook of Agricultural Economics. Vol. 1B: Marketing, Distribution and Consumers. The Netherlands: Elsevier, pp. 1083-1136.

Astley, W.G., 1984. Towards an appreciation of collective strategy. Academy of Management Review, 9: 526-535.

Astley, W.G. and C.J. Fombrun, 1983. Collective strategy: social ecology of organizational environments. Academy of Management Review, 8: 576-587.

Bresser, R.K.F. and J.E. Harl, 1986. Collective strategy: vice or virtue? Academy of Management Review, 11: 408-427.

Bresser, R.K.F., 1988. Matching collective and competitive strategies. Strategic Management Journal, 9: 375-385.

Briscoe, R. and M. Ward, 2006. Is small both beautiful and competitive? A case study of Irish dairy cooperatives. Journal of Rural Cooperation, 34: 113-134.

Carney, M.G,. 1987. The strategy and structure of collective action. Organization Studies, 8: 341-362.

Cook, M.L., 1995. The future of U.S. agricultural cooperatives: a neo-institutional approach. American Journal of Agricultural Economics, 77: 1153-1159.

Edström, A., B. Högberg and L.E. Norbäck, 1984. Alternative explanations of interorganizational cooperation: the case of joint programmes and joint ventures in Sweden. Organization Studies, 5: 147-168.

Eilers, C. and C.-H. Hanf, 1999. Contracts between farmers and farmers – processing co-operatives: a principal-agent approach for the potato starch industry. In: G. Galizzi and L.Venturini (Eds.), Vertical relationships and coordination in the food system. Heidelberg, pp. 267-284.

Fulponi, L., 2006. Private voluntary standards in the food system: the perspective of major food retailers in OECD countries. Food Policy, 31: 1-13.

Fulton, M., 1995. The future of Canadian agricultural cooperatives: a property rights approach. American Journal of Agricultural Economics, 77: 1144-1152.

Fulton, M. and K. Giannakas, 2001. Organizational commitment in a mixed oligopoly: agricultural cooperatives and investor-owned firms. American Journal of Agricultural Economics, 83: 1258-1265.

Gulati, R., P.R. Lawrence and P. Puranam, 2005. Adaptation in vertical relationships: beyond incentive conflicts. Strategic Management Journal, 26: 415-440.

Hanf, J.H. and E. Schweickert, 2003. Co-operative success by forming a strategic member group. Paper presented at the Conference Vertical Markets and Co-operative Hierarchies, Bad Herrenalb, Germany.

Hanf, J. and C.-H. Hanf, 2005. Does food quality management create a competitive advantage? Paper prepared for the 92nd EAAE seminar on Quality Management and Quality Assurance in Food Chains, March 2-4, Göttingen, Germany.

Hanf, J. and A. Pieniadz, 2006. Quality management in strategic networks – Is there any relevance in the Polish dairy sector? In: M. Fritz, U. Rickert and G. Schiefer (Eds), Trust and risk in business networks. University Bonn-ILB Press, pp. 459-467.

Hanf, J.H. and K. Dautzenberg, 2006. A theoretical framework of chain management. Journal on Chain and Network Science, 6: 79-94.

Hanf, J.H. and R. Kühl, 2005. Supply chain networks in the agri-food business - challenges and threats for co-operatives. In: Theurl (Ed.), Strategies for cooperation. Aachen: Shaker Verlag.

Hendrikse, G.W.J. and J. Bijman, 2002. Ownership structure in agrifood chains: the marketing cooperative. Americam Journal of Agricultural Economics, 84: 104-119.

Hockmann, H. and A. Pieniadz, 2006. Is a full diffusion of EU standards optimal for the development of the food sectors in the CEEC? The case of the Polish dairy sector. In: K. Mattas and E. Tsakiridou (Eds.). Food quality products in the advent of the 21st century: production, demand and public policy. Cahiers Options Méditerranéennes, 64: 179-196.

IERiGZ-PIB, 2007. Rynek mleka. Stan i perspektywy, Raporty rynkowe [Report on the Polish dairy market: performance and developments], Institute of Agricultural and Food Economics - National Research Institute, No. 32, Warsaw.

Jarillo, J.C., 1988. On strategic networks. Strategic Management Journal, 9: 31-41.

Karantininis, K. and A. Zago, 2001. Endogenous membership in mixed duopsonies. American Journal of Agricultural Economics, 83: 1266-1272.

KPMG, 2000. Vertikalisierung im Handel. Published Consulting Study.

Mentzer J.T., S. Min and Z.G. Zacharia, 2000. The nature of inter-firm partnering in supply chain management. Journal of Retailing, 76: 549-568.

Patton, M.Q., 2002. Qualitative research and evaluation methods, 3rd ed.. Thousand Oaks: Calif. Sage Publications.

Schramm, M., A. Spiller and T. Staack, 2006. Zur Markenlücke genossenschaftlicher Industrieunternehmen in der Ernährungswirtschaft – Eine empirische Untersuchung. Zeitschrift für das gesamte Genossenschaftswesen, 56: 229-242.

Sjurts, I., 2000. Kollektive Unternehmensstrategie. Grundfragen einer Theorie kollektiven strategischen Handelns. Habilitation, Wiesbaden.

Spencer, H. and T. Reardon, 2005. Private agri-food standards: implications for food policy and agri-food systems. Food Policy, 30: 241-253.

Swinnen, J.F.M. (Ed.), 2006. Global supply chains standards and the poor: how the globalisation of food system and the standards affects rural development and poverty. Cambridge, MA: CABI, p.cm.

Sykuta, M.E. and M.L. Cook, 2001. A new institutional approach to contracts and cooperatives. American Journal of Agricultural Economics, 83: 1273-1279.

Ugland, T. and F. Veggeland, 2006. Experiments in food safety policy integration in the European Union, Journal of Common Market Studies, 44: 607-624.

Unnevehr, L.J., G.Y. Miller and M.I. Gomez, 1999. Ensuring food safety and quality in farm-level production: emerging lessons from the pork industry. American Journal of Agricultural Economics, 81: 1096-1101.

Webster, F.E. Jr., 1992. The changing role of marketing in the corporation. Journal of Marketing, 56: 1-17.

Wildemann, H., 1997. Koordination von Unternehmensnetzwerken. Zeitschrift für Betriebswirtschaft, 67: 417-439.

The roles of geographical indications in the internationalisation process of agri-food products[36]

G. Belletti, T. Burgassi, E. Manco, A. Marescotti, A. Pacciani and S. Scaramuzzi

> We heartily remember with great affection and esteem Elisabetta, her joy, her grace, her lost presence among us

Abstract

The purpose of this paper is to analyse the role PDO and PGI (Geographical Indications) can play in the internationalisation process of agri-food firms producing Origin products. The hypotheses we discuss are twofold. Firstly, the role of PDO/PGI as a tool for defending the geographical protected name from imitations on markets is only one of the roles firms assign to PDO/PGI. Secondly, the PDO/PGI protection transforms the geographical name into a 'club good', which normally has the consequence of reinforcing the collective organisation between firms. The study pointed out the high average impact PDO/PGI has in capturing the benefits from the reputation of a geographical name. But the average satisfaction in using PDO/PGI derives also from its many 'marketing oriented' functions, such as satisfying specific consumer requests, increasing turnover in existing channels or opening new ones, and guaranteeing consumers by means of the EU logo. Furthermore, the results of the survey support the hypothesis that the collective dimension, linked to the use of PDO/PGI, has important positive effects in supporting exports of small-medium enterprises. Collective organisations and appropriate public local policies, aiming at supplying firms with the necessary capabilities to face the international market, are the two pillars with which PDO/PGI allows origin product systems to benefit from the internationalisation of food markets.

Keywords: quality food products (QFPs), PDO and PGI, small and medium enterprises (SMEs), marketing strategies

1. Introduction

In the face of mass produced food, the market success of Origin Products (OP), that is agri-food products whose characteristics and reputation are linked to *terroir,* derives from their suitability to respond to consumers' needs in terms of genuinity and authenticity. Such success has pointed out the usefulness of agri-food products' territorial origin, highlighted by the products' name and label, which has become a strategic tool for differentiation.

The EEC Reg. 2081/92 (substituted by EC Reg. 510/2006) introduced the Protected Designation of Origin (PDO) and Protected Geographical Indication (PGI), two quality signs pointing out the link between the production process of an agri-food product and its territorial origin by regulating the use of geographical names.

---

[36] The paper is a product of the research project PRIN 2006, funded by the Italian Ministry of University and Research (MIUR) entitled 'The internationalisation of agro-food firms: strategy analysis and role of geographical indications', coordinated by prof. A. Pacciani. We thank the referees for kindly reading the paper and providing us with their useful suggestions and comments to improve it; all the responsibility of what is stated in the paper remains of the authors.

The use of suitable marketing strategies for OPs and the role PDO and PGI is an issue which has only recently been considered in economic literature. To date, there is not much evidence of the effects the recognition and use of PDO and PGI may have on the internationalisation of agri-food firms. Indeed, because firms producing PDOs and PGIs are mainly Small and Medium Enterprises (SMEs), marketing strategies to orient consumer demand are often difficult to implement, both in terms of costs and supply availability.

This paper aims at verifying the roles PDOs and PGIs play within the internationalisation strategies of agri - food firms. Our research questions will thus be: when and why do firms use PDO/PGI to sell their products on the international market? Are PDO/PGI a valuable strategic tool to firms?

The paper starts by analysing the problems faced by OPs on international markets and the contribution the protection of Geographical Indications (here PDO/PGI) can give to support firms' internationalisation strategies (Section 2). It then describes the aims and the methodology followed in investigating the factors of success and limitation in the use of PDOs and PGIs in internationalisation, and the approach followed in the selection of the case-studies (Section 3). Moreover, a short description of the production system is provided in the case-studies (Section 4). Finally, the results of the survey are presented, at an aggregate level (Section 5), at product-type level (Section 6) and at single firm level (Section 7). The paper ends with some concluding remarks (Section 8).

2. Expected contributions of geographical indications (PDO/PGI) to the internationalisation of agri-food firms

The effect of globalisation on consumer preferences and the availability of new technologies in production, communication and information, have offered agri-food firms, producing OPs, new opportunities on the international markets although delocalising one or more production process phases (Galavotti, 2005) is usually not easy for OPs. This is due to their specificity (and the rationale for protecting them) which is based on traditional production practices localised in well defined geographical boundaries and on the use of local raw materials and ingredients. Consequently, internationalising their commercial channels is the most relevant way by which agro-food firms, producing OPs, can face international markets.

However, firms producing OPs are often Small and Medium Enterprises (SMEs). This often prevents them from implementing 'traditional' marketing strategies due to their lack of capabilities and financial resources for promotion (Albisu, 2002; Rangnekar, 2004). Consequently, agri-food SMEs need suitable tools to take advantage of internationalisation, as for instance, physical factors and above all non–physical factors such as know–how, organisational and managing capabilities (Barjolle *et al.*, 1998; Galizzi, 1995).

Furthermore, the internationalisation of marketing channels implies several complex transactions and actions to increase consumer awareness and information on a product's specific qualities, in order to lower informative problems and asymmetries of 'distant' consumers. Indeed, misleading information and frauds are quite common, especially for more reputed OPs (Marette *et al.*, 1999)[37].

---

[37] As documented by Nomisma (Nomisma, 2005), usurpation and misuses resulting from non–Italian made products which imitate and/or use Italy's notoriety on the US market is widespread, provoking a huge economic damage and causing many difficulties for firms exporting the 'original' OP.

To this point, EC Reg. 510/2006 provides PDO and PGI, quality labels indicating not only the link between product quality and its territorial origin but also guaranteeing protection against misleading uses of the geographical name. As stated in EC Reg. 510/2006, 'in view of the wide variety of products marketed and the abundance of product information provided, the consumer should, in order to be able to make the best choices, be given clear and succinct information regarding the product origin' (EC Reg. 510/2006: 4th 'whereas'). Both PDO and PGI share the same normative system and registration procedures, and grant the same guarantees to producers and consumers. Differences between these two signs depend on how closely the quality specificities of the products are linked to the geographical area of which they bear the name[38].

Agri-food firms are paying growing attention to PDO and PGI, and have great expectations on the positive effects these labels can exert on international markets. Studies (Belletti, 2000; Anania and Nistico, 2004; Menard, 1996; Pacciani *et al.*, 2003) have underlined the fact that PDO and PGI labels may not only support the reputation of a product, thus having a market-cleaning effect (exclusion of non-authentic OPs) or facilitating the creation of notoriety around a geographical name, but they can also act as a 'key' to access modern markets and/or long distance commercial channels (Belletti *et al.*, 2006). As a matter of fact, PDO and PGI labels exert a 'reassurance' effect on consumers (Tregear *et al.*, 2007) and on intermediate customers. The reassurance comes both from the Code of Practice, which establishes shared rules on production and product quality, and from the presence of an independent and qualified third party control (Anania and Nistico, 2004). In this sense PDOs and PGIs can act as important competitive levers for agri-food firms which want to penetrate long-distance markets. Indeed, the opening of international markets and trade lead to both a growing demand of information from consumers, and the use of 'halo country effects' (for reputed geographical names) by sellers (Almonte *et al.*, 1996; Belletti and Marescotti, 2006).

Furthermore, the need of the Code of Practices requires agri–food firms to codify management procedures and increase internal coordination between different stages of the supply chain (Belletti *et al.*, 2007). This can ease the implementation of internationalisation strategies either in terms of producers' improved confidence toward management functions (planning, organising, leading and controlling), and collective marketing initiatives on foreign markets, as a result of the collective dimension of PDO and PGI, which is often amplified by the presence of collective organisations such as consortia (Canada and Vazquez, 2005).

Most studies have focussed on the effects PDOs and PGIs have on producers, on consumers' behaviour, on the supply chain, whereas little attention has been granted to the impact PDOs and PGIs have on local development, rural development and environment. By contrast, the use of suitable marketing strategies for PDO and PGI products is an issue which has only recently been discussed and to date there is little scientific evidence of the effects the application of PDO and PGI can have on the internationalisation of agri-food products.

---

[38] The Protected Designation of Origin (PDO) is meant for those products which show an objective and very close link between their quality and the geographical area specificities (including both human and natural factors, such as climate, soil quality and local know – how); the Protected Geographical Indication (PGI) also designates products linked to the area of which they bear the name but with a looser objective link.

3. Objectives, data and methodology

As mentioned, the purpose of this paper is to analyse the role PDO and PGI can play on the internationalisation of agri-food firms producing Origin products. The hypothesis we discuss is twofold.

First, the role of PDO/PGI as a tool for defending a geographical protected name from imitations on markets is only one of the multiple roles firms assign to PDO/PGI, in addition to the othe PDO/PGI firms can use in their (international) marketing strategies.

Second, the PDO/PGI recognition transforms the protected name into a 'club good', which normally has the consequence of reinforcing the collective organisation between firms, thus exerting important positive effects in supporting export of SME production systems.

Specifically, the main objectives of the paper are:
• to determine the motivations of agri-food firms in using PDOs and PGIs, with special reference to export markets;
• to assess the level of usage of these signs;
• to describe the marketing channels mostly used;
• to assess the satisfaction of those firms which use the denominations, matching them with their declared motivations;
• to identify the role of collective organisations managing the denomination.

To reach these objectives, a survey was carried out on four products from Tuscany (Italy): Olio Toscano PGI (Tuscan extra-virgin olive-oil), Olio Chianti Classico PDO (extra-virgin olive-oil of Classic Chianti), Pecorino Toscano PDO (Tuscan sheep-milk cheese), and Prosciutto Toscano PDO (Tuscan ham). These products represent the most exported PDO/PGI[39] Tuscan products, although each with a different export orientation. Average data on the four denominations show how two thirds of the two PDO/PGI extra-virgin olive-oils (Toscano and Chianti) are exported, while Pecorino and Prosciutto production systems export only a small portion of certified volumes (Table 1).

These products bear a well-known geographical name (*Toscana* and *Chianti*) which has a particular appeal to foreign markets, thus exerting a positive 'halo country effect' on food products. The motivations behind the local actors' application for a PDO/PGI are quite diverse in the four cases.

---

[39] The EC Reg. 510/2006 does not apply to wines and vinegars.

Table 1. PDO/PGI exported quantities on total PDO/PGI production, average years 2004 and 2005.

Product	% Export on total PDO/PGI certified volumes
Olio Chianti Classico DOP	57.0%
Olio Toscano IGP	65.0%
Pecorino Toscano DOP	16.0%
Prosciutto Toscano DOP	6.0%

Toscano extra-virgin olive oil is well-known on international markets and suffers from imitations. PGI has been motivated mainly by the need to protect the name 'Toscano' from imitations and other unfair uses that were highly widespread on the market, hence the strong support from olive growers associations and regional administrations. By contrast, Pecorino Toscano and Prosciutto Toscano have both a long tradition in national denomination of origin but they are less known on international markets. Firms involved in these PDO are multiproduct, and therefore less dependent on the use of PDO signs. The role played by processors associations was stronger than the breeders associations in supporting this PDO, and the PDO was perceived by processing firms as a marketing tool to improve their distribution on Italian and foreign markets. Chianti Classico olive oil is a high quality niche product and imitations were not so widespread. The motivation behind applying for a PDO was to gain the EU quality label as recognition of the product's prestige and of the Chianti name. Pre-existing DOCG (a Denomination of Origin which applies to the wine sector) for Chianti wine used by the same olive growers stimulated the demand for an olive oil PDO recognition.

The empirical analysis of the above cases was carried out in spring 2006 and was articulated in three phases: a general overview of PDO/PGI products supply chains, an in-depth analysis on a group of exporting firms in each supply chain, and an analysis of collective organisations supporting the PDO/PGI (Consorzi di Tutela or Producers' associations).

The first phase aimed at highlighting the characteristics and the dynamics of the supply chains, with special reference to the use of PDO-PGI signs in marketing strategies on export channels. Information was collected by means of official databases, previous researches results, and interviews of the relevant actors of each supply chain (consortia and/or producers' associations representatives, local administrations, certification bodies, etc.).

The second phase aimed at focussing on the role PDO/PGI played in the marketing strategies of the firms. An in-depth analysis was carried out of a selected group of firms which take the decision of using the PDO/PGI label within the supply chain (mainly bottlers for extra-virgin olive-oils, cheesemakers for Pecorino Toscano PDO, and seasoners for Prosciutto Toscano PDO). In the case of production systems with a high number of export-oriented producers, through a *critical case sampling* methodology we identified the firms to include in our survey, in order to cover the typology of firms involved in the supply chain. The firms selected are key-elements to understand the matters covered by the study or have a particular capacity of influencing the development of the internationalisation process. Therefore, our aim was not to build a statistically representative sample, but to maximise information on a specific subject.

In the third phase the representatives of collective organisations (the four consortia of the PDO/PGI products) were interviewed, with the aim of evaluating their role in promotion and more in general in responding to the needs of their associated firms.

We adopted mainly open qualitative interviews, a choice motivated by the fact that our main goal was not to gather quantitative information but rather to explore the range of motivations and the producers' level of satisfaction in using PDOs or PGIs in their internationalisation processes. Furthermore, we aimed at characterising the problems and opportunities of the different marketing channels and the role of collective organisations. The tool used for interviewing producers and other actors was a semi-structured questionnaire to gather information related to each of the main research objectives. The questions proposed were partly open-ended, especially when focussing on more qualitative aspects of the survey, while the remaining questions were the multiple choice type. A total of 16 firms belonging to the 4 PDO/PGIs and the 4 consortia were interviewed.

4. Characteristics of PDO/PGI production systems and internationalisation

The four production systems analysed are very different one from another, since firms are very diversified in terms of typology (dimension, degree of specialisation in the origin products analysed), marketing channels, and level of internationalisation.

The next paragraphs give an overview of the production systems of these products with particular reference to the marketing channels used to trade the PDO/PGI products.

4.1 Tuscan extra: virgin olive oil PGI

The 'Toscano' (Tuscan) extra-virgin olive oil is a product which enjoys great, longstanding 'renown' in Italy and abroad. The PGI was obtained in 1998, in order to face misleading problems related to the renowned name 'Tuscan' for olive oil.

The olive oil production system is very articulated, managed by a variety of professional and non–professional actors (olive growers, olive pickers, olive mills, merchants, both small and industrialised mixing and bottling firms) and is characterised by a very strong fragmentation. Bottlers, that is firms (both farms, millers, cooperatives and local traders) specialised in the bottling and the final certification of the olive-oil as PGI, are 360. Approximately 15% of the total olive-oil production in Tuscany is certified as PGI. The PGI olive-oil production is directed toward heterogeneous marketing channels, from very short (direct sale from olive growers) to long channels (large bottling firms selling to supermarket chains). More than 60% of the total PGI production is sold through large retailers, followed by middle-men and wholesalers, while just 4% of the production is distributed through direct channels (direct sale to final consumers). Internationally, the main channels used to export olio Toscano PGI, are buyers of foreign large retailers and middle-men (as foreign importers and/or national exporters), although direct sales to foreign consumers are growing in importance too.

Tuscan extra-virgin olive oil PGI is the most export oriented product among Italian recognised PDOs and PGIs. In 2006, sales relied on foreign markets for more than 66% of the whole production, gaining, during the last five years, an increasing importance in terms of quantity and value from international trading. The main foreign markets for Tuscan olive oil are non-European countries, in particular the United States which accounts for more than 60% of the total exported volume of protected product.

Within the supply-chain, owing also to the fragmentation of the production system, the role of the consortium (Consorzio di tutela dell'Olio di Oliva toscano) is particularly relevant for the services offered to its members, which represent all the Olio Toscano PGI producers.

4.2 Chianti Classico extra: virgin olive oil PDO

Chianti Classico extra virgin Olive Oil is a new denomination that obtained the PDO in 2000, but enjoys the high reputation of its geographical name thanks to the renowned wine production.

The first phase of production (olive growing) is very fragmented among many family run farms. Some farms are medium-sized but because olive oil production is not often the principal activity (which is generally wine production, as well as saffron and lavender, in some cases), the degree of specialisation is very low.

The PDO Chianti Classico Olive Oil supply chain counts more than 250 olive growers (119 of them are also olive millers and/or bottlers), 30 olive mills and 100 bottlers, that is, firms (farmers, millers and traders) allowed to apply for the final PDO certification. However, producers do not certify all the olive oil production as PDO. This depends on the requirements of the marketing channels used to sell the product. Overall, about 20% of the local olive oil is certified production.

However, exports account for approximately 60% of PDO certified product. Their main commercial channels are direct sale through agritourist farms owned by olive oil producers and through import wholesalers. In particular, 65% of the exported Chianti Classico olive oil PDO goes to the EU markets (Germany, France, England, Austria and Belgium), whereas 35% is destined to the USA, Canada, Switzerland, Vietnam and Norway.

Some internationalisation strategies had already been adopted by Chianti Classico PDO producers before PDO had been approved. Most farms were already on foreign markets with their olive oil, thanks to Chianti Classico wine sales, which had achieved consolidated commercial channels.

4.3 Pecorino Toscano PDO

Pecorino Toscano is a sheep cheese produced in a wide geographical area which includes the whole of Tuscany, part of Umbria and Lazio. It has a defined and structured production process which results in a product marketed on structured commercial channels.

Before obtaining the PDO, Pecorino Toscano was (since 1954) under a national protection system of local production defined Denomination of Origin (DO). Sheep cheese has been considered for long a product needing protection against misuses and frauds. Therefore, after the D.O. protection, producers applied for the European recognition of their geographical indication and obtained PDO in 1996. Due to the existence of different typologies of pecorino (sheep cheese) traditionally produced in Tuscany, a PDO code of practice that could include and protect all these heterogeneous products was established. Pecorino Toscano PDO is produced by non-specialised dairy factories, which also produce other kinds of cheese. However, the peculiarity of Pecorino is its milk only from sheep bred in the PDO area. The number of dairy factories producing Pecorino Toscano PDO in 2006 were 15. In the Pecorino Toscano PDO cheese production system, some of the actors of the supply chain are associated to a consortium whose activities include technical assistance to its members, recording and management of the data related to milk, dairy production and the marketing phase with all the promotional activities. The commercial channels for Pecorino Toscano PDO cheese on foreign markets are various. These are mainly indirect selling through wholesalers and/or direct selling to tourists, even if a high percentage of the product is sold to buyers of large retailers. The sale of Pecorino Toscano PDO cheese on foreign markets is about 16% of the total PDO production. Specifically, 80% is destined to EU countries (England, France, Germany, Sweden, Austria, Spain, Greece, Denmark, Belgium and Finland) while the rest is exported to the USA, Canada, Australia, Japan, China, Switzerland and Luxemburg.

4.4 Prosciutto Toscano PDO

'Prosciutto Toscano PDO' is raw ham from pigs of specific breeds reared in a determined area (Tuscany and other Italian regions including those recognised for PDOs Prosciutto di Parma and S. Daniele), which undergo a dry salting process carried out only in Tuscany. The product is linked to tradition and to the geographic Tuscan area due to the type of seasoning, which involves specific know-how of the process. The relatively low breeding of pig in the Tuscan area and the higher costs of raw materials, that satisfy the standards established in the Code of Rules, constrain producers from increasing their PDO production volumes.

The PDO Prosciutto Toscano production system in 2006 included 25 producers who were involved in the production process by butchering, sectioning and seasoning. Other kinds of hams (for example ham obtained by non Italian pigs) and other types of meat products (salami, sausages, fresh meat, etc.) are added to the PDO ham production. Most producers using the PDO have joined the consortium since it was set in 1990, in order to protect the origin product from unfair national and foreign competition, which started to threaten Tuscan ham by selling inexpensive similar products made of pork of 'unknown' origin. After requesting the recognition of the Italian denomination of origin, the consortium applied for EU PDO, obtaining it in 1996.

PDO Prosciutto Toscano is one of the growing leaders of the Italian ham market, after the most popular PDOs Prosciutto di Parma and S. Daniele and on the domestic market it does not seem to be suffering the direct competition of other Italian raw hams, given its particular features. Nevertheless, many imitations of the PDO product strongly mislead the consumer, both on local and foreign markets.

The majority of PDO ham is sold on domestic and local markets, either as a result of its characteristics and quality attributes, or because firms do not produce sufficient volumes of certified ham product to sell it overseas. This implies a low export rate which impacts negatively on the turnover of the Prosciutto Toscano PDO system.

The firms which export are very few and are generally the biggest ones, those which produce from 50,000 to 150,000 PDO hams per year. The PDO Tuscan ham registers an average export rate of 4% of value and 6% of quantity of total PDO production. Moreover, exports of PDO ham on total firms' turnover have a lower impact than other products sold overseas. Prosciutto Toscano PDO firms mainly export their product through importers-exporters, whereas the direct channel is less important as they belong to a more industrialised sector.

4.5 An overview of export performance and the role of PDO/PGI

In all the production systems analysed, the use of PDO/PGI in labelling origin products has varied among firms, since it often represents a minor part of the total production of the origin product which depends on the structure and strategies of the firms, on the marketing channels used and on the 'strength' of the PDO-PGI itself in attracting consumers' and customers' attention.

As to export channels, the cases selected show how the exports of PDO/PGI products represent a small percentage of each firm's total turnover, and a limited share of the total turnover on foreign markets, with the only exception of Tuscan Extra Virgin Olive Oil PGI (Table 2). This indicates that the specialisation level of the firms both in Origin Products and its denomination is generally very limited. For example, if we consider the Chianti Classico olive oil firms, their oil production is often marginal compared to the production of wine (Chianti Classico D.O.C.G.). Instead, in the case of Tuscan Ham, production is often sided by the production of other items based on meat; in the case of Pecorino Toscano (Tuscan Sheep Cheese) cheese makers produce also other dairy products. On the contrary in the case of Tuscan Extra Virgin Olive Oil there are higher levels of specialisation of the firms in the production of Olive Oil and its certification as PGI.

Table 2. Export share of firms turnover and of total interviewed firms certified production, average years 2004/2005.

Denominations of origin	PDO-PGI export / firms turnover	PDO-PGI export / total PDO-PGI production value
Olio Chianti Classico DOP	0.35%	20.00%
Olio Toscano IGP	13.21%	61.05%
Pecorino Toscano DOP	3.00%	18.00%
Prosciutto Toscano DOP	1.00%	6.00%

The export channels of PDO/PGI products are mainly indirect: the products are marketed by specialised import-export firms that, besides the products analysed, often market other agro-food products (Table 3). In the production area direct sale to foreign customers is widespread and, though it concerns only limited volumes, it is interesting for the higher margins that firms can achieve.

As far as destination markets are concerned, the majority of the firms analysed export Pecorino Toscano and Olio Chianti Classico to EU Countries, whereas Olio Toscano and Prosciutto Toscano (Table 4) are mainly exported to extra-EU countries.

Table 3. Distribution channels for the export of PDO/PGI products, share of firms on total answering ones, average years 2004/2005.

Distribution channels	Olio Chianti Classico PDO	Olio Toscano PGI	Pecorino Toscano PDO	Prosciutto Toscano PDO
Importers/exporters	100%	80%	75%	100%
Foreign large retailers buyers	-	60%	50%	-
Own network of agents	40%	-	50%	67%
Own branches in foreign countries	-	-	-	-
International E-commerce	-	-	-	-
Direct sales to foreign customers	80%	60%	75%	33%
Other	20%	-	25%	-

Note: multiple answers were allowed, so the total per column may be over 100%.

Table 4. Distribution of the exported product between the EU and the extra-EU market, average years 2004/2005.

Geographical indications	Extra-EU countries	European Union
Olio Chianti Classico DOP	35%	65%
Olio Toscano IGP	62%	38%
Pecorino Toscano DOP	19%	81%
Prosciutto Toscano DOP	53%	47%

5. The use of Geographical Indications (PDO/PGIs) on international markets: motivations and satisfaction

The analysis of the roles Geographical Indications (PDO/PGIs) play on international markets started from the motivations producers have in the use of PDOs and PGIs both on national and international markets.

The literature analysis and previous research projects carried out by the authors have allowed to identify some of the key functions PDO and PGIs have in:
- defending the geographical name of protected products against misuses and abuses. This phenomenon is largely present for most origin products from Tuscany, given the strong reputation the name of the region and some sub-zones (as Chianti) have in Italy and abroad;
- benefiting from renown and reputation of the geographical name of PDO/PGI products;
- responding to specific customers' requests (middle-men, importers-exporters, buyers for foreign large based retailers, etc.);
- gaining a higher premium price than those products not bearing PDO or PGI;
- increasing firms' turnover on the same international commercial channels;
- facilitating the opening of new commercial channels and/or new geographical markets;
- steadying commercial relations and reducing (volume and/or price) uncertainty on the intermediate and final markets;
- benefiting from collective product promotion thanks to the presence of a collective organisation;
- offering more guarantees to final consumers by means of the EU PDO/PGI logo;
- differentiation from competing firms.

These PDO/PGI functions may partly overlap, or be consequential (some of them are a means to join other functions), but are certainly not self-excluding. They can be referred to three main strategies that originate them. In a *defensive strategy* PDO/PGI is viewed as a tool for defending a geographical name from abuses. In an *offensive strategy* PDO/PGI is viewed as a tool to differentiate a product and more generally to redesign relationships within the supply chain (both at horizontal and vertical levels). In a strictly *commercial strategy* PDO/PGI is viewed as a tool that guarantees an increase in prices, turnover and/or revenues.

During the field analysis, in order to identify the roles PDO/PGIs play for each firm, we started from a face-to-face open discussion of the problems encountered in marketing the origin product on domestic and international markets. We then identified the motivations in the use of PDO/PGIs in general and for export. Moreover, firms were asked to express a level of importance for each possible motivation for the use of PDO/PGIs, according to the above list. The interviews proceeded in analysing the firms' internationalisation strategies, markcting channels, promotional activities and performance. At the end of the interviews the firms were asked to give an evaluation, in terms of personal satisfaction, of the effective impact PDO/PGIs have on their internationalisation.

The ranking of functions in terms of motivation varied from a minimum ('not important') to a maximum ('very important') level of assessment, while for effective satisfaction the ranking varied from 'unsatisfied' to 'very satisfied'. These motivation and satisfaction assessments were later scored[40]. This allowed the calculation of an average index both related to each function for motivation and for satisfaction, as a means of the assessment expressed by each firm for each function. Moreover, the Average Motivation Index for function j was calculated as $(\sum i=1,n\ Ii)/n$, where i are the firms which answered to the question and Ii is the Importance score given by a generic firm. The same applies to the Average Satisfaction Index, calculated as $(\sum i=1,n\ Si)/n$, where Si stays for the Satisfaction score. These indexes can vary from 1 (minimum value) to 100 (maximum value).

To better understand the role PDOs and PGIs have on international markets, the analysis started from a comparison between the general motivations, which led producers to use the geographical indication, and the specific motivations for international markets (Figure 1).

Overall the motivations for the use of PDO/PGIs are stronger on international markets, except for the defence against abuses of the geographical name which is stronger on domestic markets.

On the contrary, the importance of PDO/PGI on international markets is stronger than average for the functions 'Benefit from reputation of geographical name' and 'Guarantee to consumers represented by the European Logo'. This different attitude may be explained by the fact that, on foreign markets, the use of reputation of territory origin (in this specific case Tuscany and Chianti) has a stronger effect on the foreign consumer than, for example, on the Italian one, who already knows the characteristics of the product and the territory it originates from. Firms perceive the European logo as being better known in export countries (mainly in Northern Europe, Northern America and Japan) than on the domestic market. This may originate from customers being professional buyers with a higher level of information also about the meaning, implications and functioning of EU PDO/PGI than the individual consumer.

---

[40] The 4 levels of importance of motivations were scored as follows: 'not important'= score 1, 'less important' = score 33, 'important'= score 66, 'very important' = score 100. For the satisfaction: 'unsatisfied'= score 1, 'less satisfied'= score 33, 'satisfied'= score 66 and 'very satisfied'= score 100.

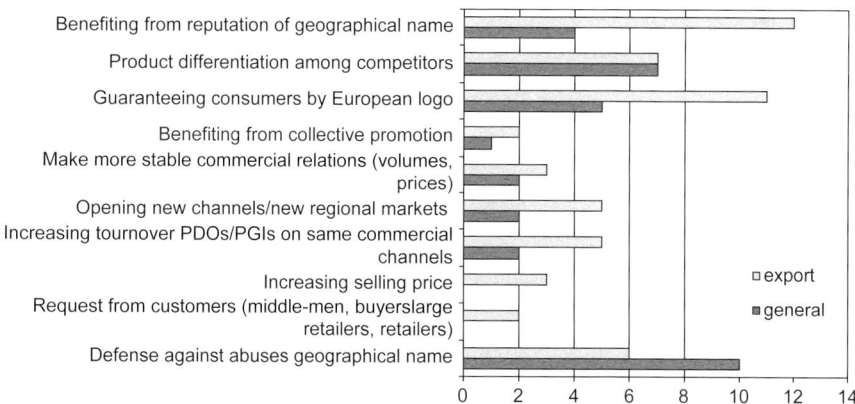

Figure 1. Motivation for the use of PDO/PGI: comparison between general usage of PDO/PGI and specific usage on international markets.
Note: the values are referred to the number of firms answering 'very important' on each motivation; multiple answers were allowed (total responses: 16).

The motivation index on foreign markets shows what PDO/PGI mean to firms in both an offensive and in a defensive logic (dark bars in Figure 2). In fact, on the one hand, producers state that they use the geographical indication in order to defend products, with a high reputation, from abuses; on the other, they take advantage of the reputation the names 'Tuscan' or 'Chianti' have *per se*.

The aggregate analysis of producers' satisfaction, deriving from PDOs and PGIs performance on international markets (referring to the same specific functions used for exploring driving forces), shows an overall good satisfaction for the use of the reputation of the geographical name as well as for the European logo (light bars in Figure 2).

From a comparison of motivation and satisfaction indexes some unexpected functions of the PDO/PGIs on international markets emerge, that is, functions for which the importance level attributed by interviewees to the use of PDO or PGI on international markets (motivation) is lower than the satisfaction grade associated with the same functions if the results reached are considered. These unexpected functions mainly refer to the marketing field, in particular the 'Request from customers' in order to guarantee not only the origin and the characteristics of the product, but, more in general, the reliability of the process and of the firm producing the PDO/PGI product. Even the 'Benefit of collective promotion' has a satisfaction which is significantly stronger than the motivations.

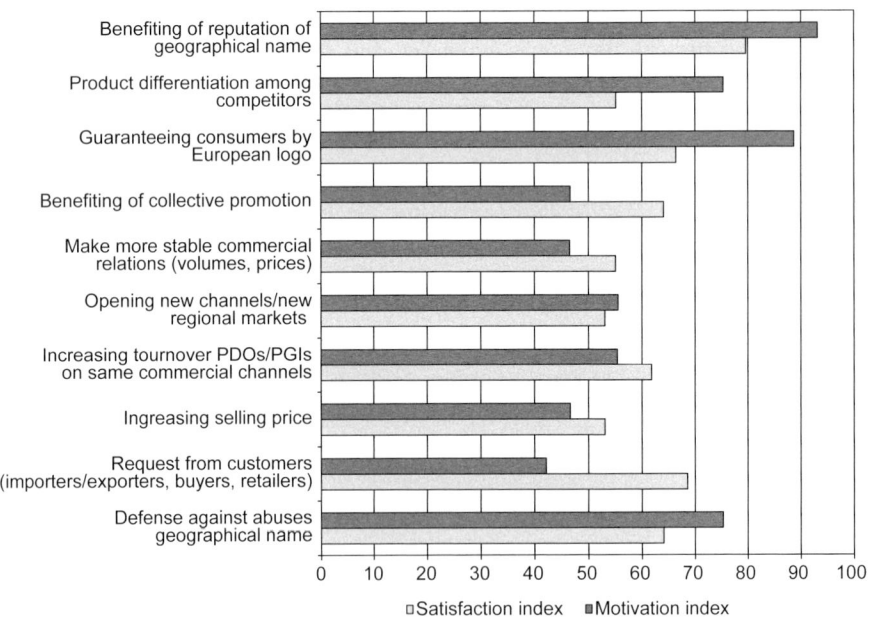

Figure 2. Motivation and satisfaction indexes for the use of PDO/PGIs on the international market.

6. Motivations and satisfaction in different production systems

Different origin product systems highlight different motivations for the use of PDO/PGI, and different satisfaction levels (Figure 3 and 4).

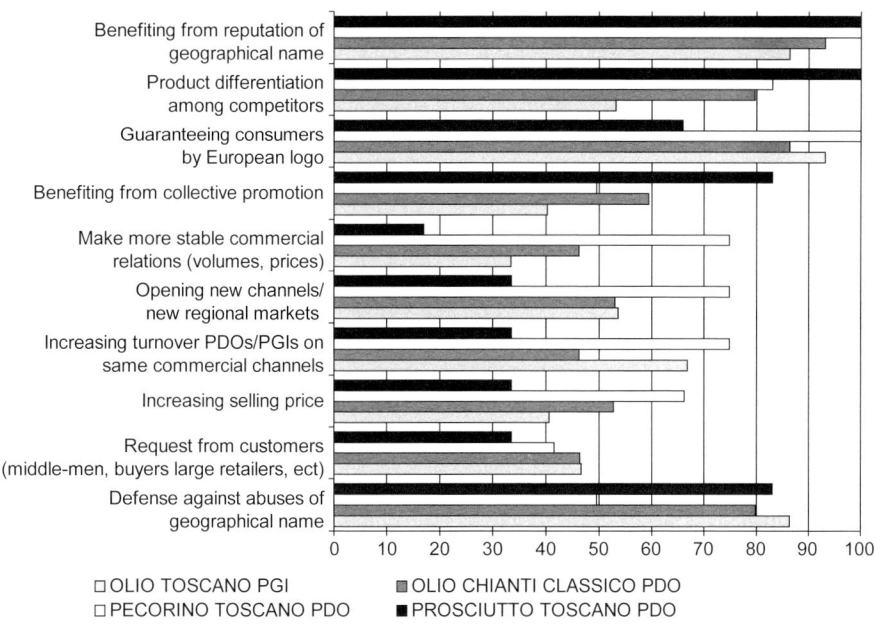

Figure 3. Motivation index for PDO/PGI functions for some PDO/PGI products.

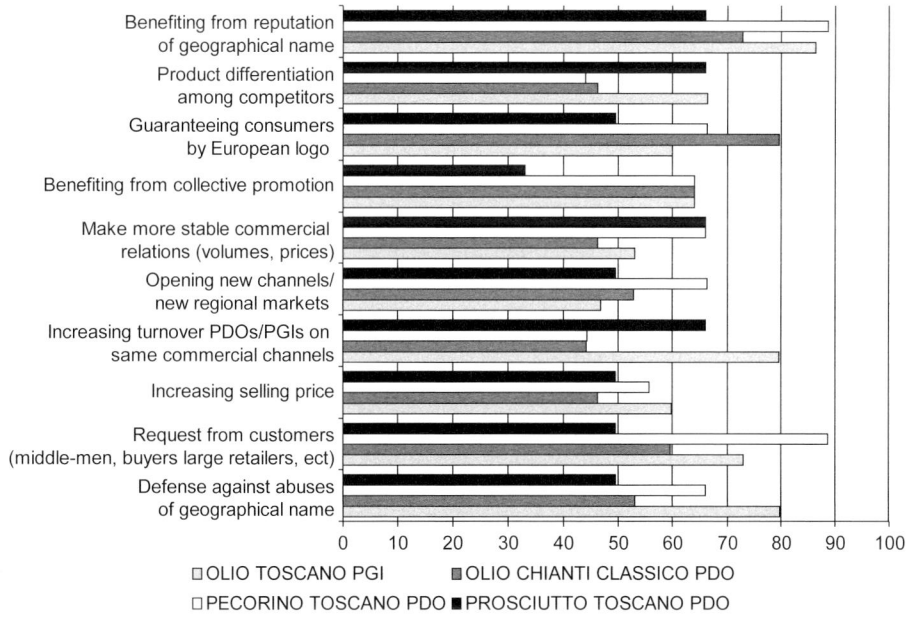

Figure 4. Satisfaction index for PDO/PGI functions for some PDO/PGI products.

For all four products 'Benefiting from reputation of geographical name' appears as a very important matter, whereas 'Defence against abuses' is less important for Pecorino Toscano PDO which is the least reputed product. Firms producing Pecorino Toscano PDO attribute a stronger offensive role to PDO, in terms of stabilisation of commercial relations, increasing turnover and prices, and opening new channels and markets.

The satisfaction in the use of PDO/PGI (Figure 4) is different among the four products and according to each PDO/PGI potential function. It is on average higher for Olio Toscano and Pecorino Toscano. 'Benefit from reputation of geographical name' is the most satisfactory PDO/PGI function for all products. The defence against abuses of a geographical name is performed mainly by Olio Toscano PGI, thanks to a stronger commitment of Tuscan Oil Consorzio in the field. The most marketing oriented functions are the least satisfactory, with a partial exception of Pecorino Toscano PDO and Tuscan Olive oil.

The effective impact in terms of producers' satisfaction with regard to the PDO/PGI role on international markets for many functions does not match firms' expectations. The comparison of the motivation and satisfaction indexes for each product and for each PDO/PGI function by the difference of their values (Figure 5) allows to evaluate the coherence between expectations and results (though these evaluations are based of firms' perceptions). In Figure 5 the positive values highlight results that are higher than expected, while negative ones indicate a certain disappointment of firms (even if the satisfaction may be positive, but in any case below expectations).

This analysis underlines a widespread incoherence between expectations and satisfaction for all products for the functions 'Benefiting of reputation of geographical name' (J in the graph) and 'Guaranteeing consumers by European logo' (H), and for almost all products for 'Defense against geographical name abuses' (function A in the graph) and 'Product differentiation among competitors' (function I). On the contrary, 'Answering to an explicit request by customers' is the only function where a high level of coherence is reached for all products.

As a result, the situation between the four products analysed is highly differentiated.

In the case of Tuscan olive oil, a product with a high export impact on total sales, the presence of PGI strongly satisfies producers in order to benefit from the reputation of the name Tuscany as well as for the guarantee offered by the EU logo (Figure 4), even if less than expected (Figure 5). The role of PGI as a defensive tool against unfair competition is not satisfactory compared to the importance producers give to the function. This probably depends on the efficacy of control systems and mechanisms of managing and treating abuses and misleading products, in particular on non-EU markets to which the biggest part of exports is addressed. In general, the analysis indicates a good level of satisfaction and coherence for many functions oriented towards a commercial logic. In fact most customers of PGI olive-oil are professional buyers often acting for supermarket chains, who have a good knowledge (better than final consumers) of the functioning of 510/2006 EU Regulation, and in particular of the contents of the Code of practice and of product and process certification schemes guaranteed by a third party control. Tuscan olive-oil PGI has given firms a strong contribution for opening new marketing channels and new geographical markets.

For the Chianti Classico Olive Oil firms, the use of the Geographical Indication (PDO) on foreign markets is basically prompted by the opportunity of benefiting of product geographical reputation, but the functions of 'Defence from abuses' and the linked 'Benefiting from reputation' are not

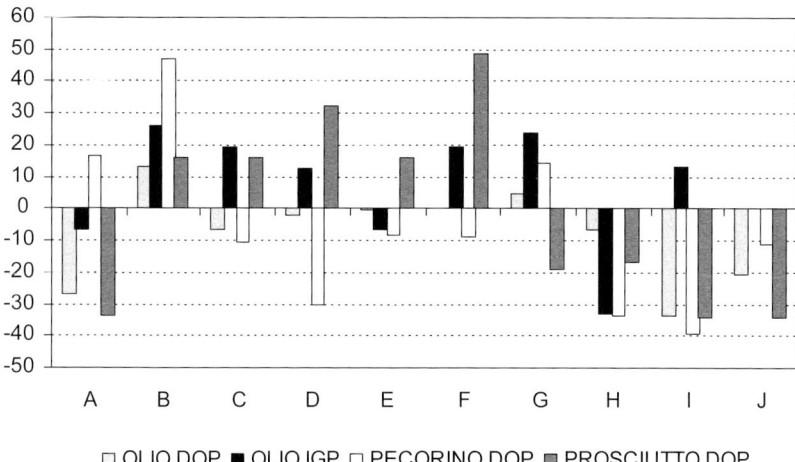

Figure 5. Difference between motivation and satisfaction index by functions and products.
Legenda:
A. Defense against abuses geographical name
B. Request from customers (middle-men, buyerslarge retailers, retailers)
C. Increasing selling price
D. Increasing tournover PDOs/PGIs on same commercial channels
E. Opening new channels/new regional markets
F. Make more stable commercial relations (volumes, prices)
G. Benefiting from collective promotion
H. Guaranteeing consumers by European logo
I. Product differentiation among competitors
J. Benefiting from reputation of geographical name

fully satisfactory. According to the firms interviewed, PDO does not significantly contribute to differentiate and signal the product to all producers, and in particular to producers who already enjoy their own brand reputation, as in the case of wine firms. By contrast, PDO allows smaller Chianti firms (not well reputed for wine production) to participate in a collective system and represents an important quality standard for buyers, middle-men and importers, due to the guarantee of the link between the territory of production and the third – party certified production process. Promotional activities could increase the informative power of PDO on final markets, in particular by creating synergies between Chianti Classico olive oil and wine, supporting the 'halo effect' from the latter that influences the reputation of olive oil.

The Pecorino Toscano cheese case study highlighted how PDO can be an important marketing tool, that allows the stabilisation of commercial relationships along the supply chain, the penetration of foreign markets, thanks to the role of standards generally accepted by retailers of foreign markets and the protection against abuses of the geographical name of the product. However, the satisfaction index in general does not correspond to motivation, except for the 'Request from customers'.

As far as Prosciutto Toscano PDO is concerned, from the comparison between motivation and satisfaction we observe a disappointment both in functions linked to the use of the name on the market ('Defense against abuses geographical name', 'Benefiting from reputation of geographical name', 'Guaranteeing consumers by European logo'), due to a not fully developed collective promotion. The good evaluation given by firms to some commercial functions of PDO, that is, the rewarding increases in price and turnover of PDO exports on the same commercial channels, facilitations in opening new commercial channels and regional markets, and the steadying effect on commercial relations, can be motivated by the acceptance of PDO as a requested standard by middle-men such as importers/exporters, buyers of foreign large based retailers who seem to prefer the PDO product because it satisfies a precise production process by presenting stable characteristics and product attributes.

7. The role of geographical indications (PDO-/PGI) at single firm level

The average results discussed in the previous paragraphs are a synthesis of heterogeneous situations at single firm level also within each case study. The aim of this paragraph is to analyse more in depth, at single firm level, the existing relation between motivations and satisfaction related to the most important functions identified by the interviewees as motivations: the benefit from the reputation of the geographical name, the defence of the reputation of the product from abuses and misleading of the geographical name, the guarantee of consumers through the EU logo, the differentiation of the product, the explicit request of the PDO/PGI by customers (middle-man, importers-exporters, buyers for foreign large retailers).

Scatter graphs are built for each of the main detected functions, in order to position each firm by considering expressed motivation and satisfaction. The first quadrant of each figure represents a *coherence area*, where the motivation of the firm for the specific function has been fulfilled through a satisfactory performance; the second quadrant is the *disillusioned area*, where firms which had high motivation for that function, have been disillusioned by unsatisfactory results, the third quadrant is the *indifference area*, where firms had a low motivation for the function and no or little satisfaction has resulted; the fourth quadrant is the *unexpected results area*, where firms had low motivation for that function and registered satisfactory or very satisfactory results (see Figure 6).

Considering the benefit from the reputation of the geographical name (Figure 7), all firms are positioned in the coherence area. PDO/PGI is hence a very effective tool in conveying reputation on the market, in particular to products bearing names (as Toscana and Chianti) with a strong

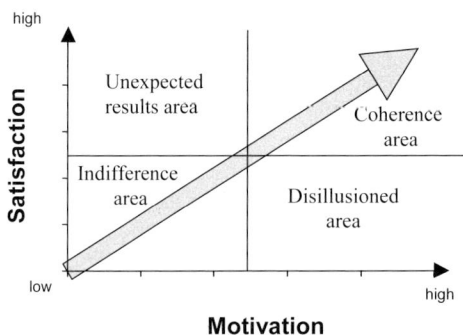

Figure 6. Comparing levels of motivation and satisfaction, a general interpretation of the different combinations.

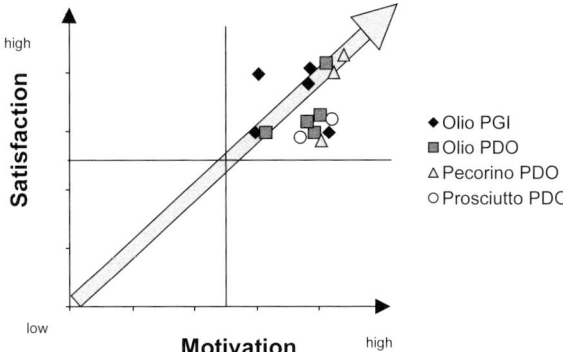

Figure 7. 'Benefiting from the reputation of geographical name', comparing importance levels of motivation and satisfaction.

evocative value and well known to many consumers. PDO/PGI allow a collective action in this field, because in all the 4 products analysed, Consortia have developed a collective trademark incorporating the geographical denomination. This trademark facilitates consumers' identification of the product and it can be supported by means of communication and promotion.

Considering the 'Defence of the product from unfair competition in the usage of geographical name' (Figure 8), most firms are in the coherence area, as they expressed a level of satisfaction in line with the importance assigned to the function with regard to motivation. However, some of the firms expected a much stronger effect than the effective satisfaction they achieved (disillusioned area), mainly in PDO Olio Chianti Classico. The interviewed firms motivated their low level of satisfaction through the low efficacy of the existing control and sanction system, particularly in non-EU countries. In general, less satisfaction is expressed by those firms who are already well positioned with their own brand on niche markets.

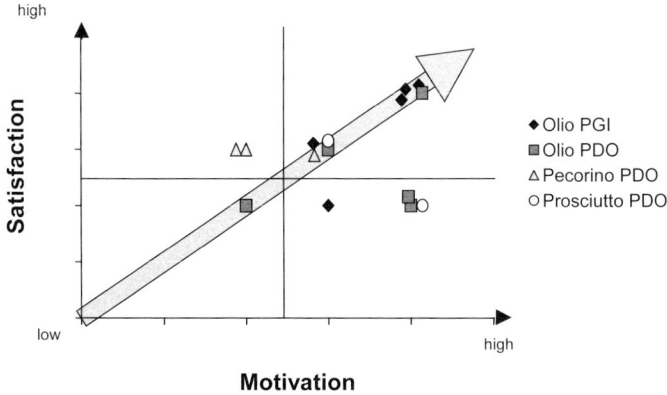

Figure 8. 'Defence of the product from abuses of geographical name and misleading', comparing levels of motivation and satisfaction.

The PDO/PGI guarantee to final consumers, with the presence of the EU logo, is an important or very important function for all firms. A part of them are disillusioned by the effective performance obtained (area with high motivation and low satisfaction, Figure 9). This is due to little information about the meaning of the logo (product traceability, coded production process), besides the consideration that the meaning of the logo is still recognised only in Europe.

Regarding the role of PDO/PGI as a product differentiation tool, most firms appear not to be very satisfied considering also that this motivation was graded as a medium to a maximum important factor (Figure 10). Therefore, PDOs and PGIs do not appear as a strong differentiation lever on foreign markets, except for Olio Toscano PGI producers who are mainly located in the high satisfaction area of the graph. In the case of Olio Toscano most satisfied producers are the two main cooperative firms selling overseas to big buyers possessing their own trademark, while firms located in the low motivation/satisfaction area are those which mainly operate on markets by promoting their own brand.

The usefulness of PDO/PGI as a differentiation tool depends also on the presence of stronger competing PDOs. This is the case of PDO Tuscan ham, whose performance is hindranced by the

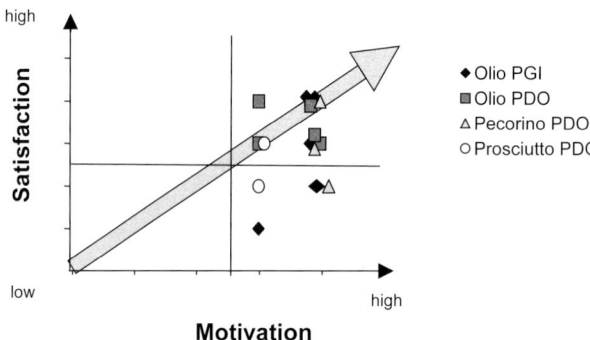

Figure 9. 'Guaranteeing final consumer by EU logo', comparing importance levels of motivation and satisfaction.

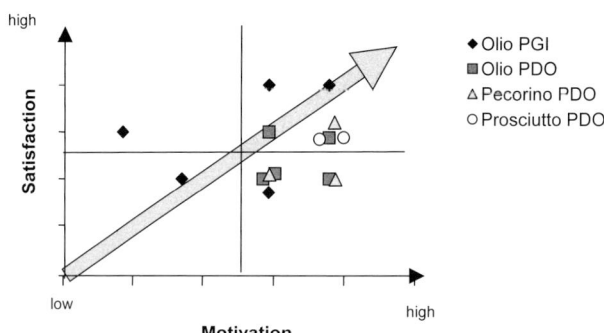

Figure 10. 'Product differentiation', comparing importance level of motivation and satisfaction grade of interviewed firms.

presence of Parma PDO and San Daniele PDO hams. These hams have been on international markets for a longer period, and have consequently gained a stronger reputation and a higher renown, thus creating difficulties in the promotion of the Tuscan ham specificities.

A relevant finding emerging from the survey is the role of PDO as a commercialisation standard requested by customers (middle-men, importers/exporters, buyers of foreign large retailers) (Figure 11). This is an unexpected function for many firms. Pecorino Toscano PDO producers are all located in the area of 'unexpectation', as well as the main exporter of Prosciutto Toscano PDO production system.

In the case of Olio Toscano PGI the most satisfied firms are two big cooperatives, where many small and very small producers converge and are provided with additional services (for instance bottling on behalf of buyers according to their specific requests). This allows to respond to the needs of some customers, for example, supermarkets chains, and to access new channels.

In the case of Olio Chianti Classico PDO, some producers gained an extraordinary result on the trade market by specifically responding to customers' requests thanks to PDO. It is the case of firms specialised in the olive oil sector, while other firms (those which do not expect/gain great effect in terms of requested standard from the distribution sector), mainly wine producers, do not show the same satisfaction, having already implemented strong strategies of international marketing.

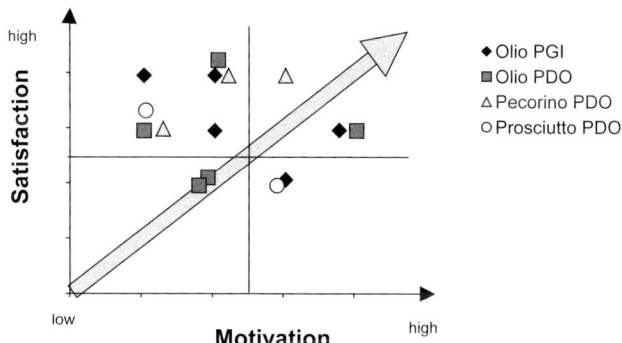

Figure 11. 'Explicit request from customers', comparing importance levels of motivation and satisfaction.

8. Final remarks

Geographical Indications (PDO and PGI) play multiple roles for the internationalisation of small-medium scale agri-food products, but their effectiveness depends on several factors: the reputation of the product, the reputation of the territory of origin (halo country effect), the importance given by customers to the guarantee offered by PDO-PGI certification (linked to the structure of marketing channels and especially modern channels), the capabilities of firms to implement PDO-PGI marketing strategies.

The case-studies analysed concern products with well known geographical names (Tuscany, Chianti) with positive evocative power also on foreign markets, even when the product itself is not well known by consumers. For this reason imitations are quite common and firms use PDO/PGI as a defensive tool with a quite high degree of satisfaction.

But small firms are interested in PDO and PGI as marketing devices too, since they can prevent the development of other marketing initiatives on an individual basis.

Regards satisfaction of producers related to the motivations expressed, a complex situation that cannot be explained only through a 'key product', has been highlighted. The study underlined a differentiation in the impact of PDO/PGI among different production systems (inter-diversity) and among firms producing the same agri - food protected product (intra-diversity). This inter-diversity depends on the organisation of the supply chain, and mainly on the acquired reputation of the product, instead the intra-diversity is related to the heterogeneity of firms in terms of dimension, organisation, strategy, production volumes, availability of capabilities to operate with foreign markets through professionals, management skills, knowledge of exporting rules, etc. Overall, firms trading on foreign markets with their own brands show a lower interest in PDO or PGI, in order to avoid a conflict between (collective) PDO/PGI and firms' brand name, while more specialised producers and smaller firms for PDOs and PGIs present high levels of motivation/satisfaction.

On the whole, this research pointed out the high impact of PDO/PGI in capturing the benefits from the reputation of a geographical name. But the satisfaction in using PDO/PGIs is positive also for many 'marketing oriented' functions, such as satisfying specific requests from customers (middle-men, importers/exporters and buyers for large retailers), increasing turnover in existing channels or opening new ones, and guaranteeing consumers by means of the EU logo. Therefore PDOs and PGIs act as a quality standard, being the product and processing characteristics codified and certified by a third-party control body. Along export marketing channels, PDO/PGIs carry out a standard function primarily for professional operators, besides certifying the effective origin of the product and its production process.

The results of the survey support also the second hypothesis of the work, that is, the collective dimension linked to the use of PDO/PGI which has important positive effects in supporting the exports of small-medium enterprises. As a matter of fact, the survey stresses that the performance of PDO/PGI depends not only on the characteristics of a product and its producer, but also on the collective organisation of the production system. Many firms state that the PDO/PGI label alone is not sufficient to communicate product characteristics on relevant markets, and that it should be accompanied by promotion and support activities. A fundamental role is therefore played by the consortia, which contribute to the realisation of collective initiatives, support the firms' management of the European certification systems, and support the individual international activities of their associated firms.

Specifically, we refer to those activities which are able to influence the internationalisation process of PDOs and PGIs either in a direct (collective promotion internationally) or in an indirect way (supporting firms in the management of production activities). Consortia support the empowerment of firms in the internationalisation processes by integrating their lack of structural capabilities to operate on foreign markets. In addition, they help firms to comply with traceability and control systems and provide information on rules in export markets to individual firms.

Collective organisations along with appropriate public local policies, which supply firms with all the necessary capabilities to face the international market, are the two pillars that with the PDO/PGI scheme allow origin product systems to benefit from the internationalisation of food markets.

References

Almonte, J., M. Cardenas, C. Falk and R. Skaggs, 1996. Product-country image and international food marketing: relationships and research needs. Agribusiness, 12: 593-600.

Albisu, L.M., 2002. Link between origin labelled products and local production systems, supply chain analysis - Work Package 2 Final Report, July 2002. DOLPHINS - Concerted Action, Contract QLK5-2000-0593, European Commission.

Anania, G., and R. Nisticò, 2004. Public regulation as a substitute for trust in quality food markets. What if the trust substitute cannot be fully trusted? Journal of Institutional and Theoretical Economics, 160: 681-701.

Barjolle, D., J.M. Chappuis and B. Sylvander, 1998. From individual competitiveness to collective effectiveness: a study on cheese with Protected Designations of Origin. EAAE-ISHS Seminar, Competitiveness: does economic theory contribute to a better understanding of competitiveness?, The Hague (NL).

Belletti, G., 2000. Origin labelled products, reputation, and heterogeneity of firms. In: Sylvander, B., Barjolle, D. and Arfini, F. (eds.), The socio-economics of origin labelled products in agri-food supply chains. INRA Actes et Communications, 17: 239-260.

Belletti, G., T. Burgassi, E. Manco, A. Marescotti, A. Pacciani and S. Scaramuzzi, 2006. La valorizzazione dei prodotti tipici: problemi e opportunità nell'impiego delle denominazioni geografiche. In: C. Ciappei (Ed.), La valorizzazione economica delle tipicità locali tra localismo e globalizzazione, Firenze (IT): Florence University Press, pp. 189-265.

Belletti, G., T. Burgassi, A. Marescotti and S. Scaramuzzi, 2007. The effects of certification costs on the success of a PDO/PGI. In: L. Theuvsen, A. Spiller, M. Peupert and G. Jahn, (Eds.), Quality management in food chains. Wageningen, the Netherlands: Wageningen Academic Publishers.

Belletti, G. and A. Marescotti, 2006. GI social and economic issues, D2 - WP2 Report Theoretical frame, SINER-GI project.

Canada, J.S. and A.M. Vazquez, 2005. Quality certification, institutions and innovation in local agro-food systems: Protected designations of origin of olive oil in Spain. Journal of Rural Studies, 21: 475-486.

Galavotti, S., 2005. L'internazionalizzazione produttiva dell'Italia nel settore agroalimentare. In: Nomisma - INDICOD (eds.), Originale Italiano. Milano (IT): Agra editrice.

Galizzi, G. (Ed.), 1995. Il commercio internazionale dei prodotti agroalimentari. Milano (IT): Franco Angeli.

Marette, S., J.M. Crespi and A. Schiavina, 1999. The role of common labelling in a context of asymmetric information. European Review of Agricultural Economics, 26: 167-178.

Ménard, C., 1996. On clusters, hybrids, and other strange forms: the case of the French poultry industry. Journal of Institutional and Theoretical Economics, 152: 154-183.

Nomisma, 2005. Originale Italiano. Rapporto Indicod - Ecr - Promozione e Tutela dell'Agroalimentare di Qualità. Milano (IT): Agra Editrice.

Pacciani, A., G. Belletti, A. Marescotti and S. Scaramuzzi, 2003. Strategie di valorizzazione dei prodotti tipici e sviluppo rurale: il ruolo delle denominazioni geografiche. In: A. Arzeni, R. Esposti and F. Sotte, (Eds.), Politiche di sviluppo rurale tra programmazione e valutazione. Milano (IT): Franco Angeli, pp. 235-264.

Rangnekar, D., 2004. The socio-economics of geographical indications - a review of empirical evidence from Europe. UNCTAD - ICTSD Project on Intellectual Property Rights and Sustainable Development, Issue Paper n. 8, United Kingdom May 2004.

Tregear, A., F. Arfini, G. Belletti and A. Marescotti, 2007. Regional foods and rural development: the role of product qualification. Journal of Rural Studies, 23: 12-22.

Value propositions and Fair Trade supply chain organisation and performance: the case of Italian Alternative Trade Organisations (ATOs)[41]

C. Zanasi and L. Paluan

Abstract

Fair Trade products import from developing countries is quickly growing worldwide although it still represents a small share of the total import. Its influence on the development of rural areas in developing countries is related to both quantitative growth and the respect of its ethical code. Large food multi-national companies are increasingly interested in Fair Trade; part of the Fair Trade movement considers the risk of a related loss in the products' identity; others consider the refusal of a more 'professional' approach to Fair Trade management as a constraint to its growth. This debate is particularly felt in Italy. The goal of this paper is to evaluate how the most important Italian Fair Trade importers (ATOs) business models influenced their growth strategies; transaction costs analyisis and logistics performance indicators were adopted to measure the supply chain coordination efficiency and performance. The results showed that the ATOs growth strategies, and logistics performance, seemed more influenced by their value propositions, than the lack of managerial skill.

Keywords: Fair Trade, logistics, alternative trade organisations

1. Introduction

The Fair Trade objectives, as defined by FINE[42], state that: 'Fair Trade is a trading partnership based on dialogue, transparency and respect that seeks greater equity in international trade. It contributes to sustainable development by offering better trading conditions to, and securing the rights of, marginalised producers and workers - especially in the South. Fair Trade organisations (backed by consumers) are engaged actively in supporting producers, awareness raising and in campaigning for changes in the rules and practice of conventional international trade.'

The introduction of Fair Trade in the agricultural products international trade can thus positively influence the developing countries socio-economic conditions. Fair Trade increases the level of welfare and reduces economic and social inequalities (Becchetti and Costantino, 2006; Becchetti and Paganetto, 2003; Ronchi, 2002; Castro, 2001). The negative impact for the less competitive countries (e.g. sub- saharian Africa) and producers (small producers in remote rural areas) due to the reduction in agricultural international trade barriers (Conforti and Velazquez, 2004; Robbins, 1999) can also be reduced. Fair trade contributes in fact to the products' differentiation, positively affecting their demand. The Fair trade products and services characteristics are designed to meet the expectations of an ethical consumer. Fair Trade 'incorporates values in their products'.

---

[41] The authors jointly prepared this paper; Lorenzo Paluan particularly contributed to paragraphs 2 and 6, Cesare Zanasi particularly contributed to paragraphs 1, 3, 4 and 5.

[42] FINE is an informal network that involves the Fair Trade Labelling Organisations International (FLO), the International Federation for Alternative Trade (IFAT), the Network of European Worldshops (NEWS!) and the European Fair Trade Association (EFTA).

Promoting a Fair Trade related international agricultural trade is also a way to increase the consumers awareness of their role in defining a more sustainable model of development under many aspects: social, environmental and economic[43].

2. Background

The consumption of Fair Trade products is quickly growing in the industrialised countries (Krier, 2005). The size of the market, and its impact on the rural communities in developing countries, is still quite small. Total Fair Trade sales in 2005, amounted to approximately US$ 1.4 billion worldwide, it represents a 37% year-to-year increase over 2004 (FLO, 2006), but a mere 0.01% of the total world trade (US$ 10,511 billions (WTO, 2006).

The interest in Fair Trade is growing in large national and trans-national food companies (supermarket chains, fast-food chains among others). It is not a case that the only countries where Fair Trade products (coffee and bananas), gain a significant market share are Switzerland and United Kingdom, where the supermarket chains play a key role in Fair Trade distribution (Krier, 2005). As a consequence Fair Trade is facing a 'growth crisis' which can be summarised by the debate between the 'product certification' supported by FLO and its FairTrade Mark, and 'organisations certification' supported by IFAT, (International Fair Trade Organisations) and its Fair Trade Organisation Mark. IFAT is critical on the inclusion of supermarket chains and other non specifically Fair Trade oriented organisations in the distribution, unlike FLO which stresses the importance of a relevant quantitative growth of the Fair Trade market size to be obtained also with the contribution of the supermarket chains. This debate is particularly strong among the Italian Alternative Trade Organisations (ATOs) (Liberomondo, 2004), (Commercio Alternativo, 2005).

The Italian Fair Trade supply chain (Figure 1) is in fact basically defined by two main channels: supermarkets and/or *World Shops* oriented, partially overlapping; they mirror the two different strategic approaches to Fair Trade above examined. The ATOs are an important node in the products and information flow along the chain.

Fair Trade products are 'credence goods' and also 'merit goods' which heavily rely upon an effective communication, and certification, of the benefits that their consumption generates for the whole of the society. The demand for these products can consequently be affected if a conflict among different organisations arises, reducing the consumer trust in the positive externalities for the society related to Fair Trade.

The interaction between market growth and the Fair Trade ethical approach of IFAT has therefore become a major concern for the Fair Trade movement. This problem is particularly felt in Italy, where the large agro-food companies involvement in the Fair Trade market is increasing, while the growth of the ATOs and *World Shops* market, still bigger than the supermarket chains', is slowing down [44](Figure 2).

---

[43] For more information on the relationship between Fair trade and the economic, social and environmental aspect see the certification standards principles defined by FLO (Fairtrade Labelling Organisations International), the main international organisation defining the Fair Trade certification standards (FLO, 2007).

[44] The data on the ATOs turnover growth was related to a negative performance for CTM, by far the main player among the Italian ATOs; the other ATOs growth is impressive in relative terms but their size remains very small.

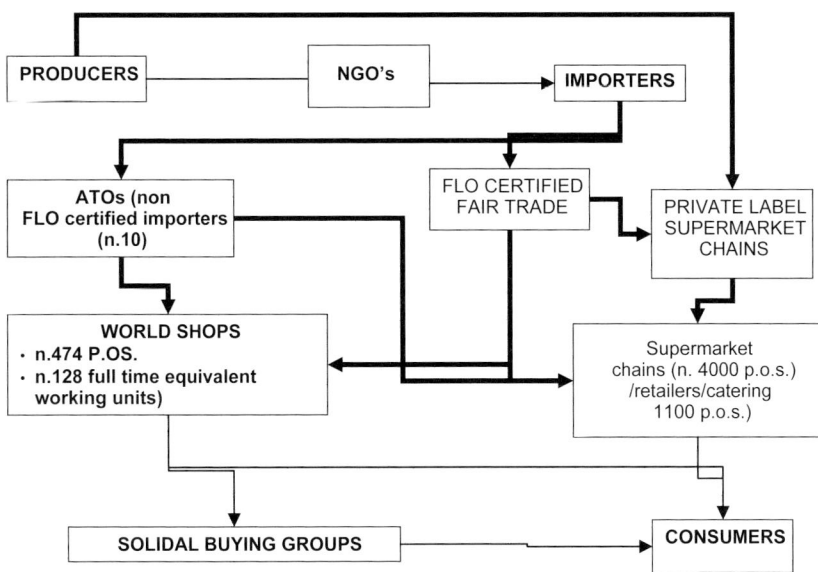

Figure 1. The Fair Trade supply chain in Italy. Source: the Fair Trade advocacy office, our interviews.

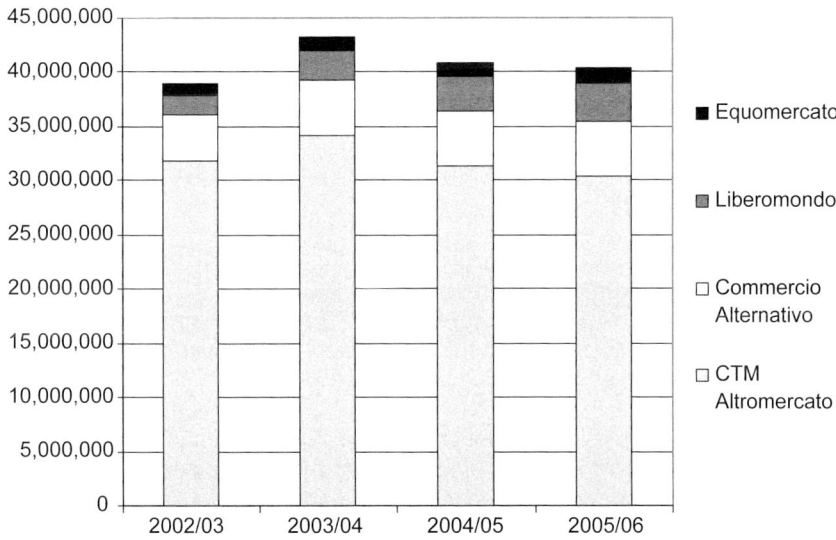

Figure 2. Main Italian ATO's turnover (€ current).

A study from Barbetta (2006) provides an insight on the main problems the Italian *World Shops* and Fair Trade importers are facing:
• small economic size;

- relative scarcity of financial and human resources with respect to the range of professional skills required and variety of products sold;
- little transparency in the price formation mechanism and in the supply chain economic and technical relationship;
- need to increase external economies of scale and collaboration between Fair Trade companies, both vertically and horizontally.

A good starting point to overcome part of problems above listed, and at the same time fulfil the Fair Trade objectives, should be the encouragement of collaboration and transparency to reduce the obstacles to the different supply chain agents communication.

To this end logistics are of paramount importance in influencing an efficient management of the physical goods and information flows along the supply chain (Pinna, 2005). This will positively affect the competitiveness of the supply chain not only in cost reduction terms but also by increasing the product value for the consumer through, among others, an efficient communication of the values differentiating Fair Trade products. A more efficient information flow reduces information asymmetry and the risk of opportunistic behaviour, supporting the reduction of inequalities in the income distribution along the supply chain, mainly for rural communities.

The information flow efficiency depends also from its compliance with the different supply chains coordination needs: the costs of coordination should not exceed its benefits. The relationship between the ATOs values and business models, with the Fair Trade logistics and supply chain organisation, and performance, should then be investigated.

3. Objectives

The goal of this paper is to determine how the impact of the ATOs business models on the supply chain organisation and performance, influenced the growth strategies and the ATOs ethical goals implementation.

4. Data and methodology

The most important Italian ATOs, almost exclusively operating in the Fair Trade market, have been analysed. The variables have been collected through the ATOs websites[45], and by interviewing the ATOs management and other agents in the supply chain. The interviews were finalised to collect quantitative and qualitative information on the companies structure, performance, and to an in depth analysis of their strategies, organisation and management.

Six interviews were conducted. The respondents for the ATOs are: the *CTM* and *Commercio Alternativo* logistics manager, an *Equomercato* management board member and the customer relationship manager for *Liberomondo*; given the relatively small size of the companies, the persons interviewed, although with different roles, were chosen for their specific competence on the subject and overall knowledge of the company. The other two Fair Trade supply chain agents interviewed are the buyers from two of the most important processing companies related to Fair Trade (*CONAPI* and *Coind*). The interviews were divided into two parts: the first regarded a broad discussion on the company business model: vision, strategies and relationship organisation along the Supply chain; a second part involved, only for the ATOs, the adoption of a questionnaire

---

[45] http://www.altromercato.it/; http://www.commercioalternativo.it/; http://www.liberomondo.org/; http://www.equomercato.it/.

with a static set of questions concerning the logistics and customer relationship related indicators. The interviews provided the bulk of the information but were integrated with other information on the company structure, budget and supply chain relationship with their suppliers and retailers, available from the ATOs websites[46], as suggested by the respondents.

The list of 'business model building blocks' defined by Osterwalder (Osterwalder, 2004) and the logistics indicators reported by Vignati (Vignati, 2002), provided a useful tool to organically collect and describe the variables influencing the analysis.

The analysis was carried out in three different steps.
- Step 1. Assessment of the different degrees of vertical coordination needed by the ATOs, for an efficient supply chain management (overall costs reduction and better services for the customer). It provides a benchmark for the following step.
- Step 2. Analysis of the supply chain organisation. The more the ATOs supply chain organisation matches its vertical coordination needs the higher the efficiency.
- Step 3. ATOs Supply chain logistics performance indicators and strategic goals analysis. It shows to which extent the ATOs managed to actually implement the vertical coordination needed and how their strategic goals influenced the logistics management.

The analysis indicates the relevant areas (organisation and/or management) where interventions should be made to increase the harmonisation between the Fair Trade companies values with the necessity to compete, and grow, in a market economy. If step 2 shows a difference between the actual and desirable supply chain coordination level the ATOs should try and modify the supply chain organisation (that is the contractual arrangements along the chain). Step 3, on the other hand, shows how efficiently the supply chain is managed in terms of products and information flow. The role of their strategic goals in influencing both their organisation and management is also considered.

Step 1. Assessment of the different degrees of vertical coordination

The Hobbs and Young approach to the analysis of the relationship between transaction cost and vertical coordination was adopted (Hobbs and Young, 2000). The authors consider that transaction costs are influenced by the level of uncertainty related to the relationship between the characteristics of the products traded and the type of transaction (Table 1).

Higher uncertainty leads to higher transaction costs and, consequently, to the need of a stronger vertical coordination within the supply chain. Vertical coordination is an extension of the vertical integration concepts as it considers: 'all means of harmonising vertically interdependent production and distribution activities' (Frank and Henderson, 1992). The authors consider four different types of progressively intensive vertical coordination, or 'harmonisation': (1) spot markets, (2) market specification contracts, (3) strategic alliances[47], and (4) full vertical integration.

The Fair Trade products characteristics and the following ATOs structural variables were examined, in order to assess their different levels of vertical coordination need:

---

[46] The websites contain many updated information as a consequence of their transparency policy.

[47] In a strategic alliance different firm join forces sharing their resources (personell, informations, capital) and the access to the market, to reach a specific goal. The firms collaborate but remain independent.

Table 1. Relationship between product characteristics and type of transaction (Hobbs and Young, 2000).

	Transaction characteristics						
	Uncertainty for buyer: quality	Uncertainty for buyer: reliable supply timeliness and quantity	Uncertainty for buyer and seller: price	Uncertainty for seller: finding a buyer	Frequency of transaction	Relationship specific investment	Complexity of transaction (variety of outcomes)
Product characteristics							
Perishability	V	V		V	V		V
Product differentiation	V	V	V	V		V	V
Quality variable and visible		V	V	V			V
Quality variable and invisible	V	V	V				V
New characteristics of importance to consumers	V	sometimes	V	V		V	V
Regulatory drivers							
Liability	V			V		sometimes	V
Traceability				V		V	V
Technology drivers							
Company-specific technology						V	sometimes

- the share of products sold by categories: fresh food, other food, non-food; this allows to evaluate their perishability, quality and regulation related characteristics;
- the turnover by type of distribution channel; it influences the complexity of the transaction: the higher the share of products sold through the supermarkets the more complex the transaction, the stronger the coordination need.

Step 2. Analysis of the supply chain organisation

In a second part the ATOs supply chain contractual relationships were analysed to evaluate how they match the supply chain coordination level required. The partnership relations (contractual arrangements) along the supply chain were examined by interviewing the ATOs management and from the websites' 'company presentation'. The variables considered, following the approach to vertical coordination measurement above mentioned, are the prevailing types of contracts occurring at each supply chain stage, according to the Hobbs and Young classification.

Step 3. ATOs Supply chain logistics performance indicators and value proposition analysis

Comparing Step 1 and 2 with Step 3 allowed appreciating how the ATOs managed to harmonise the consideration of the Fair Trade values with the necessity to compete, and grow, in a market

economy. The capacity of the ATOs to efficiently implement an adequate level of coordination was evaluated analysing their customer relationship:
- no. of clients, destinations and deliveries;
- orders management, in particular the communication technologies adopted for the orders transmission.

The logistics performance indicators reported by Vignati (2002) were also considered; they are mainly oriented towards evaluating the reliability, efficiency, timeliness and productivity of the internal and external logistics of the ATOs:
- Delivery reliability
  - R1: Stock breaking index: it indicates the frequency with which the demand cannot be satisfied from the stock.
  - R2: % of orders fulfilled/ total orders.
  - R3: % of orders unfulfilled/total orders.
  - R4: % orders fulfilled with multiple deliveries.
  - R5: % of orders fulfilled within the customer's requested date (delivery performance to request date).
- Inventory management performance
  - E1: Inventory Turnover by product category: is the ratio of the cost of annual sales to the average inventory level. The higher the inventory turns, the better the firm uses its inventory assets. Measuring the index by products categories allows a better evaluation of the different products supply chain management performance (Cost of Sales / Average Inventory Level).
  - E2: Days of supply by products category: n.of days the demand for the different products categories can be satisfied by the existing inventory (Average inventory / cost of a day's sales).
  - E3: Use of available warehouses areas and volumes: (e.g. % pallet racks area/ total warehouse area). Indicates the warehouse operative efficiency. E.g. pallet racks can be used from 45-50% up to 90% of the warehouse area and volume. The higher the area covered the higher the space utilisation efficiency (Vignati, 2002).
- Order fulfilment timeliness
  - T1: Delivery Lead Time: the total time that elapses between an order's placement and the products delivery.
  - T2: Actual delivery date/ agreed delivery date ratio.
- Warehouse operational efficiency
  - P1: Reception operational time (e.g.: handling units[48] received/time).
  - P2: Placement operational time (e.g.: handling unit placed/time).
  - P3: Picking operational time (e.g. no. of daily deliveries; no. of pallets picked and delivered).
  - P4: Wrapping and packing operational time.
  - P5: Vehicle uploading operational time.

Other information were collected regarding the ATOs value propositions. They are necessary to understand how the supply chain organisation, and management performance, matches the companies ethical values system; it also shows their influence on the ATOs capacity to attain a positive relationship between ethics and growth.

---

[48] Handling units: pallet, cartons etc.

5. Results

The analysis gave the following results.

Step 1. Vertical coordination need

One of the factors involved, the New characteristics of importance to consumers[49] (the Fair Trade values) is ubiquitous, affecting to the same extent the level of transaction costs. Other factors influence with different intensity the need for coordination among the ATOs. *CTM Altromercato* and *Commercio Alternativo* show higher transaction costs, according to the Hobbs approach, related to their:

- Larger share of fresh (perishable) products traded: it positively affects the share of quality variability, both visible and invisible and the traceability costs associated to food trade (Table 2).
- Higher share of transactions with supermarket chains, coffee shops, restaurants, etc. *CTM Altromercato* and *Commercio Alternativo* (Table 3) involve more demanding and complex 'business to business' relationships, related to the more complex administrative procedures, contracts specifications and the need for a right timing, quantity and quality in the orders management. The risks associated to these types of transactions are consequently higher.

---

[49] The 'New characteristics of importance to consumers' are those products' characteristics deriving from technical innovations or newly introduced *acceptable* production practices that consumer would like to be sure not to find if negatively perceived (GMO food) or to be present if positively perceived (animal welfare practices, Omega-3 fatty acids and Fair Trade rules compliance). These characteristics are not 'visible' increasing the buyer uncertainity over the quality (and availability) of these products' supply.

Table 2. Turnover by products categories in % (no. references) during 2004/05.

	Food	Fresh food	Non-food
CTM Altromercato	59 (230)	25	16
Commercio Alternativo	70 (300)	30	
Liberomondo	70	30	
Equomercato	10		90

Table 3. Turnover by distribution channels in % during 2004/2005 (2005/2006).

	World shops	Supermarket chains	Retailers	Catering	Export	Other
CTM Altromercato	56 (60)	21 (13)	7 (6)	5 (6)	11 (11)	(4)
Commercio Alternativo	70	30				
Liberomondo	90		10			
Equomercato	90		10			

The same conditions partially apply to *Liberomondo*, oriented towards fresh food but related to a relatively high number of small clients (*World Shops* and small traditional retailers) (Table 2 and 3). The need for coordination is however less demanding when compared to more 'conventional' clients like supermarket chains or catering.

Equomercato has a relatively lower need for coordination, given the prevailing trade in non food products (Table 2). On the other hand *Liberomondo* and *Equomercato* trade involve only the IFAT Fair Trade organisation Mark. The higher number of supply chain agents involved in the Fair trade organisation Mark control implies a stronger influence on vertical coordination, in terms of transparency and information sharing[50], when compared to the FLO certification related ATOs. The analysis shows a general strong vertical coordination need for the whole of the companies involved in Fair Trade. Higher transaction costs apply to the larger companies (*CTM Altromercato* and Commercio Alternativo).

Step 2. Actual coordination

The level of vertical coordination with the different agents is the following (Table 4).

ATOs / producers anl local processors

When a producer enters the Fair Trade supply chain the start up phase is nearly always supported directly, or through NGO's, by international cooperation projects involving the different ATOs. Fair trade producers agree on a fair price with the ATOs; the price covers the costs of production and cannot be lower than the world price; they are also given an advance payment to finance their activity. A price premium is also granted by the Fair Trade labelling system to fund social and environmental projects for the local community development. Fair trade privileges long contractual arrangements with the producers. The producers are generally organised in associations or cooperatives and in some cases take care of all (or part of) the processing stage. The prevailing contractual arrangement is thus the same for each ATO and consists in a a strategic alliance where the ATOs provide investments, know-how and make their distribution channels available to the producers.

ATOs / processors in importing countries

The vertical coordination with the processors in the importing countries is also generally high. It can be defined as a strategic alliance as processors are involved in international cooperation projects with the ATOs or share resources and/or information, sometimes also horizontally (processors to processors).

All the ATOs examined have developed common projects both with producing and consuming countries processors. A smaller but not marginal share of products processed is less coordinated and is generally due to a lack of available industries Fair Trade oriented. In this case a market specification contract takes place.

---

[50] Information sharing and transparency are necessary to reduce the uncertainty related to the respect of the Fair Trade principles and certification rules.

Table 4. ATOs Supply chain vertical coordination.

ATO	Prevailing contractual relationship / level of coordination		
	ATO / local producers and processors	ATO / industrial countries processors	ATO / retailers
CTM Altromercato	Strategic alliances[1]: collaborative management / price stability / long contractual relationship / transparency in price formation and information flows	Strategic alliances / market specification contracts	Vertical integration (with associated World Shops)[1] / market specification contracts with supermarkets and other retailers. Strategic alliances with other World Shops
Commercio Alternativo	Strategic alliances[1]: collaborative management / price stability / long contractual relationship / transparency in price formation and information flows	Strategic alliances / market specification contracts	Vertical integration (with associated World Shops) market specification contracts with supermarkets and other retailers. Strategic alliances with other World Shops
Liberomondo	Strategic alliances[1]: collaborative management / price stability / long contractual relationship / transparency in price formation and information flows	Strategic alliances / market specification contracts	Market specification contracts: long lasting contractual relationship / transparency in price formation and information flows
Equomercato	Strategic alliances: collaborative management / price stability / long contractual relationship / transparency in price formation and information flows	Strategic alliances / market specification contracts	Market specification contracts: long lasting contractual relationship / transparency in price formation and information flows

[1] See organisations websites.

ATOs / retailers

The distribution involves a complex relationship between importers (ATOs) and retailers. A network of 474 *World Shops* operates in Italy, sometimes related to solidal buying groups[51]; these shops are managed by 347 different organisations representing non profit-associations or social cooperatives; food represents 40-50% of their total turnover (Musso and Perna, 2005). Fair Trade products are also sold via the traditional retailing sector: shops and supermarkets; they buy their products either from FLO licensed importers and in few occasions from the ATOs, not necessarily FLO certified, or directly from the producers (acting as importers).

---

[51] Solidal buying groups - 'Gruppi di acquisto Solidale' (GAS) - are informal groups of consumers ethically oriented. Their principles are strictly related to the organic and Fair Trade values. For more information see www.retegas.org (English page).

The ATOs are mostly supplying the *World Shops* network but also sell their products, sometimes as private labels, to supermarket chains; part of the *World Shops* (about 50) sell also FairTraded Mark products following the FLO standards.

The contractual arrangements between retailers and the ATOs examined varies; in the case of *CTM Altromercato* they range from full vertical integration, with 350 associated World Shop, to a market specification contract when dealing with supermarkets or other traditional retailers. The same apply, to a lesser extent, for *Commercio Alternativo* (Table 4).

Liberomondo and *Equomercato* are relatively less coordinated with the *World Shops* when control is considered; the information flow and collaboration is however very strong as these ATOs sell only to World Shop. A certain amount of networking among ATOs also exists; in case they cannot meet the demand with their stock they buy products from each other.

A high level of coordination between the ATOs and the other supply chain agents resulted. The ATOs play a central role in the supply chain management. They provide not only resources in terms of services and products but also very often create, coordinate and, in some cases, are vertically integrated with part of the supply chain.

The analysis showed a correspondence between the coordination needs and the client/supplier relationship structure, along the different ATOs supply chains, and also horizontally, between ATOs.

Step 3. Efficient management of the supply chain relations (customer and suppliers trading relationship and logistics)

Customers relatonship data relate to *Commercio Alternativo* and partially to *Equomercato*. The *CTM Altromercato*, *Liberomondo* and *Equomercato* managers interviewed were not able to provide quantitiative data but confirmed the data reported for *Commercio Alternativo* are representative of their customer relationship structure. On the supply side only the number of producers in developing countries was considered; data on the products flow were not available. The supplier's number is taken from the ATOs websites.

The analysis of the clients, destinations and deliveries in the (Table 5) shows their average small economic size and a small number of shipments per client; it is a fragmented, and consequently expensive, customer relationship.

The number of suppliers, nearly always members of producers' cooperatives or associations, is relatively small but involves thousands of small producers; in this case the efficiency of the relationship is probably negatively influenced by internal logistics.

Another relevant indicator of the customer relationship management efficiency, in terms of information flow, is the orders transmission/management mode (Table 6). The on-line mode is largely adopted by the ATOs; only *Liberomondo* relies (at least up to the year 2004/2005) upon e-mail and other off-line modes.

Table 5. Clients, destinations, shipments during 2004/2005.

ATOs	Suppliers	Clients						
	Suppliers (n)	Clients (n.)	Destinations (n.)	Shipments (n.)	Turnover/clients (€)	Turnover/shipments (€)	Shipments/destinations (n.)	Shipments/clients (n.)
CTM Altromercato	173							
Commercio Alternativo	116	786	1580	7760	6643	673	5	10
Liberomondo	105							
Equomercato	62	571			2478			

Table 6. Order preparation/transmission mode (%) during 2004/2005.

ATOs	On line	E-mail	Other
CTM Altromercato	99		1
Commercio Alternativo	35	45	20
Liberomondo		50	50
Equomercato	50		50

The analysis showed an extensive and appropriate[52] use of ICT's (Information and Communication Technologies); The use of CRM (Customer Relationship Management) software encourages a better and more transparent orders and logistic management both inbound and outbound. This should increase the quality of the customer services and reduce their costs.

On the other hand the large share of food traded, high number of references, suppliers, clients, deliveries, orders and shipments involved, increase the complexity and costs related to the supply chain logistics and administration.

An efficient logistics monitoring could help the management reduce the constraints to the products flow along the supply chains; this will lower not only the costs but also help suppliers fully exploit the growing demand for Fair Trade products. To this end an analysis of the ATOs adoption of logistics indicators has been carried out.

---

[52] The off-line orders reaching Equomercato (50% of total orders) are generally fulfilled by an on-site (wharehouse) order transmission and/or consignment, not needing an on-line mode.

Logistics indicators

Reliability, flexibility, timeliness and productivity indicators were analysed (Table 7a and 7b). The actual values were not available due to the sensitive nature of the data and, in the case of smaller ATOs, also from the lack of a formal measurement.

This will not affect the relevance of the information; the extent of the monitoring is in fact related to the companies interest in an efficent logistics management, which is the focus of the analysis.

The results show two different approaches to logistics monitoring (and management). *CTM Altromercato* and, to a lesser extent, *Commercio Alternativo*, measure their logistics performance (Table 7a and 7b). The other ATOs adopted an informal qualitative monitoring (inventory performance and timeliness). The analysis of the ATOs value propositions provided useful indications on the factors affecting their different behaviour.

Table 7a. Warehouse management indicators adopted.

ATOs	Delivery reliability					Inventory management performance			Orders fulfilment timeliness	
	R1	R2	R3	R4	R5	E1	E2	E3	T1	T2
CTM Altromercato	X				X	X		X	X	X
Commercio Alternativo	X							X	X	X
Liberomondo								X	X	X
Equomercato								X	X	X

Table 7b. Warehouse productivity, indicators adopted.

ATOs	P1	P2	P3	P4	P5
CTM Altromercato	X	X	X	X	X
Commercio Alternativo					
Liberomondo					
Equomercato					

[1] The warehouse productivity was only measured by CTM Altromercato in the year 2002/2003, to verify the opportunity to externalise the picking.

Value propositions

The ATOs examined show important differences in their approach towards Fair Trade; in particular:
- *CTM Altromercato* and *Commercio Alternativo* stress the importance of integrating the ethical principles of solidarity, social and economic justice with a sound management and organisational capacity making Fair Trade companies competitive (Commercio Alternativo, 2007). The growth in the market size, obtained through collaboration with non Fair Trade companies (e.g. supermarket chains) also represents an important strategic difference with the other ATOs.
- *Liberomondo* and *Equomercato* are more oriented towards the social and political consequences related to Fair Trade. *Equomercato* stresses the importance of building a political alternative to the market economy supporting cooperation as opposed to competition, and promoting the Fair Trade values through the education of consumers and producers (Equomercato, 2007). *Liberomondo*, a social cooperative, is more oriented towards the social impact of Fair Trade as a tool to enter the labour market for socially marginalised people both in developing and developed countries and to create an integrated 'ethical supply chain' (Liberomondo, 2007). It links the international circuits of Fair Trade to the Italian social economy.

The values supported by the smaller ATOs influenced their attitude towards outbound logistics efficiency and internal performance monitoring. The relative importance given to timeliness and reliability when dealing with their consumers derives from the ATOs prevailing interest in the producers support; the social and ethical content of the Fair trade products should be more important to the consumer than the distribution efficiency; the warehouse productivity was not monitored in the smaller ATOs as it was considered an excess of control over the workers. The bigger ATOs, on the other hand, try and increase the volume of products sold improving their efficiency in the customer relationship.

6. Conclusions

The analysis of the relationship between the ATOs value propositions, supply chain organisation and management, showed a degree of vertical coordination compatible with the ATOs coordination needs; they were also useful to provide information on the different companies' capacity of managing a sustainable ethical and economic growth.

When considering the relationship between ethics and economic growth the analysis showed two different types of ATOs. One is more oriented towards increasing the volume of goods traded not only by promoting the consumers awareness towards the ethical content of Fair Trade products, but also competing on the more traditional ground of quality and managerial efficiency. This implies the adoption of more sophisticated supply chain and logistics management tools and procedures, a closer relationship with more traditional distribution channels also emerged. These type of ATOs partially match the definition of 'imprese sociali del Cees' (Fair Trade social enterprises), as proposed by Barbetta when analysing different types of World Shop (Barbetta, 2006). On the other hand the smaller ATOs maintain a business model based on small size, little interest in competing within the *conventional* market as a strategy to increase their turnover. The focus of their strategy is in fact the definition of an alternative way of dealing with economic activity by promoting the ethical awareness of the consumers and the social development of the whole of the agents in the Fair Trade supply chain. This led to an exclusive relationship with the World Shop circuit, whose structural characteristics are less efficient in granting access to a large number of consumers, when compared to conventional trading and retailing channels; it also implied a marketing policy 'producer oriented'; the support to the marginal rural areas in the developing world is in fact considered a priority with respect to the consumers need for reliability, timeliness and affordable prices. This influenced the logistic integration that does not seem to be encouraged to the same extent in the smaller ATOs. The risk associated to this strategy, whose goals are projected in a relatively distant future, is that more traditional trading and retailing big companies meanwhile can greatly increase their market share, also taking advantage of the Fair Trade values promotion coming from the ATOs and *World Shops*.

The ATOs growth strategies seem more influenced by their value propositions than the lack of managerial skill or unfavourable market conditions, at least in the smaller ATOs.

The analysis confirmed that the supply chain organisation is not 'neutral' but is influenced by the company values and in turn influences the company economic as well as 'ethic' sustainability. The conflict between an organisation necessity of growth (or survival) to grant its economic sustainability in the long run, and the respect of its ethical code, is far from being solved. A contribution to overcome this apparent contradiction could be related, in our opinion, to the consideration of the entire range of theoretical and operational tools available to business management, without prejudices. An efficient flow of goods and information along the Fair Trade supply chain is in fact necessary both to the more ethical and 'market oriented' ATOs; in this sense the Fair Trade values, promoting transparency and collaboration, represent a potential competitive advantage with respect to conventional supply chains. The importance of opportunistic behaviour, as an obstacle to the flows of information and to a collaborative management, should be in fact smaller.

Future research developments should therefore investigate the monetary and non-monetary variables affecting the efficiency and sustainability of Fair Trade inter firms' relationships (clusters, networks, supply chains, etc.).

A more specific field of investigation should be oriented towards the analysis of the Fair Trade flows from developing countries both in terms of goods and information. A fair income distribution and in general economic and social justice need transparency and therefore a full, fast and clearly organised access to relevant information and knowledge.

References

Barbetta, G.P., 2006. Il commercio equo e solidale in Italia, Working Paper n. 3 di Ricerca su 'Il commercio equo e solidale. Analisi e valutazione di un nuovo modello di sviluppo'. Milano: Università Cattolica del Sacro Cuore, Centro di Ricerche sulla Cooperazione.

Becchetti, L. and M. Costantino, 2006. The effects of Fair Trade on marginalised producers: an impact analysis on Kenyan farmers. Working paper CEIS 220, Rome.

Becchetti, L. and L. Paganetto, 2003. Finanza etica Commercio equo e solidale. La rivoluzione silenziosa della responsabilità sociale. Roma: Donzelli editore.

Castro, J.E., 2001. Impact assessment of Oxfam's Fair Trade activities. The case of COPAVIC'. Oxfam.

Commercio Alternativo, 2005. Verso il Forum AGICES 2005, Encuentro n.3 marzo-aprile, Ferrara.

Commercio Alternativo, 2007. La Cooperativa. Commercio Alternativo web site http://www.commercioalternativo. it/cooperativa/, Accessed on January 2007.

Conforti, P. and B.E. Velasquez, 2004. The effects of alternative proposals for agricultural export subsidies in the current WTO round. The Estey Centre Journal of International Law and Trade Policy, V: 11-42.

Equomercato, 2007. Chi siamo. Equomercato web site, http://www.equomercato.it/chisiamo.php, Accessed on January 2007.

FLO (Fairtrade Labelling Organizations International), 2006. Fairtrade FAQs; http://www.fairtrade.net/faq_links. html?&no_cache=1;

FLO (Fairtrade Labelling Organizations International), 2007. Producer standards. Available at: http://www. fairtrade.net/producer_standards.html, Accessed on October 2007.

Frank, S.D. and D.R. Henderson, 1992. Transaction costs as determinants of vertical coordination in US food indusries. American Journal of Agricultural Economics, 74: 941-950.

Hobbs, J.E. and L.M. Young, 2000. Closer vertical co-ordination in agri-food supply chains: a conceptual framework and some preliminary evidence. Supply Chain Management, 5: 131-142.

Krier, J.-M., 2005. Fair Trade in Europe 2005. Brussels: the Fair Trade Advocacy Office.

LiberoMondo, 2004. Dove va il commercio equo e solidale? Grande distribuzione e botteghe del mondo. Supplemento al n. 8 di Tempi di Fraternità.

Liberomondo, 2007. Chi siamo. Liberomondo website, http://www.liberomondo.org/chisiamo.htm, Accessed on January 2007.

Musso, D. and T. Perna, 2005. Vediamo quanto vale il popolo dell'equo. Altreconomia, 62: 17.

Osterwalder, A., 2004. The business model ontology a proposition in a design science approach. These Présentée à l'Ecole des Hautes Etudes Commerciales de l'Université de Lausanne, Lauseanne, Suisse.

Pinna, R., 2005. L'evoluzione nella dimensione organizzativa della supply chain. Milano: Franco Angeli.

Robbins, P., 1999. Review of the impact of globalisation on the agricultural sectors and rural communities of ACP countries. Technical Centre for Agricultural and Rural Cooperation (CTA), Study Report 4-1-06-211-9, CMIS, London.

Ronchi, L., 2002. The impact of Fair Trade on producers and their organisations: a case study with coffee in Costa Rica. PRUS Working paper no. 11 june. Poverty research Unit at Sussex University of Sussex, Falmer, Brighton.

Vignati, G., 2002. Manuale di Logistica. Milan, Italy: Hoepli.

WTO, 2006. International trade statistics. Available at: http://onlinebookshop.wto.org/shop/article_details.asp?Id_Article=715&lang=EN.

The structural evolution of organic farms in the USA: the international market effect

S. Grow and C. Greene

Abstract

Rapid growth of the organic agricultural sector in the U.S. and implementation of the U.S. Department of Agriculture's national organic standards in 2002 have lead to concerns that organic production could become increasingly concentrated on larger U.S. and international farms, disrupting the market access of small domestic organic producers. However, data on the U.S. organic agriculture show that the smallest-scale farms continue to hold a small but stable piece of the organic sector and that U.S. organic farm size has grown slowly. The amount of land under organic production worldwide is growing rapidly, particularly in developing countries producing commodities for export, many of which are not widely grown in the U.S. Small-scale producers using direct markets have likely been least impacted from increased organic imports, while producers of organic oilseeds and cotton have likely been most impacted. Federal and State government agencies and the private sector have launched initiatives to sustain small-farm participation in the U.S. organic sector. Programs to better serve organic producers in the U.S. and to differentiate organic and non-organic imports and exports are being developed at the federal level.

Keywords: organic agriculture, organic certification, small-scale farmers, international trade

1. Introduction

Supply- and demand-side forces have made organic farming one of the fastest growing U.S. agricultural sectors for well over a decade. Annual growth in retail sales has equaled around 20 percent or more since 1990 and U.S. sales of organic foods were estimated at almost $14 billion in 2005, with growth forecasted to $24.4 billion by 2010, according to industry estimates (Nutrition Business Journal, 2006). Retailers have responded to rising consumer demand for variety, quality, and convenience by introducing more product varieties while conventional supermarkets and mass market merchandisers have added organic products to their shelves (Oberholtzer *et al.*, 2005). Because price premiums for organics have held steady during this time and in order to meet supply-side demand, more operations and land have become certified organic in the U.S. and small and medium-sized organic companies have grown (USDA–ERS, 2006; Sligh and Christman, 2003). The U.S. Department of Agriculture's (USDA) national organic standards, which were implemented in 2002, were designed to stimulate growth of the organic industry by building consumer confidence in organic products and facilitating commerce in agricultural products that are organically produced.

The marketing pathways and farm profile of organic agriculture have changed as the sector has grown. Until the early 1990s, the largest outlet for organic products in the U.S. was independent natural foods stores. By 2005, independent natural foods stores represented less than 25 percent of organic food sales, and natural foods chains, conventional supermarkets, grocery stores, mass merchandisers, and club stores together accounted for the bulk of sales (OTA, 2006). The use of direct markets has also declined, from approximately 22 percent of total organic sales in the early 1990s to 7 percent in 2005 (USDA-AMS, 2000; OTA, 2006). The organic farm sector historically has had smaller operations and disproportionately more fruit and vegetable production than in conventional agriculture. With the industry shift toward larger, more concentrated marketing and distribution pathways during the last decade, the number and size of organic operations in the U.S is increasing and is expected to continue doing so into the future. Between 1995 and 2005, the number of certified organic operations has more than doubled and the amount of certified organic land has quadrupled to over 1.6 million hectares, with approximately 0.9 million hectares used for pasture and 0.7 million hectares used for crops (USDA–ERS, 2006).

Small producers have expressed concern that marketplace changes and regulatory measures developed to facilitate domestic and international organic trade, including the USDA National Organic Program, may negatively influence their market access. The national organic standards also refer to their potential impact on small operations, and contain several provisions to mitigate their impact on small producers (USDA-ERS, 2000). Small-scale organic producers are concerned that industry growth will increase competition from larger domestic operations and from international farming operations (Hanson *et al.*, 2004). Small-scale farmers have noted that the fees and paperwork requirements of organic certification inhibit their broader participation in the organic market. And finally, international trade may also be a threat to small organic farms by affecting their market power through increased competition.

Because there are public benefits to sustaining a diverse organic farm sector, a growing number of public and private groups have begun efforts to facilitate the participation of small farms in the U.S. organic market. For example, U.S. certifiers are developing certification programs tailored to small-scale operations and U.S. businesses are expanding opportunities for local, direct-to-consumer marketing. The objectives of this paper are to examine the structural changes in the U.S. organic sector over the last couple of decades, identify the potential impact of international trade on this sector, and assess the potential for small-scale farmers to remain an important component of this sector.

2. Methodology

To understand the impacts of international trade on the U.S. organic agricultural sector, specifically the impact on U.S. small farms, we first explore recent structural changes in the U.S. organic sector. Organic certification data from the USDA provide the most comprehensive description of the U.S. organic sector. USDA's Economic Research Service (ERS) has published estimates of certified organic farmland and livestock, by commodity and state, since 1997, along with some data on small organic farms. National-level estimates are available since 1992. This data were collected from State and private certification groups and were analysed to determine changes in the average size of operations as well as to determine whether small organic farms, under two hectares, are in decline. These national data sets were also used to analyse regional differences in the size of organic operations. The National Organic Program does not require organic growers and processors selling less than $5,000 per year in organic agricultural products to be certified and therefore those producers are excluded from these data.

Comprehensive data about the organic agricultural sector in California is also available, and because California makes up such a large percentage of the overall certified organic production in the U.S., trends and numbers there may reflect the realities of organic operations throughout the U.S. Data about the organic agricultural sector in California was obtained from the California Department of Food and Agriculture's (CDFA) registration data and the California Certified Organic Farmers (CCOF), a non-profit. California certification data is provided to the CDFA by growers and was used to identify changes in the average size of operations enrolled in organic certification programs in that State from 1992 and 2003, as well as other structural changes in the farm sector. Because the number of organic operations in California is increasing, California registration data describes the size of new operations entering the market, illustrating disproportionate certification by large farms. Data from the CCOF, the largest organic certifier in California, provide information about California's organic sector beginning in 1985.

This paper also explores the current information available on the role of international trade on the U.S. organic agricultural sector and some of the existing, planned, and potential programs and mechanisms developed to ensure the continued participation of small organic farms in the U.S. State agencies, nonprofit organisations, and U.S. food companies are all initiating programs to support small farm organic production as the organic sector grows and changes. Additionally, government agencies are developing projects that will fill information gaps about the U.S. organic sector and the role of international trade.

3. Organic farm size trends

3.1 National trends, based on USDA data

National datasets from 1992 to 2005 of total organic farmland suggest that the amount of certified organic farmland has steadily increased since the early 1990s, with certified organic cropland increasing more rapidly than pastureland during most of this period. On the other hand, the average size of certified organic operations has trended upward fairly slowly, despite rapidly growing demand during the 1990s and the implementation of organic regulations in 2002 which facilitated further growth in the market. For the decade spanning 1992-2002, the number of certified organic operations in the U.S. doubled from 3,857 to 7,323, but the average size of certified organic operations changed by less than one percent from an average size of 105.6 hectares to 106.4 hectares. Some growth in organic cropland is seen when organic pasture is excluded from these averages: from 1992-1995, operations ranged from 45.3 hectares to 56.7 hectares and then grew to an average of 74.9 hectares in 2000. By 2005, the average size of certified organic cropland operations was 82.1 hectares.

3.2 Regional Trends, based on USDA data

National-level data can be misleading, since regions specialising in field crops have significantly larger operations than regions specialising in specialty and other high-value crops. To get a better understanding of average farm size throughout the U.S., we have broken the data into ten Regions, based on USDA production categories (Table 1). Regional farm size averages in the Continental U.S. ranged from 23.9 hectares in Appalachia to 536.6 hectares in the Mountain Region in 2000 and from 28.7 hectares in the Southeast to 637 hectares in the Southern Plains in 2005.

Neither the total number of operations in the Region, nor the total amount of certified organic land in the Region was indicative of the average size of each operation (Table 2); in 2005, the Pacific Region contained the third highest amount of certified organic land in the U.S., the highest number of certified organic operations, and had the fourth smallest average size of operation. On the other hand, the Southern Plains ranked fifth in total certified organic land, sixth in the number of organic operations, and yet had the second highest average sized operation in 2005. Three Regions, the Delta, Corn Belt, and Mountain regions, experienced declines in the average size of certified organic operations between 2000 and 2005, ranging from four to 34 percent, while rates of growth in the other Regions, excluding Alaska and Hawaii, ranged from 17 percent in the Pacific to an almost tripling in size in the Southern Plains. Regional averages were not included for the non-Continental U.S. region (Alaska and Hawaii) because of the disparate size

Table 1. USDA production regions.

Appalachia	Kentucky , North Carolina, Tennessee
	Virginia, West Virginia
Mountain	Arizona, Colorado, Idaho, Montana, Nevada, New Mexico, Utah, Wyoming
Pacific	California, Oregon, Washington
Corn Belt	Illinois, Indiana, Iowa, Missouri, Ohio
Northeast	Connecticut, Delaware, Maine, Maryland, Massachusetts, New Hampshire,
	New Jersey, New York, Pennsylvania, Rhode Island, Vermont
Southeast	Alabama, Florida, Georgia, South Carolina
Delta	Arkansas, Louisiana, Mississippi
Northern Plains	Kansas, Nebraska, North Dakota, South Dakota
Southern Plains	Oklahoma, Texas
Lake States	Michigan, Minnesota, Wisconsin
Other: non-continental U.S.	Alaska, Hawaii

Table 2. Average operation size by USA region in 2000 and 2005.

Region	2000			2005			2000-2005
	Number of farms	Organic land (hectare)	Average farm size (hectare)	Number of farms	Organic land (hectare)	Average farm size (hectare)	Percent change in farm size
Southern Plains	432	42,060	234	449	138,853	637	173
Appalachian	49	7,804	24	39	5,966	38	59
Northern Plains	96	114,425	265	144	174,791	389	47
Lake States	1,322	78,342	82	1,747	119,755	102	24
Northeast	695	45,205	34	656	71,440	41	20
Southeast	327	2,603	24	157	3,968	29	19
Pacific	1,602	90,040	56	2,760	181,266	66	17
Corn Belt	957	59,078	72	1,177	69,690	69	-4
Mountain	108	373,028	537	138	262,846	401	-25
Delta	824	8,202	168	1,008	4,282	110	-34
Continental U.S.	6496	718,400	111	8349	1,047,739	125	46

and composition of their organic farm sectors. In 2005, Alaska had nearly 0.6 million hectares of organic pasture (accounting for two-thirds of U.S. total organic pasture), while Hawaii had 2,000 hectares of organic cropland, mostly for fruit and vegetable production.

3.3 Mixed vegetable summary, based on USDA data

The organic market niche has its origins in premiums that small-scale farmers derived from marketing produce directly to consumers and small health food stores, a niche particularly well-suited to maintaining the profitability of small farms. Small mixed vegetable operations are prevalent in the organic sector, and USDA has tracked those that are smaller than two hectares (five acres) in order to capture trends affecting these small farms. USDA has asked organic certifiers for information about these small operations since 1997, but differences in reporting by certifiers has affected the precision of the data and ultimately, they can only be used to examine trends.

In 1997, mixed vegetables grown on very small plots under two hectares, as reported by certifiers, comprised 5.6 percent of all land dedicated to organic vegetable production in the U.S. The number of hectares operated as small mixed-vegetable plots has continued to expand overall between 1997 and 2005, although as a percentage of total vegetable land they have declined slightly (Figure 1). These very small farms have essentially maintained a small, but relatively stable, share of the overall certified organic vegetable market. Among regions, small mixed vegetable plots were most likely to be seen in 2005 in the Pacific, Northeast, Mountain, Lake States, Hawaii, Corn Belt, and Appalachian regions.

3.4 Certifier data

California Certified Organic Farmers (CCOF) is one of the few published sources of information about organic farm size prior to the 1990s. CCOF was established in 1973 and was one of the first organisations to offer third-party organic certification services to farmers in the U.S. CCOF certifies more farmers than any other certifier in California, and is the top certifier in the U.S., as well. In 2005, CCOF certified over 60 percent of California's certified organic farmland, after certifying nearly 77 percent in 1995, and even higher percentages in the 1980s (CCOF, 2006; Greene, 1992). Because of California's dominance in U.S. organic production and CCOF's dominance in certification, CCOF's data may represent trends mirrored throughout the U.S.

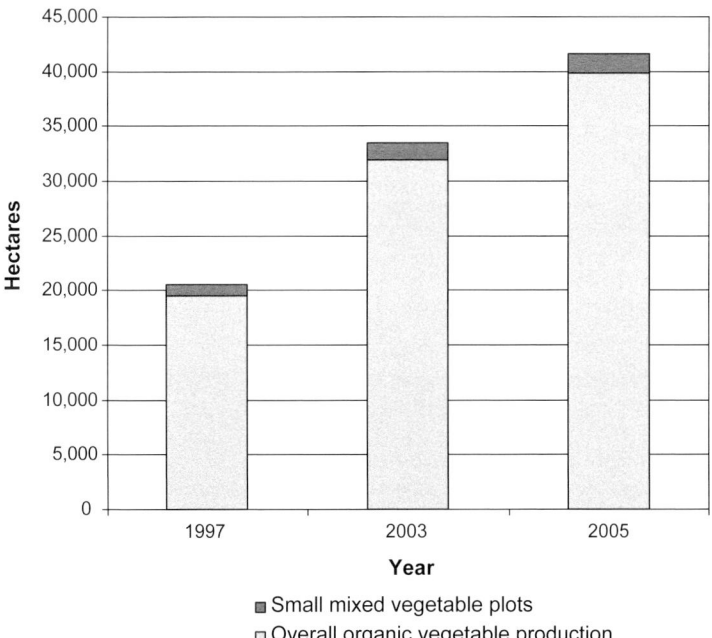

Figure 1. Small vegetable plots parallel growth in overall organic vegetable sector.

The total number of organic hectares and growers certified by CCOF rose steadily between 1985 and 2005. However, the average operation size of farmers enrolled in the CCOF certification program grew rapidly and then reached a plateau of about 61 hectares per grower in the late 1990s (Figure 2). In 1991, only five percent of CCOF's growers had organic operations larger than 405 hectares (1,000 acres) (Greene, 1992).

3.5 Summary of average organic operation size

There is no precise information on the average size of organic operations in the U.S., but analysis of a variety of data sources indicates that it is about 60 percent of the size of average U.S. farms. Trends also indicate that the average size of organic operations is generally increasing (Figure 2), but that its growth is not increasing rapidly. University of California studies suggest that over half of the registered organic operations in California were smaller than two hectares throughout the late 1990s and there is no evidence to suggest that this percentage has changed markedly (Klonsky and Richter, 2005). The certified organic livestock sector has begun to grow rapidly and it is possible that the average size of certified organic operations might also begin to grow rapidly as pasture is increasingly certified as organic.

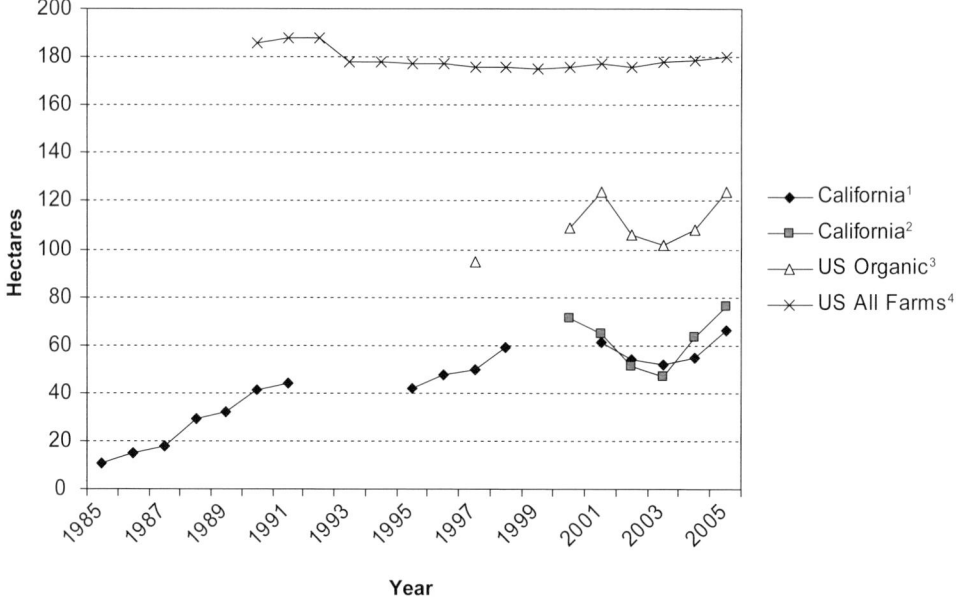

Figure 2. Average size of U.S. organic and conventional farms.
[1]California organic data as reported by California Certified Organic Farmers (CCOF).
[2]California organic data from USDA, Economic Research Service (ERS).
[3]USDA, Economic Research Service (ERS); includes all States except Alaska.
[4]USDA, National Agricultural Statistical Service (NASS) and Census of Agriculture.

4. International organic markets and production

The global market for organic products – mostly in the U.S., Europe, Canada and Japan – has more than tripled during the last decade, with retail sales reaching $30-32 billion in 2005 (Kortbech-Olesen, 2006). According to Willer and Yussefi (2006), the North American organic market is also reporting the highest growth worldwide, indicating that the region will account for much of the global revenues in the foreseeable future. The U.S. had $14 billion in organic food sales in 2005, nearly 2.5 percent of U.S. food sales and approximately 45 percent of global organic sales (OTA, 2006).

Organic production has expanded rapidly in recent years in developing countries, as well as in developed countries (Table 3). An estimated 31 million hectares of farmland are managed under organic production worldwide (Willer and Yussefi, 2006). Another 19.7 million hectares worldwide includes areas of certified forest and wild harvested plants. The U.S. has the fourth largest area under organic management in the world (Table 3), behind Australia, China and Argentina. USDA reported that in 2005 over 1.6 million hectares of U.S. farmland (0.5 percent of U.S. agricultural land) was under organic production (USDA–ERS, 2006).

4.1 U.S. organic trade

Data on organic imports and exports is incomplete because U.S. customs does not differentiate between organic and non-organic trade. USDA estimates that the value of U.S. exports was between $125 and $250 million in 2002, while the value of U.S. imports was between $1.0 and $1.5 billion, and organic imports now exceed exports by a ratio of approximately 8 to 1. (USDA–FAS, 2005). U.S. exports have stagnated as domestic demand has risen and competition for international markets has increased. However, the U.S. was likely a net exporter during part of the 1990s, with exports estimated at approximately 200 million by 1994 (Natural Foods Merchandiser, 1995), and at $200-$300 million in the late 1990s (Fuchshofen and Fuchshofen, 2000).

The U.S. National Organic Program (NOP) streamlined the certification process for international as well as domestic trade when it was implemented in 2002. Organic farmers and handlers anywhere in the world are permitted to export organic products to the U.S. if they meet NOP standards and are certified by a public or private certification body with USDA accreditation. Since 2002, USDA has accredited 40 certifiers in 19 countries outside the U.S., mostly in Latin America, Europe, and Canada, and currently has recognition agreements with six countries. In addition, nearly a dozen U.S.-based groups with USDA accreditation provided certification services in 30 countries in 2006. Among the top twenty countries with certified organic farmland, sixteen have USDA-accredited certification services available from international certifiers based in the U.S. and/or domestic certifiers located in the producing country (Table 3).

U.S. imports of organic products accounted for approximately 12 percent of the U.S. organic market in 2002 (USDA–FAS, 2005), and have likely grown substantially in the last four years. According to USDA's Agricultural Marketing Service, out of the 20,000 organic clients of USDA-accredited certifiers operating worldwide in 2006, approximately 9,000 were located outside the U.S. (C. Greene, personal communication 2006). The U.S. organic market has increased 15-20 percent a year since 2002 (OTA, 2006), and imports have increased as U.S. farmers struggle to keep pace with demand in the face of strong market competition. Organic food production is often labour-intensive, and developing countries with lower farm labour costs than those in the U.S. have a competitive advantage in organic production.

Table 3. Organic farmland is growing rapidly in most of the top 20 countries. Source: Willer and Yussefi (2006) and Yussefi and Willer (2002) reports on worldwide organic farmland, see http://www.soel.de/oekolandbau/weltweit.html for current and former editions of The World of Organic Agriculture; USA, USDA-ERS, see www.ers.usda.gov/data/organic.

| Country | Certified and transitional organic agricultural land | | | | Change 2002-2006 (%) | Availability USDA-NOP certification Services[2] |
| | 2002 survey | | 2006 survey | | | |
	(hectares)	Organic/ total (%)	(hectares)	Organic/ total (%)		
Australia	7,645,924	2	12,126,633	3	59	yes
China	40,000	<1	3,466,570	<1	>1,000	yes
Argentina	2,800,000	2	2,800,000	2	-	yes
USA	900,000	<1	1,620,350	<1	80	yes
Italy	1,040,377	6	954,361	6	(8)	yes
Brazil	803,180	<1	887,637	<1	11	yes
Germany	546,023	3	767,891	5	41	yes
Uruguay	1,300	<1	759,000	5	>1,000	yes
Spain	380,838	1	733,182	3	93	yes
UK	527,323	3	690,272	4	31	yes*
Chile	3,301	<1	639,200	4	>1,000	yes
France	371,000	1	534,037	2	44	--
Canada	340,200	<1	488,752	<1	44	yes
Bolivia	13,918	<1	364,100	<1	>1,000	yes
Austria	271,950	9	344,916	13	27	yes
Mexico	85,676	<1	295,046	<1	244	yes
Peru	27,000	<1	260,000	<1	863	yes
Greece	24,800	<1	249,488	3	906	yes
Ukraine	N/A	--	241,980	<1	N/A	--
Czech Republic	165,699	4	160,120	6	(3)	--
All countries[1]	17,156,455		31,000,000	--	81	

[1] Most estimates in the 2002 survey were as of 31.12.2000; most estimates in the 2006 survey were as of 31.12.2004.
[2] USA-National Organic Program (NOP) accredited certification services.
* The U.S. has recognition agreements with six countries, including the UK.

According to FAS, Canada is the main market for U.S. organic exports, while countries in Latin America, including Mexico, Brazil, Argentina, and Uruguay, along with China and other countries in Asia are major sources of organic imports. Among the top twenty organic countries with certified organic farmland, the countries with the fastest growth in organic production are mostly those that produce organic products for export. The amount of land under organic production systems in China, Bolivia, Chile, Uruguay, and the Ukraine for example, increased well over 1,000 percent between 2002 and 2006, while organic farmland in Europe and North America showed more modest expansion (Table 3). Worldwide, organic farmland increased approximately 81 percent between 2002 and 2006. While many developing countries were starting from a low base of certified organic farmland in 2002, several, particularly in Latin America, now manage a higher proportion of their farmland under certified organic farming systems than the U.S.

While some U.S. organic imports compete directly against similar U.S. products, many are products that are not widely grown in the U.S., such as coffee and winter produce. The impact of U.S. organic imports varies widely among commodity sectors. Small-scale farmers producing a wide variety of horticultural products – and increasingly livestock products – for sale in direct markets have likely seen the least impact from increased imports. Organic consumers at farmers markets, independent restaurants, small food shops, and other direct markets are explicitly seeking locally-grown organic products. However, some fruit and vegetable growers who marketed to natural foods grocery stores during the 1990s have reported losing some of their markets to imports as well as to larger domestic producers as these stores have expanded (Hanson *et al.*, 2004).

U.S. organic grain and oilseed producers also face market competition. U.S. organic cotton producers began losing market share in the 1990s to countries with lower labour, input, and technology costs (Greene and Kremen, 2003), and U.S. organic soybean production started declining several years ago as low-cost production began to increase in developing countries. However, U.S. cropland for wheat is still expanding, even as organic wheat production grows rapidly in the Ukraine and other parts of Eastern Europe.

5. Small farm organic initiatives

About 94 percent of all farms in the U.S. are considered small, with gross sales under $250,000 (Perry, 1998), and a survey of organic producers in California in the mid-1990s showed a similar proportion (Klonsky *et al.*, 2002). Most federal and state governments generally view organic initiatives as a mechanism to assist small producers. During the 1990s, U.S. policy on organic agriculture focused on facilitating consumer market access to a differentiated product, and national organic standards were developed during this period. More recent state and federal organic initiatives – expanding organic production and marketing research, technical assistance, and data development – are aimed at expanding market opportunities for producers.

Government research and policy initiatives often play a key role in the adoption of new farming technologies and systems. A number of federal agencies have expanded programs since the late 1990s to develop organic crop insurance, expand organic export programs and services, and broaden their intra-mural or inter-mural research on organic farming and marketing systems.

Congress also included several first-time research, conservation, and marketing assistance provisions aimed at assisting small organic farmers in the 2002 Farm Act, including cost-share funds to assist growers with the cost of organic certification, and the USDA recently proposed expanding a number of these provisions.

State support for organic farmers and handlers has also been expanding. For example, the number of States offering organic certification services – mostly at subsidised rates – has risen from 12 states in 1997 to 19 states in 2005. Several states, such as Minnesota and Iowa, began offering small subsidies for conversion to organic farming systems in the late 1990s as a way to capture the environmental benefits of these systems. The funds for these programs have mostly been from federal sources, by designating organic production as a priority for conservation cost share coverage under the federal Environmental Quality Incentives Program (EQIP) program. Additional states are now using or considering EQIP program funds for this objective. Also, at least one county – Woodbury County in Iowa – is now providing tax rebates for those who convert from conventional to organic farming practices. In 2003, the National Association of State Departments of Agriculture released a policy statement on organic agriculture expressing support for a wide range of activities that would expand public-sector organic research and education and provide technical assistance to organic and transitional farmers.

U.S. food companies are also developing innovative programs to encourage organic marketing opportunities for small farmers. For example, Whole Foods Market, the leading retailer of natural and organic foods in the U.S., announced several initiatives in 2006 to support local agriculture. The company supports weekly farmer markets in locations adjacent to their stores in many areas and developed and dedicated an annual budget of $10 million to offer long-term loans at low interest rates to support smaller scale agricultural entrepreneurs (Whole Foods Market, 2007). It is far too early to know the impact of these loans on small farms in the U.S., but Whole Foods Market has seen positive results when implementing similar loans through their Foundation in developing countries. Whole Foods Markets has encouraged individual and small groups of stores to develop on-going relationships with small, local farms for over 25 years.

6. Conclusions

During the process for implementing mandatory national organic standards for organic agriculture, the U.S. Department of Agriculture was concerned organic production could become more concentrated with larger farms if some small organic operations chose to exit the industry and others became reluctant to enter (USDA-AMS, 2000). Many U.S. organic farmers expressed similar concerns. However, since the USDA rules were implemented, data on U.S. organic agriculture shows that the smallest-scale farms continue to hold a small but stable piece of the organic sector, and organic farm size has grown, but fairly slowly. Average organic farm size is still much lower than overall farm sizes in the U.S. Overall, the U.S. organic farm sector is still steadily expanding, with cropland for fruits, vegetables, and many grains more than doubling between the late 1990s and 2005, despite rapidly increasing competition for global and domestic markets.

Gaps in the data prohibit an exhaustive description of the U.S. organic farm sector, and improved data collection is necessary to better monitor the effect of international trade and growing markets on small organic producers in the U.S. in the long run. However, progress is being made. USDA recently initiated a project to expand its annual economic survey of producers to include statistically-reliable samples of organic producers, and is working with other agencies to encourage the differentiation of organic and non-organic products as they enter and exit the country.

References

CCOF (California Certified Organic Farmers), 2006. CCOF Directory. Available at: www.ccof.org.

Fuchshofen, W.H. and S. Fuchshofen, 2000. Export study for U.S. organic products into Asia and Europe. New Lebanon, New York: Organic Insights, Inc.

Greene, C., 1992. Success steady in organic produce. Agricultural Outlook, 185: 15-17.

Greene, C. and A. Kremen, 2003. U.S. organic farming in 2000-2001: adoption of certified systems. Agriculture Information Bulletin No. 780. U.S. Department of Agriculture, Economic Research Service. February.

Hanson, J., R. Dismukes, W. Chambers, C. Greene and A. Kremen, 2004. Risk and risk management in organic agriculture: views of organic farmers. Renewable Agriculture and Food Systems, 19: 218-227.

Klonsky, K., L. Tourte, R. Kozloff and B. Shouse, 2002. A statistical picture of California's organic agriculture, 1995-1998. University of California Agricultural Issues Center. DANR Publication 3425.

Klonsky, K. and K. Richter, 2005. Statistical review of California's organic agriculture, 1998-2003. University of California Agricultural Issues Center. Available at: http://aic.ucdavis.edu/research1/organic.html.

Kortbech-Olesen, R., 2006. Demand for organic products from East Africa. CBTF Organic Agriculture Regional Workshop, Arusha, Tanzania, March.

Natural Foods Merchandiser, 1995. Organic market overview. Colorado, June: Boulder.

Nutrition Business Journal (NBJ), 2006. U.S. Organic food sales ($Mil) 1997-2010e – chart 22. Penton Media, Inc.

Oberholtzer, L., C. Dimitri and C. Greene, 2005. Price premiums hold on as U.S. organic produce market expands. Electronic Outlook Report No. VGS-308-01. U.S. Department of Agriculture, Economic Research Service. May. Available at: http://www.ers.usda.gov/Publications/vgs/may05/VGS30801/.

OTA (Organic Trade Association), 2006. U.S. organic industry overview, 2006 manufacturers survey. Available at: http://www.ota.com/pics/documents/short%20overview%20MMS.pdf.

Perry, J., 1998. Small farms in the U.S. agricultural outlook, AGO-251, Economic Research Service, U.S. Department of Agriculture. May.

Sligh, M. and C. Christman, 2003. Who owns organic? The global status, prospects, and challenges of a changing organic market. Pittsboro, NC.: Rural Advancement Foundation International – USA.

USDA–AMS (U.S. Department of Agriculture, Agricultural Marketing Service), 2000. National Organic Program. Federal Register Docket Number TDM-00-02-FR, December 21.

USDA–ERS (U.S. Department of Agriculture, Economic Research Services), 2006. Organic production 1992-2005. Data product: certified organic pasture and cropland. Available at http://www.ers.usda.gov/Data/Organic/.

USDA–FAS (U.S. Department of Agriculture, Foreign Agricultural Service), 2005. Linking U.S. agriculture to the world: U.S. market profile for organic food products. Commodity and marketing programs – Processed products division. February. Available at http://www.fas.usda.gov/agx/organics/USMarketProfileOrganicFoodFeb2005.pdf.

Whole Foods Market, 2007. Locally grown – the Whole Foods market promise. Available at: http://www.wholefoodsmarket.com/products/locallygrown/index.html; Accessed 17 January 2007.

Willer, H. and M. Yussefi, 2006. The world of organic agriculture: statistics & emerging trends 2006. International Federation of Organic Agriculture Movements (IFOAM), Germany, and Research Institute of Organic Agriculture (FiBL), Switzerland. Available at: http://orgprints.org/5161/01/yussefi-2006-overview.pdf.

Yussefi, M. and H. Willer, 2002. Organic agriculture worldwide 2002: statistics and prospects. Stiftung Oekologie & Landbau, Bad Durkheim. Available at: http://www.soel.de/inhalte/publikationsen/s_74_04.pdf.

The quality and training needs of Moroccan agri-food enterprises

M. Ismaili, M. Raggi and D. Viaggi

Abstract

A distinctive feature of present trends in international relationships is the importance of human and social capital, including training and personal growth of actors involved in the production process. The objective of this paper is to discuss the need for education and training in Moroccan agricultural and agri-food enterprises in light of the increasing connectedness with the EU economy. In particular, attention is focused on the need brought about by increased product quality requirements and by the installation of EU and US enterprises in Morocco. The study is carried out through a survey of 98 Moroccan enterprises. The results show a high degree of awareness about gaps and problems related to meeting EU consumer expectations. However, they also emphasise the need for training and information. Many of the respondents already show a clear positive strategy toward meeting such needs, and many enterprises are already compliant with many quality requirements. The main focus of the training required is on the interface between technical and marketing activities. The most preferred, and required, competences were: marketing managers, quality managers, administration and finance and food safety experts.

Keywords: education, management, quality requirements, enterprises, Morocco

1. Introduction and objectives

The Mediterranean region is experiencing increasing activity involving the exchange of goods and people. The perspective of a free trade area and the strong historical tradition as a zone of North-South and East-West interaction makes this area a particularly important node for future international trade, particularly for Italy.

A distinctive feature of present trends in international relationships is the importance of factors that are beyond the mere exchange of goods. In particular, foreign investment, the creation of international enterprises, training and personal growth of actors that are involved in the production process are key elements in the present scenario. The development of future markets and economic opportunities will depend on the structure taken by trans-national networks, and on the ability to build human and social capital able to connect different economic areas of the world.

The objective of this paper is to identify the need for education and training in Moroccan agricultural and agri-food enterprises in light of the increasing connectedness with the EU economy. In particular, attention is focused on the need brought about by the increased product quality requirements and by the installation of foreign (EU and US) enterprises in Morocco.

The study is purposefully descriptive in nature and based on empirical information derived from a survey of Moroccan enterprises. The survey was carried out as a preliminary activity of the TEMPUS project 'STRIDE 4 Développement d'un nouveau Master en Management Agricole, (2004-2007) CD_ JEP 31019-2003'.

In section 2, a short overview of the issue of food chain development and training is provided. The methodology adopted in this paper is illustrated in section 3, followed by the results in section 4. A short discussion is provided in section 5.

2. Background: human resources, training and the development of agri-food chains

The development of international relationships and markets is accompanied by a profound change in the structure of food chains. Beyond delocalisation and specialisation, an evolution of production networks can be observed.

A relevant issue in the evolution of such networks is the development of human resources. This is recognised as an important point in EU enlargement and integration with neighbouring economies, as well as in the literature on business development in the last decade. A key area of attention is the role of education and training in the creation of business networks (e.g. Butera, 1997). Different papers highlight and discuss the role of training in competitiveness (e.g. Mumma et al, 2000; Jiayanthi et al., 1996; Jatib et al. 2003). More in detail, Wang (2003) emphasises technology innovation and human resource management as determinants of organisational performance. The specific role of training in relationships with quality management is discussed by Reardon and Farina (2002).

The complexity of the issue was widely experienced and discussed in the process of transition and enlargement of Eastern Europe. On one hand, the issue of labour costs is a key driver of investment. On the other hand, local and foreign personnel require a strong learning process in order to deal with the changing working environment, and to collaborate to govern and orient such changes (EBRD, 2001a,b). The need for human capital development goes beyond the simple transfer of knowledge, and involves the whole strategy of businesses as well as the evolving role of the public administration (Viaggi, 2002, 2003).

Key topics include the use of available local knowledge, often abundant in Morocco, as well as in many Southern Mediterranean and Eastern European countries, and the life cycle of knowledge, that render education outdated and require continuous targeted training activities. In this context, the identification of focused training needs is a key factor for system success, as it fulfils the need to address scarce resources for knowledge bottlenecks and ensures timely development of competences required.

At the same time, this issue touches the complexity of the relationships between firms and training institutions. This has been the subject of a strong evolution in recent years, though the degree of collaboration does not often appear satisfactory. Different patterns of interaction between enterprises and education systems are in place, and may be adapted from case to case to deal with specific vocational training activities (Chen et al., 2004).

When enterprises already pursue growth strategies and are aware of their role as proposers of training activities, their role in training becomes somehow more complex. On one hand they can identify needs and expertise required. On the other hand they may offer up their know-how through teaching or accepting students in training internships. Finally, they may contribute by suggesting organisational structure for training activities (timing, characteristics of entrants, etc.). In a longer term perspective, the creation of stable links between enterprises and educational institutions is a strategy used to support the types of participation described above. This also includes understanding the reciprocal specificities, routines and potentialities between enterprises and training institutions.

3. The Moroccan agri-food system

Morocco is an important country in the development of agri-food chains in the Mediterranean region. Internally, agriculture plays a major role in the Moroccan economy. In 2004, agriculture accounted for 16% of the GDP, and about 40% of the labour force. Agriculture grows at approximately the same speed as the other sectors of the economy, albeit with ups and downs due to a large extent to climate conditions. In 2004 agriculture and food accounted for about 20% of exports, and 9% of imports. Exports from agriculture mainly consist of fruit and vegetables. The main export partners are France and Spain (World Bank, 2005; CIA, 2006).

Morocco is an obvious target for southern EU countries, both as a trade partner and as a place for delocalisation of agricultural production, particularly fruit and vegetables. Morocco, for its part, strongly encourages external investments.

The result is a growing network of local enterprises with commercial connections in Europe, foreign enterprises producing in Morocco, and mixed companies. In most cases, the reference market for these companies is Europe, as far as quality standards and consumer expectations are concerned.

4. Methodology

The analysis is carried out through a survey of Moroccan enterprises. The survey was carried out in 2006 on a sample of 98 enterprises. The sample includes mainly enterprises that are technologically and strategically advanced and that have already important links with markets. For the most part, the firms interviewed are production factor and agri-food enterprises, rather than farms. Small and subsistence enterprises were not considered in this survey.

The selection of the 98 enterprises was undertaken in two parts. In the region of Meknes – Tafilalet (Centre - Southern Morocco), all enterprises known to work in the area were contacted. Only a few enterprises did not collaborate. Record of the enterprises refusing to answer was not taken. However, the number can be considered negligible compared to the total population. Therefore, the sample in the region reflects almost the totality of the population in this area and the sample can be considered very representative of the region. 28 enterprises were selected in this area. In addition we selected other enterprises from other regions such as: Casablanca, Agadir, Marakech, Oujda, Fes. In these cases only a small sample was considered from each region. Enterprises were selected based on personal knowledge of the staff of the University of Meknes involved in the project. This is justified by the difficulty of entering in contact and collecting reliable answers from firms that lack a direct link with the interviewers. Within available contacts, a 'convenience' criterion was used, based on the attempt to cover the variety of business specialisations.

Depending on the size of the enterprise, we interviewed either the director, or one of the managers.

The entire process of the survey took six months and the collaboration of the entrepreneurs was positive, though in some cases it took a long time to persuade managers to accept the interview. Before being interviewed, the enterprises were sent an official letter in which we explained to them the objectives of the research, and assured them of the importance of the survey for University development and training in Morocco.

The questionnaire includes mostly multiple choice questions, though space was also provided, at the end, to openly collect the opinions of the respondents. The main parts of the questionnaire include: the description of the enterprise and its activities, the present strategy of the enterprise and its economic results, its vision regarding future opportunities and problems accessing new markets. Part of the questionnaire deals directly with enterprise activities concerning human resources. In particular, it addresses the enterprise's training policy and future training requirements in terms of preferred competences and means of interaction between educational institutions and the enterprises.

For some specific questions we found it difficult to obtain precise information and in this case we focused on a general view of competences as seen from a firm's organisational perspective.

5. Results

Those interviewed represent enterprises working, for the most part, in multiple fields of activity, with a prevailing focus on agri-food, farming and related activities (Table 1). In this study we interviewed mainly large enterprises, which employ between 50 and 500 persons (Table 2). In the majority of cases the interviewees were satisfied with the respective economic results, which also confirms the general economic trend in agri-food-related activities (Table 3).

There does not appear to be a clear relationship between size and economic results, though medium to large enterprises seem to have a higher share of answers in the range of 'satisfactory' to 'very satisfactory' (Table 4).

Table 1. Main fields of activity of the enterprises interviewed.

Activities	n	%
Research	23	23.5
Development	28	28.6
Activity related to agriculture	58	59.2
Agro-industries	72	73.5
Agricultural cooperative	28	28.6
International relationship	42	42.9

Table 2. Size of enterprises in terms of employees.

Workers	n	%
<20	17	17.3
20-50	5	5.1
50-100	23	23.5
100-500	19	19.4
>500	18	18.4
No answer	15	15.3
Total	97	99.0

Table 3. Current economic results of the enterprise.

Current results	n	%
Very satisfactory	15	15.3
Satisfactory	42	42.9
On average satisfactory	31	31.6
Not satisfactory	5	5.1

Table 4. Relationship between economic results and number of workers.

Current results	Workers						
	<20	20-50	50-100	100-500	>500	No answer	Total
Very satisfactory	2		3	5	1	4	15
Satisfactory	5	10	6	12	4	5	42
On average satisfactory	9	7	9	4		2	31
Not satisfactory		2		2		1	5

Expectations regarding increased opening of markets reveal a prevailing positive opinion, for about 50% of the interviewees (Table 5). Only 20 respondents see potential negative effects prevailing, and 10 expect no change. However, results become more complex when referring to more specific effects of market opening (Table 6).

For the specific effects of the opening markets, the data shows clear divergences, which appears in less positive expectations regarding work quality and quantity, as well as resource availability. New technologies, quality of life and training, as well as economic and policy relationships with other countries are the issues on which positive expectations appear most evident. Notably, changes in international relations are stronger for non-neighbours.

Training is perceived as important and most of the interviewees show interest in multiple aspects of training (Table 7).

Table 5. General effects of open markets.

Open market effect	n	%
Positive	48	49.0
Negative	20	20.4
None	10	10.2

Table 6. Specific effects of open markets.

Forecast (row %)	Reduction	Small reduction	Stable	Small increase	Increase
Export	2.50	2.50	32.50	15.00	47.50
Import	2.17	2.17	21.74	30.43	43.48
Work quantity	12.31	7.69	30.77	18.46	30.77
Work quality	3.03	1.52	25.76	24.24	45.45
New technology	0.00	0.00	19.12	38.24	42.65
Water availability	14.29	14.29	41.07	7.14	23.21
Life quality	1.92	7.69	40.38	23.08	26.92
Training	1.61	0.00	24.19	40.32	33.87
Economic and policy relation with EU	1.92	1.92	17.31	38.46	40.38
Economic and policy relation with Maghreb countries	10.64	2.13	40.43	23.40	23.40
Economic and policy relation with Arabian countries	10.00	2.00	34.00	30.00	24.00
Economic and policy relation with other countries	2.04	2.04	28.57	30.61	36.73

Table 7. Interest for training-related activities.

Training	n	%
Organise training course	36	36.7
Contribute to training course	57	58.2
Training partecipation list	58	59.2
Accept stageir	73	74.5
Relation with training institute	65	66.3
Free a cadre to attempt a course	61	62.2
Engage a cadre	18	18.4

In general, enterprises are interested in accepting students in training internships, establishing relationships with training institutions, or allowing employees to attend training courses. Enterprises are also willing to contribute to training courses. A number of enterprises contributed to training in Universities, by sending managers to teach courses, give seminars and conferences.

The choice is not clearly differentiated between the size of the enterprises, and accepting students for training internships tends to be consistently the main option (Table 8).

For all enterprises interviewed, the most preferred and required competences were: marketing managers, quality managers, administration and finance and food safety experts (Table 9).

The main focus of the training required is on the interface between technical fields such as soil fertility, plant pathology, water management and irrigation methods, and marketing activities

Table 8. Relationship between preferred training activities and number of employees.

Training	Workers						
	<20	20-50	50-100	100-500	>500	No answer	Total
Organise training course	7	7	5	10	1	6	36
Contribute to training course	13	11	10	12	2	9	57
Training partecipation list	11	11	12	13	1	10	58
Accept trainee	13	14	15	20	2	9	73
Relation with training institute	11	10	13	18	2	11	65
Free a cadre to attempt a course	11	8	14	16	2	10	61
Engage a cadre	3	4	4	5	1	1	18

Table 9. Most preferred competences required for trainees.

Preferred competence	n	%
Marketing manager	43	43.9
Administration and finance	35	35.7
Development expert	20	20.4
Audit expert	16	16.3
Agronomic expert	26	26.5
Environmental expert	8	8.2
Quality manager	37	37.8
Food safety expert	29	29.6

related to the quality of products and the development of new products, hence showing the need for consistent development of production and commercial activities. Training and education are expected to help in forming people through enterprise-university interaction, through research and development programs, and by making use of alternative education instruments, such as training internships and continual links between regions and universities. Training programs to upgrade the level of education of enterprise managers, and life-long learning of employees are also perceived as major needs.

Enterprises requiring quality managers are more often those who are conscious of the needs of European markets and consumers in terms of quality products, and environmental impacts of modern agriculture, and those who believe that liberalisation will bring stronger integration with the EU. This may hint at the idea that quality and quality-related training is possibly to be considered as a very biased strategy related to access to EU markets.

Also, enterprises requiring quality managers are the ones that offer to contribute to training courses as teachers and enable and encourage their employees to attend training courses, seminars and conferences. This second feature suggests that training in the fields related to quality and traceability may require more interaction and the merging of competences between enterprises and educational institutions than any other fields.

6. Discussion

This paper focused on the empirical investigation of training needs of Moroccan enterprises facing market liberalisation. Most of the results tend to confirm issues and priorities available from the literature and incorporated in policy actions concerning training and education.

The survey reveals that Morocco is characterised by an interesting business environment, where the cluster of firms included is satisfied with current results and deems that economic integration with the EU mostly involves opportunities

The enterprises show a high degree of awareness about gaps and problems in meeting EU consumer expectations. At the same time, most respondent firms already show a clear positive strategy towards meeting such needs, and many enterprises are in fact already compliant with many quality requirements.

The awareness for training needs is relatively high and supported by on-going experiences. This is confirmed to be a multifaceted issue, where firms interact with educational institutions through different combinations of traditional training and internships, providing students as well as teachers. In addition, the interviewees also highlight interest in developing network connections with educational institutions, through the exchange of pupils' records and by establishing formal connections between universities and enterprises. This opens the perspective of a challenging field of activity in which innovation in human resource development is required, and in which training directly fits into the mechanisms of interaction between firms and their environment.

The connection between training and quality issues is clearly highlighted by the high degree of priority attributed to quality issues. However, this goes together with the high priority for marketing management and administration, emphasising the need for a comprehensive development of human resources in the various fields of management.

The empirical results, however voluntarily descriptive, open the way for further research activities aimed at qualifying the role of training with respect to quality enhancement in relation to different types of enterprises, as well as the different consumer perceptions on specific issues and firms' international strategies. In particular, higher attention should be paid to the potentially varied patterns of interaction between educational institutions and firms and to innovation in this field. This could be investigated more in deep as a component of the business networks and as factor of innovation and success of such networks.

Developments from further research would be particularly useful in the area investigated, when connections between university and enterprises are still rather poor. In the background, the research shows the need for a more consistent understanding of each other's institutions and cultural settings, as well as a long-term process for learning through collaborative training, production and research.

Acknowledgments

The authors wish to thank the EU for financing this activity under the TEMPUS Stride 4 project. The authors also thank the coordinator (CEFAL, Bologna) and all the people involved in the Tempus STRIDE 4 project, in particular all of those who collaborated in the survey.

References

Butera, F., 1997. Il castello e la rete: impresa, organizzazioni e professioni nell'Europa degli anni '90. Milano: Franco Angeli.

Chen, S.H., H.T. Lin and H.T. Lee, 2004. Enterprise partner selection for vocational education: analytical network process approach. International Journal of Manpower, 25: 643-655.

CIA, 2006. The world factbook – Morocco. Available at: www.cia.gov.

EBRD, 2001a. How do foreign investors assess the quality of labour in transition economies? Results from a postal survey. Bruxelles.

EBRD, 2001b. Transition report 2001. Bruxelles.

Jatib, M.I., F. Vitella, H. Ordoñez, G. Napoletano and H. Palau, 2003. Agribusiness executive education and knowledge exchange: new mechanisms of knowledge management involving the university, private firm stakeholders and public sector. International Food and Agribusiness Management Review 5.

Jiayanthi, S., B. Kocha and K.K. Sinha, 1996. Competitive analysis of US food processing plants. Working paper 96-04, The retail food industry.

Mumma, G., A.J. Allen and W.C. Couvillion, 2000. An analysis of selected performance indicators for U.S. agribusiness sites registered to 1S0 9000 series of standards. Journal of Food Distribution Research, 31: 225-235.

Reardon, T. and E. Farina, 2002. The rise of private food quality and safety standards: illustrations from Brazil, International Food and Agribusiness Management Review, 4: 413-421.

Viaggi, D., 2002. Countries in transition and the role of education. In: Viaggi, D. (ed.) Qocun: a centre for University/enterprise co-operation in Albania. Bologna: Enterprising, pp. 33-44.

Viaggi, D. (Ed.), 2003. Institutions and development: an analysis of local public administration in Albania, STRIDE 1, Bologna.

Wang, Z., 2003. Organizational effectiveness through technology innovation and HRM strategies. International Journal of Manpower, 24: 501-516.

World Bank, 2005. Morocco at a glance. Available at: web.worldbank.org.

Part 3
Food quality and consumers

Determinants of consumer preferences for regional food products

M. Henseleit, S. Kubitzki and R. Teuber

Abstract

Over the last years there has been increasing interest in regional food, in Germany as well as in other European countries. Regression models investigating this region-of-origin effect are rare, and in most cases the region or sample size under consideration is quite small. The present study is based on a representative data set for Germany. Our objective is to identify and quantify the determining factors of consumers' preferences towards locally grown food. Therefore, a theoretical framework is proposed and tested empirically using a binary logit model. The results indicate that cognitive and normative factors are the main determinants, whereas affective and sociodemographic variables do not have a big impact in determining the preference towards local food products. If consumers are of the opinion that originating from the surrounding region is an extrinsic cue for food quality and safety, they will show strong preferences for locally grown food. The same is true for the idea to support the domestic agriculture by purchasing locally grown food. No significant influence could be examined for most of the sociodemographic variables, like gender, education, presence of children in the household and degree of urbanisation.

Keywords: consumer preferences, region-of-origin, regional food, binary logit model

1. Introduction

Regional food is defined as food, which is grown in the surrounding region, and, which is usually unprocessed (Dorandt, 2005)[53]. In Germany most of the consumers define their home federal state as their home region (ZMP, 2003: 9ff.).

Over the last years there has been an increase in preference for regional food, in Germany as well as in other European countries. Several studies have already been carried out on this phenomenon. However, in most surveys either the study region is relatively small or the sample size is rather limited. Consequently, the results are seldom statistically representative. In addition, only few researchers applied causal analytic methods like regression analyses to investigate the so-called region-of-origin (ROO) effect. Thus, the level of knowledge about the main reasons and the magnitude of preferences for regional food is still quite low.

In Germany as well as in many other European countries, regional cooperatives had been established to promote the sale of regional food. It is important for them to understand determinants of preferences for locally produced food in order to promote regional products successfully. The aim of our research is to shed some light on the discussion regarding the factors for the preference for regional food. Based on a German-wide data set we quantify determinants by means of a binary logit model.

The paper is structured as follows. Section two provides a literature review of studies investigating the preference towards regional food by applying regression analyses. In section three we explain the theoretical framework of our research and in section four the data set used is described. A summary of our empirical results is given in section five and the last section contains our conclusions and recommendations for further research.

---

[53] Local food and locally grown food are used as synonyms for regional food products.

2. Literature review

Many researchers have already tried to characterise a consumer segment with strong preferences for food from their home area (e.g. Dorandt, 2005, Schroeder *et al.*, 2005). However, only a few studies have applied advanced econometric methods to estimate the determinants of preferences towards local food. Most of these have been conducted in the United States and the majority focus on either psychographic or sociodemographic factors. Only few studies have considered a broad range of possible determinants. The following section presents an overview of the empirical findings from these studies.

2.1 Psychographic determinants of preferences for local food

Table 1 presents a review of studies considering psychographic indicators[54]. It is quite difficult to compare studies due to differences in research subjects like location and kind of product as well as in methodological aspects like the choice of measures[55]. Usually the studies examined either impact factors linked to quality and food safety, or social norms that should be accomplished, or emotional aspects of pride and regional identity.

The majority surveyed whether consumers perceive the regional product origin as a cue for product quality, food safety and health. There seems to be a common theme in that consumers' perceive regional food to be linked to higher food safety as well as to higher quality and therefore local food is preferred to other products. The indicator *quality* is significant in six out of nine studies (e.g. Jekanowski, 2000; Van Ittersum *et al.*, 2003; Lobb *et al.*, 2006), and the indicator *food safety* in seven out of nine studies (e.g. Schupp and Gillespie, 2001; Roosen *et al.*, 2003; Mabiso *et al.*, 2005), respectively.

Only two studies (Wirthgen *et al.*, 1999; Wirthgen, 2003) consider social norms. In both of them, *environmental concerns* and the willingness to *support the local economy* are the main determinants of the preference towards regional food. However, it has to be mentioned that both studies did not include characteristics of quality in their estimations. Emotional aspects had been identified as impact factors of attitudes and purchasing behaviour in a number of studies (e.g. Van Ittersum, 1999; Wirthgen, 2003; Schroeder *et al.*, 2005). But the estimated impacts could be biased due to the fact that no quality indicator was included in the regression analysis. According to Table 1 it can be concluded that, so far, no survey has been conducted that includes all the mentioned psychographic factors. Moreover, since each study considered only some aspects in the regression models, a quantitative comparison of the determinants of preferences towards regional food was not possible.

2.2 Sociodemographic determinants of preferences for local food

Table 2 presents an overview of studies considering sociodemographic indicators. *Age, sex, income, education* and the *number of children per household* are the most frequently surveyed factors. Both the impact of the time that the respondent has been a resident of the region (*lifetime*)

---

[54] Of course there are a number of more consumer country (region)-of-origin (C(R)OO) studies related to food. However, we selected the mentioned studies according to our primary focus that lies on investigations that fulfil three requirements: Firstly they have to apply advanced econometric estimation techniques. Secondly, they have to deal with own COO/ROO and, thirdly they have to focus on the explanation of the determining factors instead of simply measuring the extent of the preference towards, or willingness-to-pay for, C(R)OO.

[55] A detailed description of the design of the studies is available upon request.

Table 1. Psychographic determinants of the preference towards regional food - review of empirical studies.[1]

Author (year)	Cognitive				Normative		Affective
	Quality		Food safety	Health, nutrition	Environment-friendliness	Support of economy	Sympathy, image
	In general	Freshness					
Van Ittersum (1999)							+/+
Wirthgen et al. (1999)							+
Jekanowski et al. (2000)	+					+	
Schupp et al. (2001)	n.s.		+				
Loureiro and Hine (2002)		n.s.		+			
Loureiro and Umberger (2003)			+/n.s.				
Wirthgen (2003)			n.s.	+	+	+	
Van Ittersum et al. (2003)	+/+		n.s.				n.s./+
Roosen et al. (2003)			+				
Umberger et al. (2003)		+	+				
Loureiro and Umberger (2005)			n.s./n.s./+				
Schroeder et al. (2005)	n.s.	n.s.	+		n.s.		n.s.
Mabiso et al. (2005)[a]	+		n.s.				
Mabiso et al. (2005)[b]	+		+				
Lobb et al. (2006)[c]	+	+				+	

Notes:

[1] A description of the study designs is available upon request.

(+;-) positive and negative estimates refer to significance level of at least 0.10; (n.s.) if found to be not significant; If nothing is specified this variable was not included in the study. If several results are listed for one study this is due to different products under consideration.

[a] probit model; [b] logit model; [c] ordered probit model.

Table 2. Sociodemographic determinants of the preference towards regional food: review of empirical studies.

Author (year)	Age	Lifetime	Women	Income	Education	HH	Kids	Urbanisation
Patterson et al. (1999)	n.s.	n.s.	n.s.	n.s.	n.s.		+	
Jekanowski et al. (2000)		+	+	+	-	n.s.		n.s.
Schupp et al. (2001)	-		+	n.s.	n.s.	-[a]	-	-
Loureiro and Hine (2002)	n.s.		n.s.				n.s.	
Wirthgen (2003)[b]	+	n.s.			n.s.	n.s.		
Loureiro and Umberger (2003)			+/+	-	+/n.s.		+/n.s.	
Umberger et al. (2003)	n.s.		n.s.	-	n.s.		n.s.	
Mabiso et al. (2005)[c]	n.s.		n.s.	n.s.	n.s.		n.s.	
Mabiso et al. (2005)[d]	-		n.s.	-	n.s.		n.s.	
Loureiro and Umberger (2005)	-/n.s./n.s.		+/+/+	+/+/n.s.	-/-/n.s.		n.s./ n.s./-	
Lobb et al. (2006)[e]	+							
Lobb et al. (2006)[f]	+/+		n.s./n.s.	+/+	n.s./-		+/n.s.	-/n.s.

Notes: lifetime= lifetime in the local region; HH= household size; Kids= presence of children in the household; (+; -) positive and negative estimates refer to significance level of at least 0.10; (n.s.) if found to be not significant. If nothing is specified, this variable was not included in the study. If several results are listed for one study, this is due to different products under consideration.
[a] 1 = single household head; 0 = otherwise.
[b] Wirthgen (2003) also estimates product specific models besides the general regression. In some regressions the variable 'lifetime in the region' instead of 'age' is significant. Both factors are strongly correlated.
[c] probit model; [d] tobit model; [e] ordered probit model; [f] conditional logit model.

and the degree of urbanisation (urban versus rural areas) have not been considered as factors of preference in most of the studies. Correlation analyses and non-parametric methods have shown significant relationships between sociodemographic variables and preferences for food products from their own region (e.g. Wirthgen *et al.*, 1999; Dorandt, 2005). However, causal analyses have rarely shown statistically significant impacts. Furthermore, there is no consistency among causal analyses regarding the direction of influence of *age, income, education,* and the *number of children* on preferences. There are only consistent results regarding to the influence of *sex* on the preference towards locally grown food: women have been shown to have a higher preference for regional food than men do. All in all, the results concerning the influence of sociodemographic factors on the preference for regional food are not consistent across different studies. Moreover, the results confirm observations of Mabiso *et al.* (2005), that sociodemographic factors have only a marginal effect on the preference towards regional food.

3. Theoretical framework

3.1 Psychographic determinants

As mentioned above, the studies presented in the former section do not consider the full range of possible determinants for consumer preference for regional food in their causal analyses. Obermiller and Spangenberg (1989: 456ff) propose a theoretical framework, which gives an overview of the plurality of the factors that influence the effects of country-of-origin labels on consumer behaviour. Von Alvensleben (2000a: 6ff.) applied this concept to the region-of-origin-effect and grouped the determinants into cognitive, normative and affective processes. Figure 1 presents the theoretical framework of the psychographic determinants of the preference for regional food.

Cognitive factors

Consumers who are unsure about the quality of a product might use the geographical origin as a quality cue. This effect may result from two processes. First, the region of origin is a 'signal' for the general product quality (Verlegh *et al.*, 1999). Based on this, there might be a positive bias in the consumer's perception of other attributes that are not necessarily linked to the region-of-origin. Second, locally grown food is perceived to be fresher, healthier and more environment-friendly (Darby *et al.*, 2006: 2ff.).

Normative factors

Regional food can also be preferred due to norms and values. Both societal and personal norms, resulting from environmental values, patriotism and the aim to support local businesses, may influence the demand for regional food. Norms can cause a purchase decision independently of cognitive and affective processes. Van Ittersum, (1999: 46ff.) specifies this theory by the assumption that the demand for regional food is influenced by 'consumer ethnocentrism' which is defined as the beliefs consumers hold about the moral appropriateness to favour domestic products (Shimp and Sharma, 1987: 280ff.). Consequently, consumers feel constrained to support the local economy by their selective purchase decision.

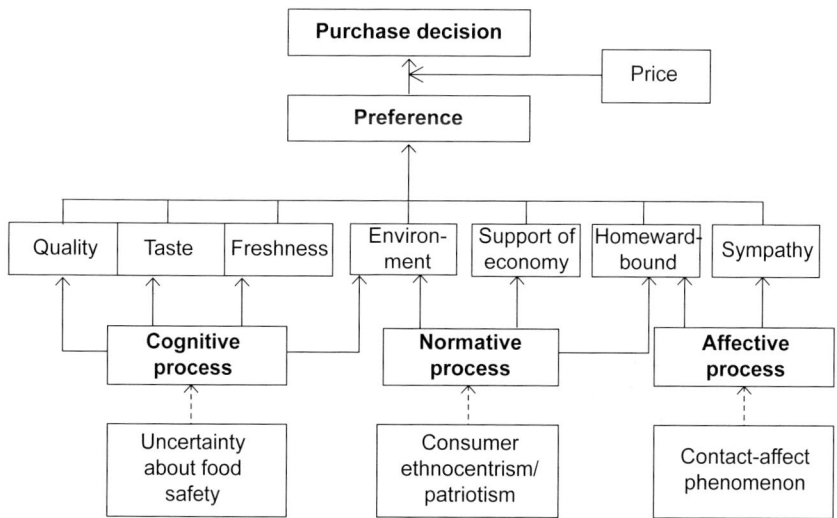

Figure 1. *Theoretical framework of the psychographic determinants of the preference towards regional food.*

Affective factors

In addition to norms and values emotional aspects might influence the demand for regional food as they are interconnected in some way with ethnocentric and patriotic issues. Emotions like pride and sympathy towards the own region may be transferred directly to the product. Von Alvensleben (2000a) assumes that sympathy to the region leads to a positive bias in the perception of the product and its attributes. The contact-affect-phenomenon is discussed as the cause of this positive image transfer from the region to the product. The mere contact to an object leads to familiarity and finally to sympathy to the object (Von Alvensleben, 2000b: 401).

The three described processes can hardly be regarded separately since they are overlapping and interacting. They are affected by the individual perception of quality indicators, personal confidence in the source of the information and by situational conditions like the heterogeneity of products and the general availability of other information (Obermiller and Spangenberg, 1989: 455ff.). Furthermore, there is a strong interdependence with demographic factors.

3.2 Sociodemographic determinants

There seems to be no consensus about the influence of sociodemographic factors on the preference towards local food and its psychographic indicators so far (see section 2b). Therefore, in the following paragraph the influences of different sociodemographic factors on the preference towards regional food are derived theoretically.

Age may have a positive impact on the preference. On the one hand, elderly consumers usually tend to be more closely connected to their home region (Dorandt, 2005), have more time for purchasing and preparing food, and are more concerned about health issues. Furthermore, age is often correlated with time spent in the home region, which in turn encourages emotional ties to the region (Wirthgen, 2003). On the other hand, elderly consumers tend to be less flexible in the food items they accept (Schupp and Gillespie, 2001: 38) and usually they are less concerned about environmental issues and the impact of pesticides on food (Loureiro and Hine, 2002: 484). The latter considerations give reason to expect a negative impact of age on the preference for local food.

Males are considered to be less interested in nutrition and health issues than females (Patterson *et al.*, 1999; Schupp and Gillespie, 2001). This leads the hypothesis that women tend to prefer food from the own region more than men do.

Consumers with high income tend to desire a larger variety of food in the marketplace, whereas regional products can enhance the variety (Schupp and Gillespie, 2001: 38ff.). However, some authors like Umberger *et al.* (2003: 111ff.) found a significant negative sign for the income coefficient. It is assumed that wealthier consumers usually buy more expensive food since they normally expect it to be of higher quality. In this case, the price is more important as a quality cue than the origin of the product. Moreover, wealthier consumers buy foreign delicacies more often, and, therefore products from the home region are not always their first choice.

Consumers with higher levels of education are expected to evaluate products rather by personal experience and by the price, rather than by brand names or labels of origin. Thus, a negative impact of education on the preference for regional food is expected. Opposite to this, higher education could lead to an increased awareness of the external effects of food consumption, which could positively influence the demand for regional products.

The presence of children in a household can have both positive and negative effects on the preference for local food products. On the one hand, parents are concerned about the safety and quality of food for their children, and thus they are more interested in food quality and safety (Patterson *et al.*, 1999: 187). On the other hand, families have to deal with time and budgetary constraints. This could reduce the efforts to buy locally produced food (Schupp and Gillespie, 2001: 38).

Further, the geographical location and the degree of urbanisation are supposed to explain the preference for regional food to some extent. Consumers living in urban residences may spend less attention to food from the own region, because they are less connected to local agriculture. Besides the supply of locally grown food is more constrained in urban than in rural areas (Lobb *et al.*, 2006). Consumers in rural areas may be more appreciative of locally produced food (Jekanowski *et al.*, 2000: 47ff.). It is hypothesised that the degree of urbanisation has a negative impact on the preference for regional food. Additionally, we assume that consumers in the southern and eastern states of Germany have higher preference for regional food than consumers in other parts of Germany. This assumption is based on two reasons. First, the agricultural sector in southern Germany is mainly small scaled, and thus a closer connection between farmers and non-farmers is expected. Second, in southern Germany more fruits and vegetables are produced, which can be sold without further steps of processing. Thus, they are usually sold close to the production area. In the north of Germany, there are comparatively more arable farms which are more industrialised. Third, a return to local products, which have been popular in the former German Democratic Republic, can be observed in eastern Germany (Ahbe, 2005).

4. Data and methodology

The Official Marketing Board of the German Agricultural and Food Industry funded a German-wide consumer survey which was conducted in October/November 2002[56]. The sampling frame is households with telephone services. The respondent should be the person in charge of food purchasing and the survey was carried out via a telephone interview. Respondents were selected using a random stratified sampling strategy. The population was sub-divided according to the federal state the respondents live in, and separate random samples were drawn from each state using random-digit dialing procedures. Small states were over-sampled, but the cases were weighted to reflect the actual population in the federal states. 3000 questionnaires were completed. During the data collection the sample was controlled automatically in terms of the representative distribution of the parameters, i.e. location of residence, age, and gender.

The questionnaire consisted of two parts; the first part aimed to identify the determinants of preferences towards local food, whereas the second part focused on specialty food products. We used the data of the first part, which contained questions about:
- the respondents' personal understanding of the meaning of the term 'home region';
- the respondents' sympathy towards the own region of residence;
- the respondents' purchasing habits of food in general and locally grown food products in particular;
- the respondents' motives and barriers of purchasing locally grown food products.

As shown in Table 3, the actual sample is somewhat biased towards female, middle-educated, and employed categories of the German population. The gender imbalance exists because

---

[56] The Official Marketing Board reported descriptive results of the survey in 2003 (ZMP, 2003) and provided the data set for advanced scientific purposes.

Table 3. Descriptive statistics of the demographics of the sample (N=3,000).

Category	Percent	
	Sample[a]	German population
Sex		
Female	78.4	51.1
Male	21.6	48.9
Age		
<20	1.3	4.2
20–39	34.4	32.9
40–59	37.3	33.3
60–79	25.5	24.9
≥80	1.4	4.6
Household size		
1	17.4	36.7
≥2	82.6	63.3
Children		
No children	44.4	43.4
Children	55.6	56.6
Education		
No formal education	0.3	7.9
Lower secondary school I (age 14-16)	27.6	45.3
Lower secondary school II (age 15-16)	40.7	26.7
Higher secondary school (age 18-20)	22.6	20.1
University degree	8.8	11.2
Employment status		
Employed full time and part time	56.5	46.0
Unemployed (including economically inactive population)	43.5	54.0
Household Income		
≤3,000 €/month	66.3	42.4[b]
>3,000 €/month	8.6	53.1[b]
refused	25.1	4.6[b]

[a] The data are weighted according to the regional distribution of the population in the federal states of Germany.
[b] Data are related to year 2000.

the respondent was the person responsible for carrying out food purchasing. Single person households are substantial underrepresented in the sample. The same is true for the households with a monthly income of 3,000 € and more, whereas one-quarter of the respondents refused to answer the income question. There is no bias in the age categories and in the presence of children in the household. The level of higher-educated respondents is also approximately equal to the level in the German population.

4.1 Measuring the preference for local food products

Within the interviews, the preference for locally grown food products was taken by a seven-point Likert scale ranging from '*I completely disagree*' (1) to '*I completely agree*' (7) to two alternative statements. The first statement '*If possible, I try to buy local products*' was coded as Preference 1 (P1) and the second statement as Preference 2 (P2): '*I am willing to pay a price premium for local products*'. Figure 2 presents the frequencies in the response categories of the two statements. The dispersion of the responses in the categories is quite uneven, because the majority of respondents (P1: 88%, P2: 80%) rather agreed with the statements. The hypothetical formulation of the statements without any real consequences for the respondent seems to cause a significant 'warm-glow' effect. Warm-glow specifies the moral satisfaction of a certain action or behaviour. It occurs whenever people get involved with public affairs because of the feeling of being a good citizen rather than due to the matter itself (Henseleit, 2006: 41). Social desirability bias is the inclination to present oneself in a manner that will be viewed favourably by others. We decided to transform the statements expressing the preference for regional food into binary variables coding the first two values (top-two-values) of the Likert scale as 1 and the remaining values as 0. This transformation should separate respondents with strong preference for locally grown food from the remainder.

Since the dependent variable is dichotomous, standard multiple regression is not applicable. Therefore, we applied binary logit regression analysis as an appropriate technique to handle the dichotomous nature of the dependent variable. Table 4 presents descriptive statistics of the items measuring the preference for local food products.

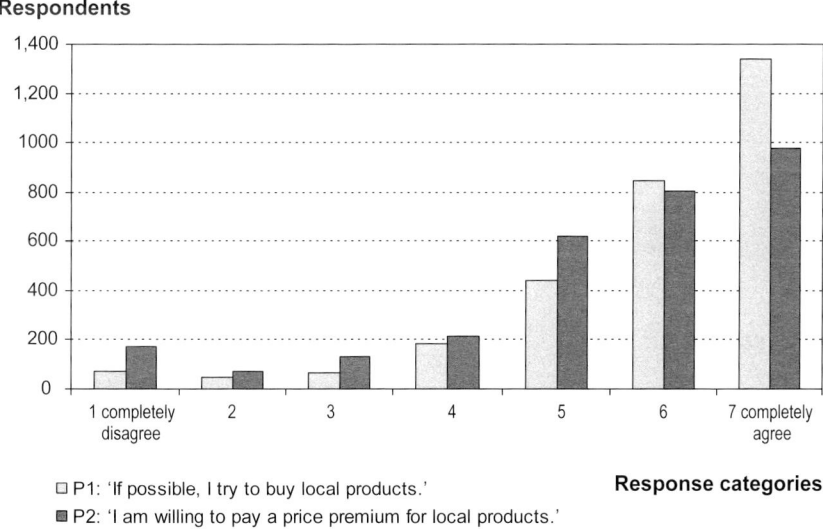

Figure 2. Response frequencies towards Preference 1 and Preference 2.

4.2 Estimating parameters of the preference for local food

Psychographic (cognitive, normative and affective) factors

Based on cognitive processes consumers may use the products' origin as a quality indicator. Thus, several items expressing the perception of product attributes and food safety were included to represent cognitive factors. Further on, affective processes can influence consumers' product evaluation. Sympathy to the own region is directly transferred to the food product. In our analysis, items, which express the sympathy to the own region and to the local food supply, are defined as affective factors. Normative aspects can also influence the preference for local food products. Statements, which express the environmental friendliness and the support of the local economy by purchasing local food, were used to define normative factors. Descriptive statistics of the items measuring the psychographic determinants of the preference for local food are presented in Table 4. The statements were measured on a seven-point Likert scale. In the logit analysis we transformed the scale into binary dummy variables as we did for the preference items.

Consumption and shopping habits

Besides psychographic factors, purchasing habits might influence the preference towards local food products. It is hypothesised that organic shoppers also prefer locally grown food products due to environmental and health reasons. The shopping frequency of organic food, formulated as a dummy variable with 'regular' and 'occasional' coded as 1 and 'seldom' and 'never' as the reference category coded as 0, is included into the analysis. Furthermore, consumers who prefer convenience (ready-to-eat) products may not buy regional food, because it is usually non-processed, and therefore needs more time for preparation. Thus, we also considered statements that express shopping habits related to organic and convenience food products in our analysis. Items regarding the preference of supermarkets versus other kinds of shopping places were included for the same reason. It is hypothesised that consumers who usually buy in supermarkets because of convenience aspects do not have a strong preference for local food products. The same is expected for consumers, who classify taste as far more important than the origin of food. Descriptive statistics of the statements measuring the consumption and shopping habits are presented in Table 4. In all cases the seven-point Likert scale was used for measuring the shopping habits, except for the shopping frequency of organic food. For the same reasons given for the transformation of the dependent variables we transformed the agreement to the above mentioned items into binary dummy variables.

Sociodemographic factors

Several variables control for demographics. We include dummy variables for *gender* and the level of *education* as described in Table 3. Other variables include *household income* (0 = less than 3,000 €/month, 1 = 3,000 €/month and more), and respondents' *age* (0 = younger than average, 1 = older than average). A binary indicator controls for the employment status of the respondent (0 = unemployed, 1 = employed), whereas 0 also includes persons who are not engaged in economic activity (e.g. pensioners, students). Furthermore, respondents were asked to characterise the area they live in. We apply a dummy variable with 1 = rural and 0 = provincial and metropolitan area. Finally, we include the geographical location of respondents' home by aggregating the sixteen federal states of Germany into four dummy variables. The former states of the GDR in the eastern part of Germany are the reference category. Descriptive statistics of the demographic variables included in the model are presented in Table 4.

5. Empirical results

The binary character of the preference variables requires the application of a nonlinear model analysing the relationship between psychographic and sociodemographic indicators and the preference for local food products.

Logit analysis calculates the probability of belonging to a certain category of the dependent variable by using the cumulative logistic distribution for each individual with personal characteristics. The degree of impact of the independent variables is reported by so-called effect-coefficients exp (b), which indicate the change of the odds[57] ratio when the independent value increases for one unit. It is defined as the ratio of the odds of an event occurring in one group to the odds of it occurring in another group, or to a sample-based estimate of that ratio (Menard, 1995: 6, 12f, 49f).

The model is estimated by the stepwise forward logistic regression analysis using the maximum likelihood function in the SPSS package.

We already mentioned the interdependences between psychographic and sociodemographic factors in section 3. Hence, multicollinearity has to be considered in the modelling strategy and estimations of the correlation between the independent variables were carried out. The highest Pearson correlation coefficient is 0.54 between the statements '*Local food is of higher quality*' and '*Local food is tastier*'. This coefficient lies under the magnitude 0.7, mentioned by Bryman and Cramer (1994) to be critical regarding multicollinearity problems in regression analysis. Moreover, the standard errors are just marginal raised and the regression coefficients are stable through several model specifications.

Two models were estimated for which results are presented in Table 5. The first model describes the relationship between P 1 '*If possible I try to buy local products*' and both psychographic and sociodemographic variables, respectively. The second model includes the alternative preference statement 2 '*I am willing to pay a price premium for local products*' as dependent variable.

The R-squared values indicate that a remarkable part of the variance of the preference variables can be predicted by the independent variables. All included explanatory variables show the expected signs. Hence, the results confirm the theoretical framework of impact factors. Not surprisingly, there are more significant variables in Model 1 than in Model 2. The effect of 'yeah-saying' seems to be higher for P1 ('*If possible, I try to buy local products*') than for P2 ('*I'm willing to pay a premium for local products*') due to the less binding character of the first statement. While in Model 1 the location dummies are significant, there is no significant difference between regions in Model 2. The same is true for the income dummy that shows that the respondents, who belong to the highest class of income are more likely to buy local products. *Age* of the respondent appears to be the only relevant sociodemographic variable in determining the preference towards locally grown food in both models. Elderly people tend to show a higher preference for regional food than younger people. This may be a result of having a closer emotional connection to the home region, and having more time to purchase and prepare unprocessed food products.

Respondents, who agreed to the statements that they prefer shopping in supermarkets and that taste is more important to them than origin, show a significantly lower preference for regional food. Not surprisingly, there is a significant positive relationship between the frequency of buying organic products and the preference for local food. As expected, the statements indicating *cognitive*

---

[57] Odds (Y=1) = P(Y=1) / 1 – P(Y=1).

Table 4. Descriptive statistics of variables included in the binary logit analysis (N=3,000).

Variables	
Preference towards local food products	
If possible, I try to buy local products.	
I am willing to pay a premium for local products.	
Psychographic factors	
cognitive	Local food is fresher.
	Local food is of higher quality.
	Local food is tastier.
	Local food is healthier.
	Legal requirements are stronger for local foods.
	Caused by the food scares in the last years I lost confidence in products from supermarkets.
	Quality is much more important for me than the price when I buy food.
	Food, which I buy directly from the farmer, is free of any pollutants.
	I spend a lot of time eating healthy.
affective	Individual sympathy to the home region.
	Individual assessment of food supply of the home region.
normative	Local products have short transportation ways.
	Local products are naturally and eco-friendly produced.
	I support local farmers when I buy local food.
Consumption and shopping habits	
Taste is more important than the origin of food.	
I prefer food, which is quickly prepared.	
I prefer supermarkets because I can buy everything at a single blow.	
Shopping frequency of organic food	'regular' and 'occasional'
	'seldom' and 'never' (=reference)
Demographics	
Geographical location of residence in Germany	
Northern states of Germany	
Southern states of Germany	
States in the middle of Germany	
Eastern states of Germany	
Male	
Age (average = 46.5)	
Younger than average (=reference)	
Older than average	
High Education (=Higher secondary school and university degree)	
Residence in rural area	
High Income (household income: 3,000 €/month and more)	
Employed full time and part time	

Code	Mean	Median	Std. Dev.	Top-two-respondents
P1	5.93	6	1.37	72.0%
P2	5.46	6	1.66	59.4%
Local_Fresh	6.31	7	1.02	84.0%
Local_Quality	5.66	6	1.25	60.0%
Local_Taste	5.78	6	1.25	65.0%
Local_Health	5.36	6	1.43	49.3%
Local_Law	5.61	6	1.39	57.3%
Scare	4.59	5	1.77	31.3%
Quality	5.61	6	1.36	59.0%
Pollutants	4.87	5	1.57	35.9%
Time	5.07	5	1.64	43.0%
Sympathy	6.24	7	1.18	80.8%
Supply	5.68	6	1.09	63.2%
Transport	6.62	7	0.88	92.9%
Nature	5.45	6	1.40	50.4%
Support	6.43	7	1.05	87.0%
H.taste	4.70	5	1.78	35.6%
H.quick	4.64	5	1.84	35.8%
H.shop	5.09	5	1.72	46.3%
H.organic	0.54	1	0.49	

Code	Percent
North	16.2
South	27.0
Middle	35.4
East	21.3
Male	21.6
Under mean	50.3
Above mean	49.7
High education	31.6
Village	42.1
High income	9.2
Employed	56.5

Table 5. Effect coefficients of the binary logit models (N=3,000).

	Model 1			Model 2		
Constant	0.21	***	(29.89)	0.05	***	(108.22)
Sociodemographic factors						
Germany (ref. East)						
North	0.47	***	(20.88)	1.23		(1.83)
South	0.58	**	(11.84)	1.15		(1.08)
Middle	0.42	***	(36.87)	1.00		(0.00)
Male	1.02		(0.02)	1.20		(2.69)
Age (ref. <mean)	1.67	***	(22.99)	1.79	***	(34.21)
High Education	1.12		(1.04)	0.99		(0.01)
Village	1.16		(2.08)	1.11		(1.14)
High Income	1.52	*	(5.55)	0.90		(0.47)
Employed	0.85		(2.43)	1.14		(1.75)
Shopping habits						
H.shop	0.59	***	(27.57)	0.68	***	(16.74)
H.taste	0.68	***	(14.06)	0.75	**	(8.36)
H.organic (ref. rarely/never)	1.17		(2.26)	1.50	***	(18.06)
Cognitive factors						
Local_Quality	1.75	***	(23.38)	1.30	*	(5.58)
Local_Taste	1.50	***	(12.52)	1.57	***	(17.08)
Local_Health	1.62	***	(17.29)	1.29	**	(5.94)
Local_Law	1.00		(0.00)	1.21	*	(3.92)
Scare	1.42	**	(9.21)	1.53	***	(16.90)
Quality	1.44	***	(12.65)	3.48	***	(181.05)
Pollutants	1.36	*	(6.63)	1.57	***	(18.60)
Time	1.58	***	(16.81)	1.59	***	(22.08)
Affective factors						
Sympathy	1.50	**	(11.95)	0.88		(1.27)
Supply	1.54	***	(18.48)	1.08		(0.62)
Normative factors						
Transport	1.57	*	(6.54)	1.24		(1.39)
Nature	1.36	**	(7.30)	1.43	***	(12.13)
Support	2.60	***	(49.90)	2.38	***	(35.40)
R^2		0.36			0.39	
Correct prediction		0.79			0.75	

Wald statistics in parentheses
*, **, *** denote statistical significance at the 0.10, 0.05 and 0.01 level

factors show in almost every case a positive influence in both of the two models. Especially the remarkable exp (b) of the item *'Quality is much more important to me than the price when I buy food'* indicates that quality and safety are important factors for the preference for regional food. *Affective aspects* determine the preference for local food significantly only in Model 1. Emotional processes do not effect the statement *'I am willing to pay a price premium for local products'*. Both logit models indicate an obvious impact of *normative indicators* on the preference variables. The two most important normative aspects are the support of local farmers and environmental considerations. By evaluating the importance of normative indicators, it is essential to consider the *warm glow effect* in the interviews. The true importance of the desire to support local farmers might be smaller than the observed and estimated levels in the models.

In comparison to the results of the studies described in section 2, our estimations confirm the importance of cognitive factors determining consumers' preference for locally grown food. However, if normative aspects are not considered in the models, an overestimation of the coefficients of cognitive factors can occur. Affective indicators play only a marginal role when cognitive and normative processes are also included. The German studies of Wirthgen *et al.* (1999) and Wirthgen (2003) seem to overvalue emotional factors in their estimations.

6. Final remarks

The results of the study indicate that cognitive and normative processes are the most important factors in determining the preference for regional food in Germany. Sociodemographic factors and affective processes are not satisfactory in explaining the variance in the preference for locally grown food. From the consumer's point of view, the origin of food is an important indicator of quality and safety. Another important factor are social norms, especially the desire to support the local economy by the purchase of local food. However, in comparing our results with other consumer country (region)-of-origin studies the results indicate that in former studies affective aspects were partly overvalued and normative processes mostly neglected. Future research on the product specific nature of the effect of products' origin on consumers' food evaluation may provide further relevant results (see also Van Ittersum *et al.*, 2003). Representative studies need to clarify impact differences according to different food products. Furthermore, cross-national studies should be undertaken in order to examine cross-cultural differences regarding the preference for regional food.

References

Ahbe, T., 2005. Ostalgie. Zum Umgang mit der DDR-Vergangenheit in den 1990er Jahren. Landeszentrale für politische Bildung, Thüringen. Available at: http://www.thueringen.de/imperia/md/content/lzt/ostalgie_internet.pdf, 09.08.2006.

Bryman, A. and D. Cramer, 1994. Quantitative data analysis for social scientists. London: Routledge.

Darby, K., M.T. Batte, S. Ernst and B. Roe, 2006. Willingness to pay for locally produced foods: a customer intercept study of direct market and grocery store shoppers. Selected paper prepared for presentation at the AAEA Annual Meeting, Long Beach, California, July 23-26.

Dorandt, S., 2005. Analyse des Konsumenten- und Anbieterverhaltens am Beispiel von regionalen Lebensmitteln. Zugl.: Dissertation, Universität Gießen. Hamburg: Verlag Dr. Kovač.

Henseleit, M., 2006. Möglichkeiten der Berücksichtigung der Nachfrage der Bevölkerung nach Biodiversität am Beispiel von Grünland in Nordrhein-Westfalen bei der Ausgestaltung eines ergebnisorientierten Honorierungskonzepts im Rahmen des Vertragsnaturschutzes. Zugl.: Dissertation, Universität Bonn. Goettingen: Cuvillier Verlag.

Jekanowski, M., D.R. Williams II and W.A. Schiek, 2000. Consumer's willingness to purchase locally produced agricultural products: an analysis of an Indiana survey. Agricultural and Resource Economics Review, 29: 43-53.

Lobb, A., M. Arnoult, S. Chambers and R. Tiffin, 2006. Willingness to pay for, and consumers' attitudes to, local, national and imported foods: a UK survey. Unpublished Working Paper of the Department for Agricultural and Food Economics, University of Reading.

Loureiro, M.L. and S. Hine, 2002. Discovering niche markets: a comparison of consumer willingness to pay for local (Colorado grown), organic, and GMO-free products. Journal of Agricultural and Applied Economics, 34: 477-487.

Loureiro, M.L. and W.J. Umberger, 2003. Consumer response to the country-of-origin labeling program in the context of heterogeneous preferences. Paper prepared for presentation at the American Agricultural Economics Association Annual Meeting, Montreal, Canada, July 27-30.

Loureiro, M.L. and W.J. Umberger, 2005. Assessing consumer preferences for country-of-origin labeling. Journal of Agricultural and Applied Economics, 37: 49-63.

Mabiso, A., J. Sterns, L. House and A. Wysocki, 2005. Estimating consumers' willingness-to-pay for country-of-origin labels in fresh apples and tomatoes: a double-Hurdle probit analysis of American data using factor scores. Selected Paper prepared for presentation at the American Agricultural Economics Association Annual Meeting, Providence, Rhode Island, July 24-27.

Menard, S., 1995. Applied logistic regression analysis. Thousand Oaks, California: Sage Publications.

Obermiller, C. and E. Spangenberg, 1989: Exploring the effects of country of origin labels: an information processing framework. Advances in Consumer Research, 16: 454-459.

Patterson, P.M., H. Olafsson, T.J. Richards and S. Sass 1999. An empirical analysis of state agricultural product promotions: a case study on Arizona grown. Agribusiness, 15: 179-196.

Roosen, J., J.L. Lusk and J.A. Fox, 2003. Consumer demand for and attitudes toward alternative beef labeling strategies in France, Germany, and the UK. Agribusiness, 19: 77-99.

Schroeder, C., H. Burchardi and H. Thiele, 2005. Zahlungsbereitschaften für Frischmilch aus der Region: Ergebnisse einer Kontingenten Bewertung und einer experimentellen Untersuchung. German Journal of Agricultural Economics, 54: 244-257.

Schupp, A. and J. Gillespie, 2001. Consumer attitudes toward potential country-of-origin labeling fresh or frozen beef. Journal of Food Distribution Research, 32: 34-44.

Shimp, T.A. and S. Sharma, 1987. Consumer ethnocentrism: construction and validation of the CETSCALE. Journal of Marketing Research, 24: 280-289.

Umberger, W.J., D.M. Feuz, C.R. Calkins and B.M. Sitz, 2003. Country-of-origin labeling of beef products: U.S. consumers' perceptions. Journal of Food Distribution Research, 34: 103-116.

Van Ittersum, K., 1999. Consumer ethnocentrism and regional involvement as antecedents of consumer's preference for products from the own region. AIR-CAT Meeting Reports, 5 (1): October 1998 – Consumer Attitudes towards Typical Foods – The European Food Consumer. (EU project AIR-CAT, Series of Meeting Reports). Matforsk, Ås, Norway, pp. 45-51.

Van Ittersum, K., M.J.J.M. Candel and M.T.G. Meulenberg, 2003. The influence of the image of a product's region of origin on product evaluation. Journal of Business Research, 56: 215-226.

Verlegh, P. and J.-B. Steenkamp, 1999. A review and meta-analysis of country-of-origin research. Journal of Economic Psychology, 20: 521-546.

Von Alvensleben, R., 2000a. Verbraucherpräferenzen für regionale Produkte: Konsumtheoretische Grundlagen. Agrarspectrum Schriftenreihe, Band 30: Regionale Vermarktungssysteme in der Land-, Ernährungs- und Forstwirtschaft – Chancen, Probleme und Bewertung. Frankfurt am Main: DLG-Verlag, pp. 3-18.

Von Alvensleben, R., 2000b. Zur Bedeutung von Emotionen bei der Bildung von Präferenzen für regionale Produkte. German Journal of Agricultural Economics, 49: 399-402.

Wirthgen, A., 2003. Regionales- und ökologieorientiertes Marketing – Entwicklung einer Marketing-Konzeption für naturschutzgerecht erzeugte Nahrungsmittel aus dem niedersächsischen Elbetal. Zugl.: Dissertation, Universität Hannover. Hamburg: Verlag Dr. Kovač.

Wirthgen, B., H. Kuhnert, M. Altmann, J. Osterloh and A. Wirthgen, 1999. Die regionale Herkunft von Lebensmitteln und ihre Bedeutung für die Einkaufsentscheidung der Verbraucher. Berichte über Landwirtschaft, 77: 243-261.

ZMP, 2003. Nahrungsmittel aus der Region – Regionale Spezialitäten. Bonn: ZMP Zentrale Markt- und Preisberichtstelle für Erzeugnisse der Land-, Forst und Ernährungswirtschaft GmbH.

The influence of label on wine consumption: its effects on young consumers' perception of authenticity and purchasing behaviour

R. Lunardo

Abstract

The last forty years have seen a dramatic decrease in wine consumption in France. In 1965, the wine consumption per people per year was 160 liters; in 2005, people didn't drink more than 70 liters of wine in a year. Moreover, from 1980 to 1990, people over 14 years who drunk wine have decreased from 80 to 67% of the population. In 2005, only 62% of them pretended drinking wine. That is one million French people less than in 2000. This decline in wine market can be explained by the fact that young people consume less wine than older people. This article identifies authenticity as a factor explaining purchasing behaviour of young consumers. Findings suggests that the label of bottled wine influences young consumers' choice of wine. Originality and projection are two dimensions of the authenticity explaining how young consumers perceive performance risk, perceived price and purchase intentions.

Keywords: authenticity, bottled wine, label, performance risk, perceived price, purchase intention

1. Introduction

Wine has become a significant beverage in many nations around the world. For example, in 2003, over 233 million cases of wine were sold in the United States and sales totaled 21,800 million dollars (Adams Wine Handbook, 2004). However, the last forty years have seen a dramatic decrease in wine consumption in France. In 1965, the wine consumption per people per year was 160 liters; in 2005, according to the INRA, people didn't drink more than 70 liters of wine in a year. Moreover, from 1980 to 1990, people over 14 years who drunk wine have decreased from 80 to 67% of the population. In 2005, only 62% of them pretend drinking wine. That is one million French people less than in 2000.

Why the decline in wine market? Firstly, the wine market is the subject of increasing interest to new foreign producers, as Californian, Australian and other New World wine producers who become more export oriented and see their national outputs grow. Secondly, wine experts suggest that this decrease in wine consumption is not surprising when one realises how the status of wine has evolved. The status of wine seems to have transited from 'wine as an aliment' to 'wine as pleasure' (Corbeau, 1997): wine was former considered as a whole part of the meal, while today it is associated with pleasure. That transition also explains that regular wine consumers are not as numerous as before. Regular consumers represented 60% of consumers over 14 years in 1980, 40% in 1995 and only 33% in 2005, according to the Onivins Institute (www.onivins.fr) These figures highlight that people drink less, and it also seems they want to drink better. The desire for quality and the degree of expertise of consumers has increased. As a questionnaire carried out in 2005 emphasised, consumers pay more attention to signs of quality, as AOC French label. When French people were asked 'Do you know what the AOC is ?', they were 58% to answer yes, while they were only 41% ten years ago (Onivins, 2005).

Along with the issue represented by the decrease of the wine market, there is also another issue represented by young people. A dynamic analysis provides information about future wine consumption. The weak wine consumption by young people suggests that wine consumption in the future is likely to keep low. Because regular wine consumers, especially represented by old people, won't be replaced after their disappearance, a decrease of wine consumption in France is therefore unavoidable. The APC econometric model forecasts a decrease in wine consumption between 13.3% and 18.1%.

Despite this decline of per capita consumption volumes in France, French producers don't seem to be interested in marketing as a useful tool to sell wine. However, marketing practices seem to be efficient in selling wine. To provide practical solutions to actual selling problems in the area of wine, French winemakers have developed quite a wide range of new kinds of wines, with new labels. Wines with distinctive, weird and funny names and labels, such as *Fat bastard* or *Cats pee on a gooseberry bush* have lately become best sellers. This success of such wines with new labels has highlighted the major influence that the label of bottled wine had on consumer behaviour. The label has become an important cue in explaining consumers' choice of wine and a strategic imperative for wine producers. Consumers often make their choices among a large numbers of alternatives in a very short time (Britton, 1992) and, in this context, packaging becomes a fundamental marketing tool for the winery. As Rocchi and Stefani (2005) suggest, the shape of the bottle, the colour of glass, types and drawing in the label should attract the attention of the potential purchaser, distinguishing a specific wine bottle from several competitors.

One marketing concept of interest to relate to wine packaging in order to understand how to make wine sell good could be authenticity. In general terms, authenticity can be defined as the fact of being original (Mc Leod, 1999). Researchers go as far as to state that the search for authenticity is one of the cornerstones of contemporary marketing (Brown *et al.*, 2003). They have identified that authenticity is often more contrived than real, but in the case of wine, authenticity is real. So, focal questions are: can the label improve the perception of authenticity for the consumer? Do wine consumers prefer authentic wine? On the contrary, do they prefer modern wine, one that does not seem to be authentic? Does authenticity improve perceived quality, decrease perceived risk, and enhance the probability of buying, especially in young people?

The purpose of this article is twofold. Firstly, this article aims to contribute to a better understanding of authenticity as a marketing tool. Secondly, it is to highlight the relationship (1) between the label of bottles of wine and perceived authenticity and (2) between perceived authenticity and purchase behaviour of wine.

The research described in this article addresses this issue by first developing a conceptual framework for examining the concept of authenticity and highlighting the relationship between authenticity in food products and consumer behaviour. This review of literature will allow us to draw hypothesis about authenticity in wine and its relationship with consumer behaviour. Then we will explain the methodology we used to test our hypothesis. The results are reported with managerial implications considered at the end of the article.

2. Background

The following section first explains the concept of authenticity by defining its dimensions and attributes. The latter sections then integrate the issue of how authenticity will interact with the consumers' buying behaviour of bottled wine to influence.

2.1 The authenticity concept: definition, dimensions and attributes

The issue of authenticity has been identified as central in marketing research. As a result, there are as many definitions of authenticity as there are those who write about it. However, Warnier (1994) suggests that any definition of authenticity must be done with reference to any place, time or product. In the field of marketing, according to Cova and Cova (2001), when authenticity is linked to a product, it refers to a four-dimension concept. Those dimensions are history, space, socialisation and naturalisation. Two dimensions have been added to form 'six worlds of authenticity' in the consumption world. Those are the archaeological world, the spaciological world, the ritualised world, the natural world, the inspired world and the technical world (Cova and Cova, 2002). Thus, several studies provide evidence in support of a conceptualisation of authentic product as something natural, uncorrupted or of clear and known provenance (Marianna, 1997). From this point, authenticity can in general be defined as the fact of being original (Mc Leod, 1999). Early research acknowledged this correlation between authenticity and naturality or tradition; for instance, Rushdie (1991) proposed that authenticity demands that sources, forms, style, language and symbol all derive from a supposedly homogeneous and unbroken tradition. Also, Marianna (1997) defines authenticity as a declaration of belonging to, identity with, knowledge about, and respect for and responsibility towards the product. These views of authenticity posit that authenticity is intrinsic to the object and must prevent any alterations against history, quality or art (Postrel, 2003). This respect for tradition helps create an image around the product that differentiates it from mass-market products by making appearing it committed to values far from commercial considerations. For instance, the National Institute of Controlled Appellations created in 1935 developed the appellation of origin as a tag that wineries put on their label to indicate the geographic pedigree of their wine and to communicate to consumers about the use of traditional methods. This appellation of origin is seen as a sign of quality and an assurance to consumers of quality standards.

In addition to a conceptualisation as a value of respect of tradition, authenticity has been conceptualised as self-expression. Following this, a product is seen authentic because it is a genuine expression of our own personality – what Postrel (2003) define as 'I like this because I'm like that'. In this sense, authenticity relates to the image wineries want to project and how that image may be associated to the consumers' own drives.

Moreover, researchers have hold uniqueness as an important dimension of authenticity (Lewis and Bridger, 2001). The uniqueness dimension means that the product must be perceived as different from mass-manufactured products which are sold by millions all around the world. The notion of terroir is another example of the necessity to promote uniqueness. Terroir has been developed to provide products with uniqueness in order to make it difficult for competitors to replicate. However, Grayson and Martinec (2004) moderated the necessity for an authentic object to be unique. They identified two kinds of authentic marketing offerings: when an object has a spaciotemporal connection to history, it has indexical authenticity, whereas when the object is an accurate reproduction of the original, iconic authenticity is present. Thus, in the case of iconic authenticity, an object can be perceived as authentic even if it is not unique.

Given all those considerations, Beverland (2005) suggests that authenticity may be defined as a 'story that balances industrial (production, distribution and marketing) and rhetorical attributes to project sincerity through the avowal of commitments to traditions (including production methods, product styling, firm values, and/or location), passion for craft and production excellence, and the public disavowal of the role of modern industrial attributes and commercial motivations'. For our purposes, we hold the definition proposed by Camus (2004) who has posited that authenticity in the context of food products may be defined as 'a characteristic of the product which brings it to an origin, which distinguishes it because it fills up a lack, an insatisfaction, and which is reinforced since the products represents a part of the identity of the consumer'. Through this definition, authenticity contains the three dimensions which have appeared from the literature review: originality (respect for tradition and origin), projection (authenticity as self-expression) and uniqueness. These three dimensions will be helpful in providing a basis for our model.

So far, we have explained how authenticity is defined and characterised. We have emphasised that authenticity refers to something original, unique, far from merchandises, usually seen by consumers as standardised goods. The next issue is how consumers integrate authenticity as a criterion while buying food products.

2.2 Authenticity and food consumption

The quest for authenticity is a characteristic of postmodern consumption (Firat and Venkatesh, 1995). People are nostalgic about old ways of life, and they want to relive them by the way of living authentic experience. According to Fine and Speer (1997), an authentic experience involves participation in a collective ritual, where strangers get together in a cultural production to share a feeling of closeness or solidarity. Researchers use the term 'authenti-seeking' for consumers searching for authenticity in a range of products, services and experiences or looking for it within themselves. In tourism area, authenticity as a concept is nothing new; destinations such as Australia, Canada or China are promoting authentic experiences in order to attract tourists (Yeoman *et al.*, 2006). In looking for authenticity, some tourists focus on the product in terms of its uniqueness and originality, its workmanship, its cultural and historical integrity, its aesthetics, and/or its functions and use (Hugues, 1995).

Also in other areas, such as food market, one of the key areas identified by research into the future of food market focuses on this concept of authenticity. This focus on authenticity is largely a consequence of the risk consumers perceive while buying food products. Indeed as Fischler (2001) noticed, there's a real paradox in postmodern consumption: while consumers have today a maximal security when they buy food products, their fear about what they eat has never been so important. Consumers have a great consciousness of what they eat and what risk can be associated to their food. This behaviour is ruled by two universal principles.

The first one is the 'principle of incorporation', which can be defined as 'I get what I eat'. By controlling the food you eat, you control what you get, in order to maintain your self-esteem. Authenticity allows people to be sure about what they eat: you eat something natural, something original, and something unique.

The second principle is the 'principle of classification'. As anthropologists notice, people are used to classifying things in order to make rules or norms. The most fundamental classification is the one related to what can be eaten and what can not be. Another classification can be about authenticity: some things are authentic, others are not.

Those two principles of incorporation and classification can be considered as risk reductors. By being conscious of the quality of food products and by classifying, people reduce risks related to food behaviour. Many risk reduction models have been suggested in marketing literature, including word-of-mouth, warranties, brand image, a price-quality association and salesperson assurance (Hawes and Lumpkin, 1986) but to our knowledge authenticity has never been integrated into consumer behaviour research dealing with consumers' choice of wine.

2.3 Wine choice, wine label and wine authenticity

One approach to studying food choice derives from social psychological research into attitude–behaviour relationships. Referring to the Theory of Planned Behaviour (Ajzen, 1991), it is assumed that most part of the influences on food choice are mediated by the beliefs and attitudes held by an individual. Beliefs about the nutritional quality and health effects of a food may be factors more important than the actual nutritional quality and health consequences in determining an individual's choice. Concerning wine, it can be both a good friend (in moderation, providing physical and social benefits) and a cruel enemy (in excess, causing moral and physical declines). That is, one of the most prominent factors influencing consumer's wine choice has been found to be perceived quality (Hauck, 1991). Quality can be perceived by human senses, as sight: for food products, and especially for wine, that means packaging and labels are some of the sources consumers refer to in order to judge the quality of the product and to make a choice.

With respect to Olson and Jacoby's typology (1973), the label is considered as an extrinsic cue, an attribute which is not part of the physical product. Rocchi and Stefani (2005) found out that consumers seem to be affected by extrinsic cues, such as shape, size, colour and dress of the bottle, represented by the set of the other packaging elements (labels, capsules). The label is the most obvious and probably the most important part of the wine package. The label has to reflect the wine in the bottle, which is not the same thing as simply catching the eye. It signals the producers' names, the types of wines, the origin, the vintage, the level of alcohol, and the government warnings. As Halewood and Hannam (2001) suggest, the label is often placed on goods to make them seem more authentic, to add a quality assurance tag, and even explain their wider context. Such marking helps to make explicit the exchange value of the product. In this sense, authenticity becomes a source of information used by consumers to assess the quality of wine before purchase. As Marianna (1997) suggests, consumers have become clearly discerning and are demanding more information about the products they buy. People want to know what they are buying and what the product's origins are. In case of wine, the 'where' question is complex and elicits notions of classifications, appellations and the terroir. Indeed, when a winery wants to indicate the geographic pedigree of its wine, it uses a tag on its label called an appellation of origin. This appellation of origin must meet federal and state legal requirements. It is seen as a sign of quality for reputable production areas, and an assurance to consumers of quality standards. The origins carry significant weight for both producers and consumers, and so much effort goes into protecting and promoting it. For instance, the National Institute of Controlled Appellations created in 1935 made the label 'Controlled Appellation' as a sign of authenticity and singularity.

Quality is not the only factor consumers refer to in their choice. Choice is not determined only by physiological or nutritional need (Shepherd, 1999) but also by other interrelating factors. There are many factors in the context in which the choice is made that are likely to be very important, such as motivations for instance. In addition to the utilitarian (physical) and symbolic (social) motivation, a third motivation labelled 'experience' must be emphasised, in line with the evolution of consumer behaviour studies of wine consumption. People choose a bottle of wine not only for the taste or for social reasons, but also to live a unique experience (Holbrook and Hirschman, 1982).

Other factors include marketing and economic variables as well as social, cultural, religious or demographic factors (Murcott, 1989). In their summarising framework, Orth and Krska (2002) identified five factors influencing consumer's choice of bottled wine (Figure 1). They include push factors, pull factors, exogenous factors and economic restraints (time and money).

Besides these situational factors, consumer's choice can be moderated by individual ones. Wine consumption has been seen as moderated by sex: men drink more alcohol than women. It is also moderated by age. It is only between 20 and 25 years old that people begin to appreciate drinking wine (Aigrain *et al.*, 1996).

3. Objectives

One of the objectives of the paper is to identify the effects of authenticity on purchase behaviour. These effects can now be linked with the previous discussion about the buying processing of bottled wine to develop the hypotheses to be tested. So, from the review of the literature, we propose three sets of hypothesis, dealing respectively with the relationship of the three dimensions of authenticity identified by Camus (2003) and perceived risk, perceived price and purchase intention.

3.1 Authenticity and perceived risk

The study of perceived risk has a long history in the marketing literature. Risk perceptions are considered to form the basis of a heuristic framework that guides decisions about behaviour (Frewer *et al.*, 1994). Researchers generally agree that perceived risk is a combination of the perception of the likelihood that something will go wrong and the perception of the seriousness of the consequences if it does (Garbarinoa and Strahilevitz, 2004). That's why, following Stone and Gronhaug's conceptualisation (1993), we define perceived risk as the subjective expectation of a loss. While a number of risk dimensions have been suggested, only one is included, performance risk. This risk dimension can be viewed as the loss incurred when a product does not perform as expected; in case of food products, performance risk can be viewed as the loss incurred when the product is not as good as expected.

With a large range of wines available for consumers to choose from and the complex nature of the varieties and brands available along with the varying tastes of different people, consumers are interested in approaches that will lower the risk of purchase and help them make a good

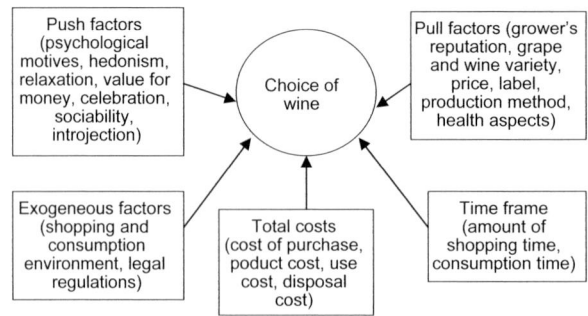

Figure 1. Factors influencing consumer's choice of bottled wine (Orth and Krska, 2002).

decision (Johnson and Bruwer, 2004). The packaging of wine can be considered as a quality cue contributing to lower the risk and define the expected quality of the product. According to Hall and Winchester (2000), the consumer uses these cues to assess alternative products with respect to his system of values following a set of subjective rules. Authenticity may be perceived from these cues.

However, the relationship between authenticity and perceived risk has not yet been deeply established in marketing literature. Only Cova and Cova (2002) suggested, without bringing any evidence, that when the product appears as not very sure, when you can see the product as physically risky, it appears as not authentic. The literature has mainly focused on the relationship between authenticity and quality.

So, we hypothesised that:
- H1: the greater the perceived authenticity, the less is the perceived risk about quality of the wine.
- H1 a: the greater the natural dimension of authenticity, the less the performance risk.
- H1 b: the greater the reflect of personality dimension of authenticity, the less the performance risk.
- H1 c: the greater the uniqueness dimension of authenticity, the less the performance risk.

3.2 Authenticity and perceived price

Literature about price has widely provided evidence about the influence of price on other variables, such as quality or risk (Roselius, 1971). Literature about risk reduction models suggests a link between quality and price, the price-quality association being viewed as a risk reductor. Concerning wine, Landon and Smith (1997) measured the absolute impact of wine quality and reputation on price and purchasing decisions for Bordeaux wines. Their results showed that reputation has a large impact on the implicit price. The label has also an influence on price. Combris *et al.* (1997) showed that the price of Bordeaux wine is essentially determined by its objective characteristics appearing on the label of the bottle year of harvest, geographical origin of grapes, and concentration of alcohol).

But the direct relationship between price and authenticity has been far less studied. To Warnier and Rosselin (1996), the value of the authentic product can not be estimated. Every master chief which is not a copy can be sold at the highest price because of its originality and uniqueness. And the lack of expertise of the consumer can be caught up by a reference to price: a low price raises the risk of a copy whereas a higher price is a sign of authenticity (Bessy and Chateauraynaud, 1995). Therefore, for any product, a decrease in price will be prejudicial to the perceived authenticity of the product. Thus, it is hypothesised that:
- H2: the greater the perceived authenticity, the more is the perceived price of the bottled wine.
- H2 a: the greater the natural dimension of authenticity, the greater the perceived price.
- H2 b: the greater the reflect of personality dimension of authenticity, the greater the perceived price.
- H2 c: the greater the uniqueness dimension of authenticity, the greater the perceived price.

3.3 Authenticity and purchase intention

The focus on the relationship between visual perceptions of the labels and purchasing process has several psychological implications that need to be taken into account. One of these psychological implications may be the perception of authenticity.

The relationship between authenticity and purchase intention has not been widely studied. On one hand, intention has often been related to confidence; Bennett and Harrell (1975) suggested that confidence plays a major role in predicting intentions to buy. On the other hand, there is evidence demonstrating that intention to buy is positively influenced by attitude (Laroche and Brisoux, 1989) and consumer's knowledge confidence (Laroche *et al.*, 1996). Concerning authenticity, we hypothesised that:

- H3: the greater the perceived authenticity, the more is the intention to buy the bottled wine.
- H3 a: the greater the natural dimension of authenticity, the more is the intention to buy the bottled wine.
- H3 b: the greater the reflect of personality dimension of authenticity, the greater the purchase intention.
- H3 c: the greater the uniqueness dimension of authenticity, the greater the purchase intention.

We do not hypothesise that perceived relative price directly reduces risk about quality. We could have, considering that Monroe (1990) regarded product quality as influenced by perceived price. Hypotheses 1-3 can be represented by the model presented in Figure 2.

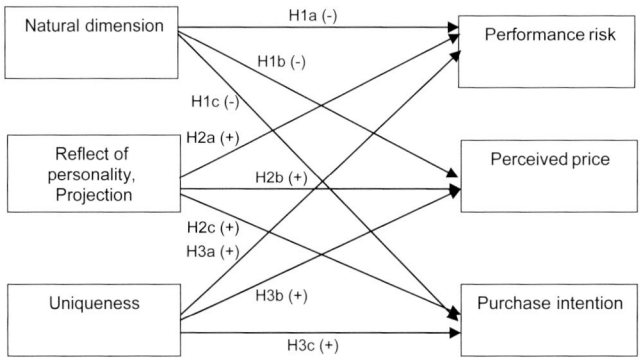

Figure 2. The conceptual model.

4. Data and methodology

As Rocchi and Stefani (2005) concluded, further developments are possible both using quantitative and qualitative approaches. Considering this conclusion, this section describes the qualitative and quantitative studies that were designed to test the propositions described in the previous section.

4.1 Qualitative study

An exploratory survey on consumers' perception of wine packaging has already been done by Rocchi and Stefani (2005). They used a repertory grid (RG) approach as a methodological framework in order to know which pattern of features is better at inducing purchase.

The purpose of our qualitative study was different. Its main objective was to have a better knowledge about what means authenticity for consumers and how they can perceive authenticity from labels of bottled wine. To do so, ten interviews of young people between 18 and 25 years were conducted. Considering the exploratory nature of the research, we needed a composition of the sample compatible with the elicitation of the broadest range of constructs. We decided to interview regular consumers and non regular consumers so that we could receipt opinions from expert and non expert consumers. Interviews were carried out with participants in French at their university, and on average lasted for twenty minutes. Questions evolved around their perception of authenticity provided by front labels on bottles of wine. All interviews were taped. Details of the sampled respondents and their responses are shown in Table 1.

As a result, we identified 7 attributes of authenticity provided by the label on the bottle: the drawing of a castle, the drawing of vine, the colour of the label, the shape of the label, the presence of a wine exhibition award, the name of the castle, the typography. This result can be compared to the six attributes of authenticity for luxury brands found out by Beverland (2006). It can also be compared to the pull factors identified by Orth and Krska (2002). It also can be compared to the traditional cues identified by Rocchi and Stefani (2005): colours, shape and size of the bottles, and labels. Another result is that authenticity seems to be linked to the structure of the wine industry. Authentic and luxury wine is seen as coming from small, family growers and not from larger producers controlling global distribution.

Table 1. Summary of case studies.

	Attributes of authenticity on labels of bottled wine	Attributes of non authenticity on labels of bottled wine
Person 1 'expert male'	Drawing of a castle or vigneyard, handwritten writing, information about the place of production, the year of production	Bright colours, non handwritten writing
Person 2 'expert male'	Parchment-looked paper, year of production, country of production, put into the bottle at the castle	Bright colours
Person 3 'non expert male'	Medals from contests, French name of the castle	Non handwritten writing
Person 4 'non expert male'	Year of production, name of the castle, drawing of vineyard	Bright colours, non handwritten writing
Person 4 'non expert female'	Pale colours, name of the castle, reputation	Non squared label
Person 6 'non expert female'	Year of production, Pale colours, name of the castle, French name	Bright colours, emptiness of the label
Person 7 'expert female'	Country of production, the year of production, the name of the castle, put into bottle at the castle	
Person 8 'expert male'	Country of production, put into the bottle at the castle, year of production, name of the castle	
Person 9 'non expert female'	Wine exhibition awards, name of the castle, year of production	Emptiness of the label, bright colours, non handwritten writing
Person 10 'non expert male'	Wine exhibition awards, name of the castle, country of production, appellation of origin	Non handwritten writing

4.2 Quantitative study and measures

The findings of the qualitative study have been used to design the questionnaire. Through the interviews, the most relevant attributes of authenticity to include in the questionnaire were determined. The second major source was the in-depth literature review. So, the categorisation of the bottles for the questionnaire emerged from the initial interviews. On the basis of the qualitative study and the literature, two bottles were selected as the target pieces for the main experiment, each of which was rated as authentic or non-authentic. In our questionnaire, we placed the pictures of these two bottles of wine. The bottle perceived as authentic provided on its label all the legal mentions plus a drawing of a castle and vines, a classic-coloured paper looking like a parchment. We decided to choose for this authentic bottle a label with a French name, a Chateau Prieuré Lalande, Côtes de Bourg 2004. On the contrary, the bottle perceived as non authentic provided on its label grey and orange colours, a non handwritten typography, an orange circle out of the label located on the bottle. For this non-authentic bottle, in order to increase the gap of perception, we chose a bottle with a French name (Art de Vivre) but with an english explanation (The art of bottling sunshine) (Figure 3). The questionnaire with these two labels inside has been administered to 94 students. In the first part of the questionnaire, people had to answer questions about the authentic bottle; in second part, they had to answer questions about the non authentic bottle. By doing so, we collected 188 data relative to the two bottles. This technique to collect data is recommended by Bowman and Gatignon (1995). The data were collected in the form of self-report questionnaires. The data were gathered from a study conducted in May 2006 on the campus of a French University in Reims.

The authenticity measure was composed of a 12-item and seven-point likert scale derived from the scale developed by Camus (2004). Participants were asked to rate the items according to how they thought the bottles of wine were original, unique, and able to reflect their personality. The items required the respondent to indicate the extent to which he or she agreed or disagreed with the statement, and ranged from 'Strongly disagree' (1) to 'Strongly Agree' (7). Traditional scale development procedures, including exploratory factor analysis and coefficient alphas were used to eliminate items that did not adequately contribute to the reliability and validity of the proposed scales. That is, we examined the dimensionality of the scale by using an exploratory factor analysis. The Bartlett' sphericity test provides good results (KMO=0.784), as does Chi-

Figure 3. The two labels for the questionnaire.

Square test (1024.982, df=66). Communalities were acceptable (>0.500). These results allowed us to factorise the data and along with Camus' results we found out by using a Varimax rotation the tridimensionality of the scale. Authenticity can be measured by the three dimensions identified by Camus (2004): originality, uniqueness, and projection dimensions. These three dimensions provide 70.376% of the overall variance. Reliability estimates (coefficient alpha) and convergent validity coefficient are acceptable (Table 2).

Consistent with Anderson and Gerbing's (1988) advice, a test of the measurement model was conducted (Table 3). The measurement model specifies a confirmatory analysis of the hypothesised relationships between manifest variables and latent constructs. Because a measurement model can be tested only if the model contains more than 3 items, only natural and projection dimensions were tested. For the measure of each dimension, the chi-square test was statistically significant. The goodness-of-fit index (GFI, adjusted=AGFI), the root mean square error of approximation (RMSEA), the normed-fit index (NFI), the Tucker-Lewis index (NNFI) and the comparative-fit index (CFI) indicated an acceptable fit (Bagozzi and Yi, 1988).

Table 2. Measures used in the study and reliabilities.

Items	Factor loadings		
	Factor 1	Factor 2	Factor 3
When you're looking at the label on the bottle number, you can say about the wine:			
it is natural	0.880		
it is made from natural stuffs only	0.862		
it is not made from natural stuffs (inversed)	0.758		
you know how it has been produced	0.712		
you know where he comes from	0.658		
it can reflect your personality		0.928	
it can define yourself		0.891	
it can help you being yourself		0.823	
it is at your style		0.745	
it is unique			0.868
it is one-of-a kind			0.854
there's not other like it			0.836
Eigen values	4.217	2.469	1.787
Cronbach's alpha	0.8484	0.8853	0.8221
Jöreskog	0.7310	0.8334	0.8185
Convergent validity	0.5616	0.6730	0.6009

Table 3. Goodness-of-fit indicators for the measurement model of authenticity.

	χ^2	ddl	χ^2/ddl	GFI	AGFI	RMSEA	NFI	NNFI	CFI
Natural dimension	14.5	5	2.91	0.97	0.91	0.04	0.96	0.95	0.97
Projection dimension	7.55	2	3.75	0.98	0.90	0.02	0.98	0.96	0.98

Concerning the other measures, the perceived risk measure was designed to assess participants' evaluation of the risk they perceive while consuming bottled wine. Perceived risk was assessed with a single-item and seven-point Likert scale, taken from the scale developed by Dandouau (1999), which was 'When you're looking at the label on the bottle, you can say about the wine that its quality may not come up to my expectations'. Perceived price was assessed with the single-item and seven-point Likert scale 'When you're looking at the label on the bottle number X, how would you rate the price of the bottle?'. Purchase intention was assessed with the single-item and seven-point Likert scale 'When you're looking at the label on the bottle number X, you can say about the wine, you would seriously consider buying the bottle'.

5. Results

To examine whether authenticity has any effect on the consumers' behaviour, all the relationships between authenticity provided by the label of bottles and consumer behaviour attributes (performance risk, perceived price and purchase intentions) have all been tested by using linear regressions.

About the relationship between authenticity and performance risk, which can be seen as the perceived quality of wine, we found a significant main effect of the natural dimension of authenticity on purchase intention (sig = 0.000, â = -0.788, t = -5.732), strongly supporting prediction 1a. Further, the R^2 was 17.0%, meaning that when the label is perceived as authentic, young consumers don't see any risk buying the wine because the presence of the label is a definitive indication of the product's authenticity. When the label is perceived as modern, they perceive a risk.

However, the linear regression made to test the influence of the projection dimension of authenticity did not bring significant results (sig = 0.175). Prediction 1b is not supported. This results means that, when the label of a bottle of wine reflects his personality, a young consumer does not perceive it as a sign of quality, as a guarantee that the quality of the bottle is good enough to buy it.

The linear regression made to test the influence of uniqueness on performance risk shows that the influence is significant at 10%. At this level of significance, we found a significant main effect of the uniqueness dimension of authenticity on performance risk (sig = 0.087, â = -0.248, t = -1.669), supporting prediction 1c at 10% only. The R^2 for the analysis was low (1.7%), reflecting a low proportion of variance in performance risk explained by the uniqueness dimension of authenticity. As far as the level of significance allows us to bring any conclusion from the analysis, this result may means that young people who perceive a wine as unique from its label may perceive it as less risky to buy.

About the relationship between authenticity and perceived price, the main result should interest producers: the only dimension of authenticity that affects perceived price of bottles is the natural dimension. This results supports prediction 2a, while prediction 2b is not supported (sig = 0.274). Wines with a label improving the natural dimension are perceived as more expensive (sig = 0.029, â = 0.128, t = 2.197).

However, the linear regression made to test the influence of uniqueness on perceived price shows that the influence is significant at 10%. At this level of significance, we found a significant main effect of the uniqueness dimension of authenticity on perceived price (sig = 0.071, â = 0.930, t = 1.820), supporting prediction 2c, meaning that young consumers perceive bottled wine as more expensive when the label provides a sign of uniqueness. The R^2 for the analysis was low, at 2.0%.

About the relationship between authenticity and purchase intention, we found a significant main effect of the natural dimension of authenticity on purchase intention (sig = 0.000, â = 0.839, t = 6.572), strongly supporting prediction 3a. Further, the R^2 for the analysis was 20.7%.

An interesting result shows that people are more intended to buy a bottle of wine when the label reflects their personality, supporting the prediction 3b claiming that the second dimension of authenticity has an influence on purchase intention (sig = 0.000, â = 0.683, t = 5.122). The R^2 for this linear regression was 13.5%. Maybe this result could mean that young people may be more attracted by bottled wine with modern labels, while elderly people may prefer authentic labels. However, given the sample composition of our study, this hypothesis cannot be tested.

Inversely, the last dimension of authenticity, uniqueness, does not improve purchase intention (sig=0.829). Prediction 3c is not supported; in contrast with our hypothesis, consumers may perceive physical risk when buying a wine that seems too different from others. That is, the challenge wineries has to cope with is the way they could promote uniqueness of wine and distancing their product from mass-production, without increasing uncertainty for the consumer about the quality of wine.

6. Final remarks

Wine marketers spend billions of dollars annually seeking to enhance consumers' perceptions of value associated with their bottles. Because of the size and the negative evolution of the market, it is critical for them to have a clear understanding of the way the labels on the bottles can influence buying behaviour, especially for young consumers. Indeed, although young consumers still account for only a small portion of total consumers, they represent the future consumers for wine producers.

This study was intended to provide a more complete understanding of the influence of the authenticity perceive from the label of bottled wine. As an attempt to extend the research on the influence label of bottled wine can have on consumers' decisions of buying, the current article shows some interesting results. Based on the use of a recently built scale measuring authenticity and of the regression results, the answer the study gives to the research questions can be summarised as follows. Our central finding is that authenticity consumers perceive from the label on bottled wine influences the performance risk they perceive while buying the product. Bottles of wine with labels perceived as authentic by young consumers are seen as less risky to buy. New kinds of labels, without any drawing of castle of vineyard for example, or with bright colours, are seen as risky.

This is not, however, the only one interesting result. Rather, our second major finding is that all the dimensions of authenticity do not affect the consumers' behaviour. As original dimension of authenticity influences performance risk, perceived price and purchase intention, reflect of personality and uniqueness dimensions do not influence all the dependant variables. For instance, the fact that the label reflects the consumers' personality does not influence perceived price, while natural dimension does.

In this context, we also show that young consumers only develop purchase intentions from two dimensions of authenticity. Natural dimension and the fact that the label reflects the consumers' personality influence purchase intentions. The fact that young people want to buy wine that reflects their personality is interesting for marketers. Wine has become a situational product, a product you consume for special times, as parties or important dinners. A young people would like to offer his guests a wine they would enjoy drinking, a wine he can be proud of, a wine he can 'you like it, you like me'. Wine can be seen here as an extended self product (Belk, 1988). Implications for producers are numerous. Producers could adopt a marketing strategy based on labels. For young people, they could make typologies in order to have a good knowledge of their customers and adapt the labels to their personality.

6.1 Limitations and future research

What is clear from these findings is the major role played by labels. However, our research holds some limitations. First of all, from an academic point of view, because our results are directly relevant only to students and young people, researchers should be interested in understanding the effects of labels of bottled wine on other kinds of targets. The middle-aged people can be considered as an important target for wine producers and the research may be replicated to know if this target is influenced by authenticity as young consumers are. Further research should clarify the extent to which the relationships we have found will broadly hold. Additionally, we only studied the influence of authenticity for red wines. Further research should clarify the extent to which the relationships we have found will be similar for white wines. Moreover, like most part of research (Wansink, 2003), we examined only front label perception. While this paper examines how front labels influence the consumers' perception of authenticity, it is also important to investigate the influence of back labels and how such labels could improve the perception of authenticity and the willingness to buy the bottle.

Moreover, of particular interest could be the study of the moderator influence of the consumption situation to explain consumer's wine choice. Consumers may buy bottled wine with funny labels if they forecast to drink it at a party or with some friends; on the contrary, they may prefer wine with authentic labels if the wine has to be drunk at a professional diner.

In concluding the article, we have not dealt, for reason of space, with some important questions and we wish to suggest possible avenues that consumer research could take. For example, do consumers expect same kinds of labels for domestic wine and foreign wines? And should we distinguish between red wine and white wine to develop authentic or funny labels? How other aspects of packaging, such as the shape of the bottle, could interact with the label in the perception of the bottle? Not only front and back labels should be presented to consumers but the complete packaging, the authenticity of which could be affected by glass colour, size, shape… Of particular interest could also be the influence of point of sale on authenticity. Future research might focus on the way the point of sale, if it is perceived as authentic or not, could influence the perceived authenticity of the wine sold in it.

From a methodological point of view, we only presented front labels in the questionnaire. While front label is usually considered for evocation, back label is expected to provide to an informative function, containing the relevant technical information about the wine. This back label could have been presented. Further research should measure its influence during the purchasing process. Another methodological limit is due to linear regressions. Structural equation modelling (SEM) could be chosen in future research because it can support simultaneously latent variables with multiple indicators, interrelated dependent variables, mediating effects, and causality hypotheses. Structural equations can measure independent variable errors while regression analysis cannot (Bollen, 1989).

All these questions are of major interest deserve much attention from wine producers and wine researchers.

6.2 Implications for market

A number of implications for research and practice flow from this line of research. An obvious implication of these findings is that, in order to increase a consumer's intention to buy a bottle of wine, a marketer needs to enhance his/her perceived authenticity. Authenticity decreases the level of performance risk, enhances perceived price and purchase intention. Enhancing authenticity can be done by making a label that makes the wine be perceived as natural and unique (the projection dimension does not significantly influence consumer behaviour). Making the wine be perceived as natural can be easy, by putting a picture of vineyard or castle on the label. Making it being perceived as unique can be done by enhancing the quality of the label for instance.

According to Seth Godwin (2005): 'Authenticity: if you can fake that, the rest will take care of itself'. As a conclusion, we emphasise the jeopardy of faking authenticity. Labelling bottled wine in a way that enhances the consumers' perception of authenticity could be doomed to failure. Consumers could perceive the wine as 'false authentic' and develop negative affect toward the producers and negative purchase intentions.

References

Adams Wine Handbook, 2004. Adams Beverage Group, Norwalk, CT.

Aigrain, P., D. Boulet, J.-B. Lalanne, J.-P. Laporte and C. Mélani, 1996. Les comportements individuels de consommation de vin en France, évolution 1980-1995. Paris: Rapport ONIVINS/INRA.

Anderson, J.C. and D.W. Gerbing, 1988. Structural equation modeling in practice: a review and recommended two-step approach. Psychological Bulletin, 103: 411-423.

Ajzen, I., 1991. The theory of planned behavior. Organizational Behavior and Decision Human Processes, 50: 179-211.

Baggozi, R. and Y. Yi, 1988. On the evaluation of structural equation models. Journal of the Academy of Marketing Science, 16: 74-94.

Belk, R.W. 1988. Possessions and the extended self. Journal of Consumer Research, 15: 139-168.

Bennett, P.D. and G.D. Harrell, 1975. The role of confidence in understanding and predicting buyers' attitudes and purchase intentions. Journal of Consumer Research, 2: 110-117.

Bessy, C. and F. Chateauraynaud (Eds.), 1995. Faussaires et Experts. Pour une Sociologie de la Perception. Paris: Métailié.

Beverland, M.B., 2005. Crafting brand authenticity: the case of luxury wines. Journal of Management Studies, 42: 1003-1029.

Beverland, M.B., 2006. The 'real thing': branding authenticity in the luxury wine trade. Journal of Business Research, 59: 251-258.

Bollen, K.A., 1989. Structural equations with latent variables. New York: Wiley.

Bowman, D. and H. Gatignon, 1995. Determinants of competitor response time to a new product introduction. Journal of Marketing Research, 32: 42-53.

Britton, P., 1992. Packaging: graphic examples of consumer seduction. Beverage Industry, 83: 21.

Brown, S., R.V. Kozinets and J.F. Sherry, 2003. Teaching old brands new tricks: retrobranding and the revival of brand meaning. Journal of Marketing, 67: 19-33.

Camus, S., 2003. L'authenticité marchande perçue et la persuasion de la communication par l'authentification: une application au domaine alimentaire. Thèse de Doctorat en Sciences de Gestion, Université de Bourgogne, Dijon, France.

Camus, S., 2004. Proposition d'échelle de mesure de l'authenticité perçue d'un produit alimentaire. Recherche et Applications en Marketing, 19: 39-63.

Combris, P., S. Lecocq and M. Visser, 1997. Estimation of a hedonic price equation for Bordeaux wine: does quality matter? The Economic Journal, 107: 390-402.

Corbeau, J.P., 1997. Identité et image du mangeur. In: Images du goût. Vol. 5. Paris, France: L'Harmattan, pp. 11-20.

Cova, V. and B. Cova, (Ed.), 2001. Alternatives marketing: réponses marketing aux évolutions récentes des consommateurs. Paris: Dunod.

Cova, V. and B. Cova, 2002. Les particules expérientielles de la quête d'authenticité du consommateur. Décisions Marketing, 28: 33-42.

Dandouau, J.C., 1999. Le besoin d'information en situation d'achat et le comportement d'information face au rayon: utilisation et effet du média de communication électronique interactive. Thèse d'Etat en Sciences de Gestion, Université de Bourgogne, Dijon, France.

Fine, E. and J. Speer, 1997. Tour guide performance as sight sacralization. Annals of Tourism Research, 12: 73-95.

Firat, A.F. and A. Venkatesh, 1995. Liberatory postmodernism and the reenchantment of consumption. Journal of Consumer Research, 22: 239-265.

Fischler, C., 2001. La peur est dans l'assiette. Revue Française du Marketing, 183/184: 7-10.

Frewer, L.J., R. Shepherd and P. Sparks, 1994. The interrelationship between perceived knowledge, control and risk associated with a range of food related hazards targeted at the individual, other people and society. Journal of Food Safety, 14: 19-40.

Garbarino, E. and M. Strahilevitz, 2004. Gender differences in the perceived risk of buying online and the effects of receiving a site recommendation. Journal of Business Research, 57: 768-775.

Godwin, S. (Ed.), 2005. All marketers are liars: The power of telling authentic stories in a low-trust world. New York: Penguin Books.

Grayson, K. and R. Martinec, 2004. Consumer perceptions of iconicity and indexicality and their influence on assessments of authentic market offerings. Journal of Consumer Research, 31: 296-312.

Halewood, C. and K. Hannam, 2001. Viking heritage tourism: authenticity and commodification. Annals of Tourism Research, 28: 565-580.

Hall, J. and M. Winchester, 2000. What's really driving wine consumers? The Australian and New Zealand Wine Industry Journal, 15: 68-72.

Hauck, R., 1991. Buying behavior and attitudes towards wine-findings of a field survey among younger consumers. Acta Horticulturae, 295: 127-132.

Hawes, J.M. and J.R. Lumpkin, 1986. Perceived risk and the selection of a retail patronage mode. Journal of the Academy of Marketing Science, 14: 37-42.

Holbrook, M.B. and E.C. Hirschman, 1982. The experiential aspects of consumption; consumer fantasies, feelings, and fun. Journal of Consumer Research, 9: 132-140.

Hugues, G., 1995. Authenticity in tourism. Annals of Tourism Research, 22: 781-803.

Johnson, T. and J. Bruwer, 2004. Generic consumer risk-reduction strategies (RRS) in wine-related lifestyle segments in the Australian wine market. International Journal of Wine Marketing 16: 5-32.

Landon, S. and C.E. Smith, 1997. The use of quality and reputation indicators by consumers: the case of Bordeaux wine. Journal of Consumer Policy, 20: 289-323.

Laroche, M. and J.E. Brisoux, 1989. Incorporating competition into consumer behavior models: the case of the attitude-intention relationship. Journal of Economic Psychology, 10: 343-362.

Laroche, M., C. Kim and L. Zhou, 1996. Brand familiarity and confidence as determinants of purchase intention: an empirical test in a multiple brand context. Journal of Business Research, 37: 115-120.

Lewis, D. and D. Bridger, 2001. The soul of the new consumer: authenticity – what we buy and why in the new economy. Naperville, IL: Nicholas Brealey Publishing.

Marianna, A., 1997. The label of authenticity: a certification trade mark for goods and services of indigenous origin. Aboriginal Law Bulletin, 3: 4-15.

Mc Leod, K., 1999. Authenticity within hip-hop and other cultures threatened with assimilation. Journal of Communication, 49: 134-149.

Monroe, K.B., (Ed.), 1990. Pricing: making profitable decisions. 2nd edition, New York: McGraw-Hill Book Company.

Murcott, A., 1989. Sociological and social anthropological approaches to food and eating. World Review of Nutrition and Dietetics, 55: 1-40.

Olson, J.C. and J. Jacoby, 1973. Cue utilization in the quality perception process. In M. Venkatesan (ed.). Proceedings of the 3rd Annual Conference of the Association for Consumer Research, Chicago: 167-179.

Onivins, 2005. Facteurs de compétitivité sur le marché mondial du vin. www.onivins.fr/pdfs/1176.pdf.

Orth, U.R. and P. Krska, 2002. Quality signals in wine marketing: the role of exhibition awards. International Food and Agribusiness Management Review, 4: 385-397.

Postrel, V., 2003. The substance of style: how the rise of aesthetic value is remaking commerce, culture, & consciousness. New York: Harper Collins Publishers.

Rocchi, B. and G. Stefani, 2005. Consumers' perception of wine packaging: a case study. International Journal of Wine Marketing, 18: 33-44.

Roselius, T., 1971. Consumer ranking of risk reduction methods. Journal of Marketing, 35: 56-61.

Rushdie, S. (Ed.), 1991. Imaginary homelands: essays and criticism. London: Granta.

Sheperd, R., 1999. Social determinants of food choice. Proceedings of the Nutrition Society, 58: 807-812.

Stone, R.N. and K. Gronhaug, 1993. Perceived risk: further considerations for the marketing discipline. European Journal of Marketing, 27: 39-50.

Wansink, B., 2003. How do front and back package labels influence beliefs about health claims. The Journal of Consumer Affairs, 37: 305-316.

Warnier, J-P. (Ed.), 1994. Le paradoxe de la marchandise authentique. Imaginaire et consommation de masse. Paris: L'Harmattan.

Warnier, J.-P. and C. Rosselin (Ed.), 1996. Authentifier la Marchandise. Paris: L'Harmattan.

Yeoman, I., D. Brass and U. McMahon-Beattie, 2007. Current issue in tourism: the authentic tourist. Tourism Management, 28: 1128-1138.

Willingness to pay for organic food in Argentina: evidence from a consumer survey

E. Rodríguez, V. Lacaze and B. Lupín

Abstract

Throughout these last years, organic agriculture has undergone a remarkable expansion due, among other things, to the greater interest shown by consumers aware of food safety concerns involving real or perceived quality risks. This paper aims to estimate consumers' willingness to pay (WTP) for organic food products available in the Argentinean domestic market, with a view to providing some useful insights to gain support and outline strategies for promotion of organic production, marketing, regulation, and labelling programs of organic food products. A Binomial Multiple Logistic Regression model is estimated with data from a food consumption survey conducted in Buenos Aires city, Argentina, in April 2005. The Contingent Valuation Method was chosen in order to calculate their WTP for five organic selected products: regular milk, leafy vegetables, whole wheat flour, fresh chicken and aromatic herbs. The empirical results reveal that consumers are willing to pay a premium for these products and that although prices play an important role, lack of store availability and of a reliable regulatory system to mitigate quality risks constraint consumption of organic products in this country.

Keywords: willingness to pay, food quality attributes, organic price premium, Argentina

1. Introduction

Throughout these last years, organic agriculture has undergone a remarkable expansion due, among other things, to the greater interest shown by consumers aware of food safety issues involving real or perceived quality risks (Henson, 1996). In Argentina, key factors such us very good agro-ecological conditions, intensive labour requirements, and increasing export perspectives for these differentiated foods, could transform organic production into a profitable activity for farmers, distributors and retailers, thereby improving the development of our regional economies.

Argentina has developed national organic regulations which have turned it into the first Third Country to adapt its national regulations to the European Union requirements (1993)[58]. It has also implemented a private certification system accredited by SENASA (National Service of Agrifood Quality and Safety) and carried out significant public research actions through certain technological institutions such as the INTA (National Institute of Agricultural Technology) and private and state universities. Still information scarcity remains a gap to be bridged as it confines supply and demand quantification and restrains potential market growth (Rodríguez, 2005).

When purchasing food, consumers make their choices based on price and quality. Such choices are certainly conditioned by the information available to them. In the Argentinean domestic market, many consumers are willing to pay higher prices for healthy products, i.e. organics, because they increase their utility level by reducing perceived health risks. But also some other reasons like lifestyles are explaining these choices.

---

[58] Having been recognised as an organic certified country has enabled Argentina to export organic products processed and certified in agreement with standards equivalent to those of the EU (IFOAM, 1998).

Information about the quality attributes of food products, i.e. safety attributes; convenience; place and manner of product production, and environmental concern, is imperfect for consumers, producers, government regulators, and researchers (Antle, 1999). This is particularly true when production process attributes cannot be readily observed or tested, and the product's health effects are difficult to determine once it has been consumed

Although 'safe products' still constitute a small part of the Argentinean food expenditure, they are considered a market niche of great potential growth. The main restrictions to domestic demand growth are the lack of information available to consumers; organic prices over those of conventional foods; and the erratic supply oriented to domestic market, as organic products' main target is the foreign market. In 2006, 96 percent of the Argentinean total organic production was destined to the foreign market. The domestic market accounted for as little as the remaining 4%[59] (SENASA, 2007).

2. Conceptual framework

2.1 Willingness to pay (WTP)

Increase in consumers' concern about food safety and food quality is driven by recent scientific discoveries, new information about the relationship between diet and health, novel food technology and mass communications (Kinsey, 1993). However, many of the scientific and economic variables related to food safety and food quality are difficult to measure. A well-used method to determine the benefit of a given improvement in food safety and food quality is the estimation of consumers' willingness to pay (WTP) for risk-reduced food (Goldberg and Roosen, 2005).

The notion of *willingness to pay* could be defined as the amount of money represented by the difference between consumers' surplus before and after adding or improving a given food product attribute.

Some previous efforts to develop a WTP model for an attribute change are found in the works by Van Ravenswaay and Wohl (1995) and Halbrendt *et al.* (1995). These models are based on Lancaster Demand Theory (1966) according to which consumers are hypothesised to derive utility not directly from goods, but from a collection of characteristics or attributes those goods possess. Van Ravenswaay and Wohl (1995), on the other hand, modelled consumer's WTP for a single product and applied that model to the analysis of the effect a change in pesticide residues has; and Halbrendt *et al.* (1995) incorporated consumer's socio-demographic characteristics in the estimated WTP function.

Four major methods are employed to measure or infer consumers' WTP for a given attribute. These techniques fall into two general categories depending on the type of data employed. The first category collects primary data directly from consumers, and it includes Contingent Valuation, Experimental Auctions and Conjoint Analysis methods. The second category, which includes the Hedonic Prices method, employs indirect sources to infer willingness to pay from the market

---

[59] The largest marketing export volumes are grains: bread wheat, rice and maize, and oilseeds. Other processed organic products such as olive oil, sugar, concentrated juices, honey and wines, notwithstanding their low production volumes, are also attractive export alternatives. The European Union imports more than 80% of Argentinean organic products; the remaining 20% is exported to the United States. Cereals and oils are also central products in the domestic market due to their high volume, and vegetables are noteworthy because of their diversity.

itself. These techniques are based on observed consumers' choices and revealed preferences (Lee and Hatcher, 2001).

2.2 Determinants of WTP for organic food

Most recent studies conducted in the potential markets for organic agriculture have tried to establish connections between the WTP for these products and a particular consumers' lifestyle (Hartman and New Hope, 1997; Gracia *et al.,* 1998). Consumers segmentation based on those variables has resulted in several profiles of potential organic consumers. Despite the notorious ambiguity of the socio-demographic profile, these consumers show a purposeful attitude towards a balanced life, eating healthy food, and decreasing agriculture impact on the environment (Thompson, 1998).

Results from empirical works carried out in countries with a significant level of organic food consumption demonstrate that the main reason why these products are acquired is health care, either because of disease suffering or disease prevention (Kuchler *et al.,* 2000). Besides, due to their low pesticide-residue content, these products are considered beneficial, mainly for produce (Weaver *et al.,* 1992; Baker, 1999). As regard meat products, e.g. chicken meat, the risks perception linked to hormone use along the production process is remarkable when conducting consumers' studies in Brazil and Argentina (Farina and De Almeida, 2003; Rodríguez and Lacaze, 2005).

Earlier studies performed in Buenos Aires city,[60] Argentina, concluded that Argentineans are worried about healthy and nutritive food, unsafe production processes and health care, which are key factors to organics consumption. Yet consumers are unaware of environmental issues. Taste and nutritive attributes are other relevant factors mentioned as well (Rodríguez *et al.,* 2005). Results from focus groups studies conducted in four different Argentinean cities (Buenos Aires, Mar del Plata, Mendoza and Córdoba) demonstrated that consumers do not trust organic certification bodies, and they recognise the lack of information available in the domestic market regarding organic food[61] (Rodríguez and Lacaze, 2005).

Argentineans place no trust in the regulatory system's ability to monitor and guarantee food safety. The better-educated consumers, who eat healthy food, and consider food control bodies 'inefficient', are more likely to buy organic products. Educated people seem to be more exposed to diet and health information sources, and can better understand and process the information. It is worth mentioning that the degree of relative satisfaction obtained from organic consumption compared to conventional food constitutes a relevant factor when explaining Argentinean consumers' choices (Rodríguez *et al.,* 2006).

Some studies have found direct associations between income and WTP either regarding risk reduction, derived from consuming healthier and safer food products, (Jordan and Elnagheeb, 1991; Blend and Van Ravenswaay, 1998) or certified quality (Misra *et al.,* 1991; Underhill and Figueroa, 1996).

With regard to educational level as a socio-economic predictor, Misra *et al.* (1991) obtained a negative correlation between education and fresh organic products consumption. Govindasamy

---

[60] Buenos Aires, the capital city of the Republic of Argentina, is the most densely populated city and also concentrates most trading activity in the country.

[61] To conduct the focus group studies (2003, 2004 and 2005), the referred cities were chosen not only for sharing consumption patterns, but also for being near production regions.

and Italia (1999) concluded, on the one hand, that the lower the educational level, the higher the risk perception; and, on the other, that the higher the educational level, the greater the confidence in production standards. In addition, Eom (1994) found that better educated people seem to understand scientific information related to food risks. Van Ravenswaay (1995) also sustained that higher education respondents can easily access to reliable information sources about food risks and benefits and, consequently, are less worried about these issues.

Several researches have focused on the obstacles hindering organic food demand expansion. Higher prices and products shortage supply in supermarkets should be mentioned in the first place (Michelsen *et al.,* 1999; Richman and Dimitri, 2000; Govindasamy and Italia, 1999; Gil *et al.,* 2000) together with the degree of relative satisfaction regarding conventional products, and the level of information about food quality consumers have access to.

3. Objective

The purpose of this paper is to estimate consumers' willingness to pay (WTP) for organic food products available in the Argentinean domestic market, with a view to providing some useful insights to gain support and outlines strategies for promotion of organic production, marketing and regulation and labelling programs of organic food products.

The following *hypotheses* are to be tested:
- Health risks perceptions linked to hormone, pesticide and preservers content in several food products affect significantly consumers' willingness to pay for organics.
- The effect of regulation programs on the willingness to pay for organic unprocessed products is lower than for organic processed products.
- Consumers' willingness to pay for organic food is lower than retail prices in the stores.

4. Data and methodology

4.1 Data

Survey design

The data in this study derives from a food consumption survey conducted in Buenos Aires city, Argentina, in April 2005, by applying a semi-structured questionnaire.

A convenience sample, in which the probability of being selected is unknown, was chosen due to the difficulty to spot the target population, i.e. individuals who usually shop for organic foods (Brewer, 1999; Chow, 2002; Schonlau *et al.,* 2002). 301 surveys were completed by trained interviewers who surveyed respondents in the largest supermarket chains and also in an important specialised organic store.[62] The sample was based on age and gender local distribution pursuant to the last National Population Census in Argentina (INDEC, 2001), for respondents aged 18 or above with a medium-high socio-economic level.[63] Table 1 provides the representativeness of the sample in terms of the demographic structure of Buenos Aires city population according to gender and age.

---

[62] Supermarket chains: Coto, Disco, Jumbo, Norte and Wall Mart. Specialised organic store: La Esquina de las Flores.

[63] As defined by the Argentinean Marketing Association (AAM). Available at: http://www.aam-ar.com

Table 1. Sample representativeness in terms of Buenos Aires city demographic structure according to gender and age (18-87 years old). Source: consumer survey, Buenos Aires City/2005 and Population Census in Argentina (INDEC, 2001).

Comparison between survey sample[1] and population census in Buenos Aires city

Demographic characteristics	Categories	Relative frequency	
		Representation in the survey sample	Representation in Buenos Aires city
Respondent's gender	Male	32%	44%
	Female	68%	56%
Respondent's age (in years)	18-24	15%	14%
	25-34	19%	20%
	35-49	26%	24%
	50-59	15%	15%
	60-87	25%	27%

Proportion of Buenos Aires city population in relation to Argentinean overall population

	Buenos Aires City	Argentina	
Population	2,174,017	23,927,108	9%

[1] N = 301.

Data collection

The semi-structured questionnaire contained both close- and open-ended questions displayed in three sections. In the first one, questions referred to organic, natural and fresh food consumption; also to purchasing frequency, and to reasons for buying these products.

The second section was designed in order to collect consumers' opinions concerning several issues linking diet and health. Questions dealt with: eating habits; reasons behind taking care in meals; risks perceptions derived from hormone, pesticide and preservers present in each of the selected products; factors of trust, such as brand, food labels, product origin, confidence in stores where respondents do their food shopping; search information, food products advertising and promotion; respondents' opinions about food control and regulatory bodies functioning; their preferences regarding private or public regulation systems; and personal beliefs about differences between organic and conventional foods.

The last section of the questionnaire collected socio-economic data, and included income ranges. Respondents had to indicate the range in which the household monthly income fell.

Sample characterisation

The socio-economic sample characterisation displayed in Table 2 shows that sixty eight percent of the respondents were female, as expected, since grocery shopping is mostly a female activity (Baker, 1999; Chen *et al.*, 2002).

The average sample age was 44, and the highest absolute frequency ranged between 35 and 49 years, and 60 years or more (26% and 25% of the total sample, respectively).

Thirty four percent of the respondents mentioned that they usually consumed organic food. These consumers were called 'organic consumers'. The remaining 66%, who stated to have never consumed organics, were called 'non-organic consumers'.

Thirty eight percent of the total sample stated that their household monthly income was US$ 500 or less per month, while the remaining 62% declared it was above US$ 500. Despite the fact that 67% of organic consumers earned above US$ 500, non-organic consumers were almost equally distributed when considering these household's income levels.

Regarding educational level, 20% of the respondents had not completed high school, and more than a half had gone into further education, even though they had not graduated. Twenty nine

Table 2. Demographic and socio-economic variables in the sample[1] for both organic and non-organic consumers. Source: consumer survey, Buenos Aires City/2005.

Variables	Total sample (100%)	Organic consumers (34%)	Non-organic consumers (66%)
Respondent's gender			
Female	68%	66%	69%
Male	32%	34%	31%
Respondent's age (in years)	15%	16%	15%
18-24	19%	19%	20%
25-34	26%	27%	26%
35-49	15%	16%	15%
50-59	25%	23%	23%
60 or +60			
Respondent's household monthly income[2]			
≤ US$ 500	38%	33%	45%
> US$ 500	62%	67%	55%
Respondent's educational level			
Unfinished High School	20%	10%	24%
Unfinished University	51%	54%	50%
University or Postgraduate degree	29%	36%	25%

[1] N = 301 - *Exchange Rate: 1 US$ = 3 Argentinean Pesos ($).*
[2] For comparative purposes, notice that in April 2005, the poverty threshold for an Argentinean citizen living in Buenos Aires city was of US$ 83.35 per month. Therefore, a 4-member family with 2 children needed US$ 285 monthly to achieve a minimum standard of living (INDEC, 2005).

percent held a university or postgraduate degree. The highest proportion of respondents who had reached a university or postgraduate degree was included in the organic consumers group (36%).

Selection of food products

Store availability was a crucial factor in the selection of these five products: regular milk, leafy vegetables, whole weat flour, fresh chicken and aromatic herbs, to which the methodology for consumers' WTP calculation was applied. The revealed consumers' perceptions regarding production processes risks, and nutritional information confidence in the organic labels also supported this selection (Rodríguez *et al.*, 2006). Table 3 below displays the description, net content and packaging of the selected products.

The organic price premiums are expressed as the percentage by which the price of any organic product is above the price of a similar conventional product (Lohr, 2001). These premiums were calculated with the current prices of both organic and conventional products collected at the stores where the survey took place.[64]

4.2 Methodology

Among the different methodological alternatives to assess consumers WTP, the Contingent Valuation (CV) approach was chosen (Hanemann, 1984, 1989; Portney, 1994). Even though CV is primarily used for the monetary evaluation of consumers' preferences for non-market goods, it is also applicable to the Argentinean organic market as it is still a small-scale niche, and organic products are not usually available in all retail stores.

CV tends to quantify the value consumers assign to products by facing a hypothetical purchasing situation in which they have to answer how much money they would be willing to pay for a given product, or if they would be willing to pay a certain price premium (Carmona-Torres and Calatrava-Requena, 2006).

In CV surveys, one of the most widely used approaches to elicit information about respondents' WTP is the so called dichotomous choice format (Hanemann, 1984). The single bound dichotomous choice format, selected herein, entails asking respondents whether they would be willing to pay a price premium for each of the selected organic products or not. It could be

---

[64] Stores mentioned in Section 4.1.

Table 3. Description of organic selected products.

Selected products	Description	Net content and packaging
Regular milk	Regular milk	1 l - carton
Fresh leafy vegetables	Chard, green onion, parsley, leeks, cabbage, rocket and chicory escarole	½ kg - plastic tray
Whole wheat flour	Whole wheat flour	1 kg - carton
Fresh chicken	Fresh chicken	1 unit - plastic tray
Aromatic herbs	Tarragon, oregano and black pepper	0.20 kg - plastic envelopes

assumed that the respondents' answer is conditioned by the organic and conventional prices they find when choosing organics instead of conventional products.

To obtain the parameters estimates for each selected product regression equation, the theoretical Model to be estimated by using a Binomial Multiple Logistic Regression is formulated as follows:

$$WTP_{ij} = \alpha + \beta_1 P_{jk} + \beta_2 Y_i + \beta_3 \pi_i + F_{(Zi)} \qquad (1)$$

Where:

WTP_{ij} Whether i respondent is willing to pay a price premium for the j selected food product or not; j = 1 Regular milk; j = 2 Leafy vegetables; j = 3 Whole wheat flour; j = 4 Fresh chicken; j = 5 Aromatic herbs.

P_j Organic price premiums charged for any of the j selected products at the k sampled stores; k = 1 Coto; k = 2 Disco; k = 3 Jumbo; k = 4 Norte; k = 5 Wal Mart; k = 6 La Esquina de las Flores.

Y_i Household income level of i respondent.

π_i Risks and quality attributes perceptions of i respondent.

Z_i Socio-economic characteristics of i respondent: respondent´s educational level and respondent´s household monthly income.

Table 4 lists the selected explanatory variables finally included in the Logit Models according to their statistical significance.

Equation 1 was estimated by maximum likelihood. The estimated parameters for each selected product equation were obtained by using the Statistical Package for Social Sciences (SPSS version 11, 2001).

After estimating the five Logit Models and in order to calculate the average consumers' WTP for each selected product, the estimated parameters were included in Equation 2. It equals the average WTP, calculated as the area below the logit functions estimated by Equation 1 truncated[65] at the maximum organic price premium, which was calculated in accordance with prices collected in the sampled stores:

$$\overline{WTP}_j = H + \frac{1}{\beta_1} \ln \left[\frac{1 + \exp\left[-(d + \beta_1 H)\right]}{1 + \exp(-d)} \right] \qquad (2)$$

Where:

$\overline{WTP}_j$ The average organic WTP calculated for the j product;[66]

β_1 Coefficient estimated for the price premium variable;

H   Maximum organic price premium (P_j) that is charged for the j selected product;

-d $= \alpha + \beta_2 Y_i + \beta_3 \pi_i + F(Z_i)$, according to Equation 1;

j   Selected food products.

---

[65] It is important to mention that truncation does not significantly affect the WTP estimates if H is large, as in this research. Also it should be mentioned that WTP were assumed to be strictly positive.

[66] The expression [2] was obtained by integrating:

$$E(WTP) = \int_0^{H} (1 + \exp\left[\alpha + \beta_1 P_j + \beta_2 Y_i + \beta_3 \pi + \beta_4 Z_i\right])^{-1} dp$$

Table 4. Description of models' variables.

Dependent variable		Categories
WTP	If the respondent is willing to pay a price premium for the organic product	1 = Yes, 0 = Otherwise
Categorical explanatory variables		Categories
CONSUMP	If organics are usually consumed in the households	1 = Yes, 0 = Otherwise
HORMONE	If the respondent perceives the high risks of hormones in conventional fresh chicken content	1 = Yes, 0 = Otherwise
PESTICIDEV	If the respondent perceives the high risks of pesticides in conventional leafy vegetables content	1 = Yes, 0 = Otherwise
PESTICIDEF	If the respondent perceives the high risks of pesticides in conventional whole wheat flour content	1 = Yes, 0 = Otherwise
RISKSCON	If the respondent believes that there are no significant risks when consuming conventional food	1 = Yes, 0 = Otherwise
AVAILABLE	If the respondent would be willing to buy organics if they were available in the market	1 = Yes, 0 = Otherwise
REGULATION	If the respondent believes that there should exist a food quality regulation system	1 = Yes, 0 = Otherwise
LABELS	If the respondent is used to reading food labels when buying	1 = Yes, 0 = Otherwise
DIFORCON	If the respondent believes that there is no difference between organic and conventional food products	1 = Yes, 0 = Otherwise
Quantitative explanatory variables		
RMPP	Organic regular milk price premium over conventional regular milk price	
LVPP	Organic leafy vegetables price premium over conventional leafy vegetables price	
WWFPP	Organic whole wheat flours price premium over conventional whole wheat flours price	
FCPP	Organic fresh chicken price premium over conventional fresh chicken price	
AHPP	Organic aromatic herbs price premium over conventional aromatic herbs price	

5. Empirical results

5.1 Binomial Logit models estimations

Table 5 displays the results from the estimated Logit Models. All the estimations were set for the higher income level (more than US$ 500) except for regular milk because the explanatory variables were also statistically significant for the lower income level (US$ 500 or less). Consequently, Model 1.a was estimated for the higher income level (more than US$ 500) and Model 1.b for the lower income level (US$ 500 or less).

Willingness to pay (WTP) for organic regular milk, is largely explained by its scarce store availability (AVAILABLE) for both income level Models. Besides, the belief that there should be a food quality regulation system (REGULATION) ranks as the second significative explanatory factor. The consumption of organics also explains the WTP for organic regular milk (CONSUMP).

Table 5. Results from the estimated Logit models and statistical models' performance. Source: consumer survey, Buenos Aires City/2005.

Variable	Model 1a: Regular milk[1] (income ≤ US$ 500)	Model 1b: Regular milk (income > US$ 500)	Model 2: Leafy vegetables (income > US$ 500)	Model 3: Whole wheat flour (income > US$ 500)	Model 4: Fresh chicken (income > US$ 500)	Model 5: Aromatic herbs (income > US$ 500)
Intercept	-2.21 (9.22)***	-3.42 (10.08)***	4.1 (2.93)*	-3.35 (8.65)***	-4.32 (11.25)***	-6.99 (10.46)***
CONSUMP	1.08 (6.42)**	1.32 (4.04)**	1.23 (7.96)***	ns	1.58 (10.33)***	1.64 (8.77)***
HORMONE					-1.30 (5.65)**	
PESTICIDEV			-0.98 (4.20)**			
PESTICIDEF				-1.61 (10.77)***		
RISKSCON	ns	ns	ns	ns	ns	1.13 (4.98)**
AVAILABLE	1.39 (10.12)***	2.45 (18.14)***	1.64 (13.39)***	1.59 (10.49)***	1.63 (11.43)***	1.32 (7.91)***
REGULATION	1.08 (4.25)**	1.54 (6.26)**	ns	1.48 (6.14)**	1.59 (7.62)***	1.58 (7.25)***
LABELS	ns	ns	ns	1.50 (4.18)**	1.28 (3.14)*	1.50 (4.39)**
DIFORCON	ns	ns	ns	ns	ns	-0.94 (3.66)*
RMPP	0.05 (3.23)*	0.08 (2.94)*				
LVPP			-0.05 (3.85)*			
WWFPP				0.23 (7.70)***		
FCPP					0.076 (5.79)**	
AHPP						0.02 (4.02)**
N	146	99	143	139	143	138
Chi-Square Statistic[2]	24.668	38.914	26.959	37.399	38.824	35.912
Cox & Snell's R[2]	0.155	0.325	0.172	0.236	0.238	0.229
Nagelkerke's R[2]	0.217	0.454	0.241	0.332	0.334	0.332
Overall Predicted Power (%)	74.7	81.8	73.4	77	75.5	76.1
Concordance index	0.72	0.84	0.74	0.80	0.80	0.78

Notes: Wald Test-value is between brackets, *** 1%, ** 5%, * 10% significance levels, Cut-off = 0.50; ns: non- significant variable. All the estimations were done for the higher income level except for (1), which was for the lower income level; (2) Chi-Square P-value = 0.000.

On the other hand, the PRESERV variable was not statistically significant (at the 0.10 level of significance) for these Models. This would be explained by the high degree of trust Argentinean consumers have in milk products quality, both organic and conventional.

65% of the respondents (n_{1a}=146) ascribed great relevance to the brands they bought, as they constitute a confidence factor when it comes to shopping choices.

Among respondents whose monthly income is above US$ 500, WTP for organic leafy vegetables is mainly explained by this product shortage in the market (AVAILABLE), since respondents would buy more organic leafy vegetables, if they were readily available. These results agree with those found by Michelsen *et al.* (1999) and by Richman and Dimitri (2000). Moreover, organic food consumption (CONSUMP) also contributes to consumers' willingness to acquire organic leafy vegetables.

Indeed, those consumers who choose these vegetables representing a highly differentiated product in terms of packaging, presentation in container, serving size, and origin have a relatively high income level. In this regard, a high proportion of the respondents (78% of n_2=143) included in this analysis, whose educational level was high, consider that knowing leafy vegetables origin gives them confidence when it comes to shopping decisions.

The perception of high health risks associated with pesticides in the conventional varieties of these products turns the PESTICIDEV variable significant. The empirical evidence of these results is consistent with those by Weaver *et al.* (1992) and Baker (1999).

WTP for organic whole wheat flour is explained mainly by regular label reading when making shopping decisions (LABELS). Besides, 78% of the respondents (n_3=139) regularly look for information about food quality, and believe that there should be a food quality regulation system (REGULATION). The scarcity of this product in the market is also worth noting (AVAILABLE). These results are in accordance with those documented by Michelsen *et al.* (1999); Richman and Dimitri (2000); Gil *et al.* (2000) and Pearson (2001).

Consumers perceive whole wheat flour as a natural and healthy product. Respondents affirm that knowing the product origin and the store where it is acquired constitute confidence factors in their shopping choices.

WTP is further explained by the high health risks perceptions associated with pesticides in the conventional products (PESTICIDEF). In addition, 68% of the respondents believe that the greater this product processing, the higher the quality distrust.

High income level respondents are willing to pay price premiums for organic fresh chicken mainly because they believe that there should be a food quality regulation system (REGULATION). This result reinforces those previously found in focus groups studies, in which consumers expressed a high degree of concern about conventional chicken production processes as well as the notorious lack of control by the regulatory bodies (Farina and De Almeida, 2003; Rodríguez and Lacaze, 2005).

On the other hand, this product shortage in the market (AVAILABLE) together with the regular label reading by consumers when making shopping decisions (LABELS) play a minor, though significant, role in WTP. Finally, consumption of some of these products (CONSUMP) as well as the perception of high health risks associated with hormones present in the conventional varieties

(HORMONE) also contributes, to a lesser extent, to WTP understanding. In this sense, 60% of the respondents (n_4=143) sustain that knowing the product's origin constitutes a confidence factor when it comes to shopping choices.

WTP for organic aromatic herbs is explained mainly by regular label reading when making shopping decisions (LABELS) as well as by the REGULATION variable.

It is also worth noticing the perception of this product shortage in the market (AVAILABLE). This is explained by the fact that most organic aromatic herbs production is exported, as export prices are more profitable. Knowledge and identification of organic food are also relevant to explain WTP, as it is evidenced in CONSUMP, RISKSCON and DIFORCON variables. In this sense, 68% of the respondents (n_5=138) sustain that knowing the product's origin constitutes a confidence factors when it comes to shopping choices.

Finally, it should be highlighted that more than 60% of the respondents included in both Model 4 (fresh chicken) and Model 5 (aromatic herbs) believe that the greater this product processing, the higher the quality distrust. This was also mentioned when explaining the explanatory variables for organic whole wheat flour.

After running the models, both the respondent's educational level and the household monthly income were not statistically significant as explanatory variables. Therefore, they were disregarded when estimating the final models.

Considering that the interviewers surveyed respondents with a medium-high socio-economic level (80% of the respondents had gone into higher education and 62% affirmed to earn more than US$ 500 monthly), it was not feasible to isolate the particular effect of both income and educational variables. In addition, as respondents had to indicate the range within which their monthly household income fell, misinformation could have been provided regarding the income range. Undoubtedly, those results should be verified in subsequent researches.

The Models' Performance was tested with Pearson's Chi-Square Statistic, which indicates that all Models fit adequately.

The alternative forms of R^2 for Binomial Logit Models are Cox & Snell's R^2 and Nagelkerke's R^2. The highest values of alternative R^2 are yielded in Model 1.b for regular milk (0.325 and 0.454 respectively) (Ryan, 1997; Menard, 2000).

The Overall Predicted Power is above 73% for all Models. The Concordance Index, which estimates the predictions and outcomes probability of concordance, yields values above 0.50 for all the estimated models, indicating that predictions are better than random guessing (Agresti, 2002).

5.2 Analysis of 2002-2005 organic price premiums trends

According to a European Union Report,[67] organic price premiums are lower for processed products (e.g. whole wheat flour and regular milk) than for unprocessed products (e.g. fresh chicken and leafy vegetables). Table 6 below shows that such trend replicated in Argentina, when analysing the organic price premiums prevailing in the domestic market when this study was

---

[67] Commission Européenne G2 EW – JK D 2005 Report.

carried out.[68] For comparative purposes, the organic price premiums calculated in a previous study, also conducted in Buenos Aires city in 2002, were included too (Rodríguez *et al.*, 2003).

Since the devaluation of the Argentinean peso in 2002, the prices of both conventional and organic food products have increased, but taking into account that Argentinean organic production has foreign markets as its main destination, the domestic prices of tradable goods rise in the country as export prices do. This has led to changes in the organic vs. conventional price relations. In this sense, the case of organic aromatic herbs is a remarkable example, with a price premium sharp increase from 62.35% in 2002 to 298.33% in 2005. Regarding all the selected products, the retail price premiums prevailing in this year vary from 6% to 298%.

In 2002 organic regular milk was cheaper than conventional milk. This could be explained by the steady increase of dairy conventional products prices due to decrease in production and also an increase in foreign demand of these products. In 2005 the opposite occurred with a 13% price premium due to the increase of organic international prices. The same applies to organic leafy vegetables, which experienced a sharp rise in 2005 (84.54%). On the other hand, the organic price premium for whole wheat flour decreased comparing 2005 to 2002.

5.3 WTP Calculation

By applying the Equation 2 described in Section 4.2, Table 7 below displays the average WTP for each selected product, i.e. the additional premium respondents are willing to pay for each organic product over the price of the conventional product. These values are expressed in %/kg or %/l. As mentioned in Section 5.1, all the estimations were made for the higher income level (more than US$ 500) except for regular milk, which was estimated for both income levels.

This table also includes the averages additional premiums charged for organic products at the stores considered in the survey. Finally, the differences between respondents' calculated WTP [A] and the real premiums [B] are presented.

---

[68] Premiums calculated based on real prices of both organic and conventional selected products and collected in the stores where the survey took place.

Table 6. 2002-2005 organic price premiums trends. Source: consumer Survey, Buenos Aires City/2005 and Rodríguez et al. (2003).

Organic over conventional products average price premiums			
Selected product	2002 price premium[1]	2005 price premium[1]	2002-2005 price premium change
Regular milk	-0.61%	13.84%	↑
Leafy vegetables	21.80%	84.54%	↑
Whole wheat flour	172.31%	5.91%	↓
Fresh chicken	[2]	24.61%	[2]
Aromatic herbs	62.35%	298.33%	↑

[1] Organic price premiums are expressed in %/kg or %/l.
[2] No data available because organic Fresh Chicken was unavailable in the domestic market in 2002.

Table 7. Average WTP estimations. Source: consumer survey, Buenos Aires City/2005.

Model	Average WTP (%/kg)	Average price premium (%/kg) [2]	% Difference
	[A]	[B]	[A] – [B]
1a. Regular milk	12.2 [3]	13.8 [3]	-1.64
1b. Regular milk [1]	11.6 [3]		-2.24
2. Leafy vegetables	87	84.5	2.46
3. Whole wheat flour	7.5	5.9	1.59
4. Fresh chicken	20	24.6	-4.61
5. Aromatic herbs	110	298.3	-188.33

[1] Estimation for the lower income level.
[2] Calculated as the percentage by which the price of the organic product is above the price of a similar conventional product. Premiums derived from price collection carried out in the stores where the survey took place.
[3] Expressed in %/l. *Exchange rate: 1 US$ = 3 Argentinean pesos ($).*

While higher income level respondents (Model 1.a) are willing to pay WTP for organic whole wheat flour is 7.5% higher if compared to the price paid for conventional whole wheat flour; this WTP being slightly above the organic whole wheat flour real price premium in as much as 1.59%.

The results yielded by Model 4 show that WTP for organic fresh chicken is 20% higher if compared to the price paid for conventional Fresh Chicken. This WTP value is below the organic Fresh Chicken real price premium in as much as 4.61%.

Finally, WTP for organic aromatic herbs is 110% higher if compared to the price paid for conventional Aromatic Herbs; this WTP being below the organic Aromatic Herbs market price premium in as much as 188%.

To sum up, it is worth mentioning that the key factors that help to explain organic WTP for the selected products are consumption of organic products, health risks perceptions linked to hormone and pesticide content, regulation concerns, perceptions of irregular organic availability in the domestic market, labels reading, and the real price premiums charged over the conventional prices. Still, the relative importance of these factors is different when WTP is explained for each case.

Health risks perceptions contributed to explaining WTP for leafy vegetables, whole wheat flour, fresh chicken and aromatic herbs, but have no relevance when trying to explain WTP for regular milk. Hence, hypothesis #1 - *Health risks perceptions linked to hormone, pesticide and preservers content in several food products affect significantly consumers' willingness to pay for organics* - has been rejected only for the regular milk estimations.

According to the results of the estimated models, the effect of the regulation program was statistically significant for both unprocessed products (like fresh chicken) and processed products (like aromatic herbs, regular milk and whole wheat flour); but had no significance for leafy vegetables. Therefore, hypothesis #2 *The effect of regulation programs on the willingness to pay for organic unprocessed products is lower than for organic processed products* - has also been rejected. This could be explained by the fact that the degree of product processing may not seem to condition the effect regulation programs have on consumers' WTP.

It should be mentioning that 74% of the respondents affirm that the regulatory bodies are inefficient, and 70% prefer a public food regulation system to a private one.

Undoubtedly, price premiums play a critical part in the applied methodology when calculating WTP. If organic market prices were slightly reduced, the differences between WTP and real price premiums would get reduced as well. Consequently, consumers would have greater access to organic regular milk and organic fresh chicken. On the other hand, organic aromatic herbs real price premiums restrict their consumption in the domestic market, which is exceedingly influenced by the high revenues obtained when exported.

Even tough WTP for organic leafy vegetables is somewhat above the real price premiums, the problem seems to be the lack of regular supply of these vegetables in the domestic market. Also WTP for organic whole wheat flour is barely above the organic price premium charged in the market.

To conclude, hypothesis #3 - *Consumers' willingness to pay for organic food is lower than retail prices in the stores* - can not be rejected for regular milk, fresh chicken and aromatic herbs. In addition, it is important to point out that the difference between observed prices and stated WTP may be caused by the hypothetical survey itself. It should be useful to test for this hypothesis by applying an alternative approach.

6. Final remarks

The results of WTP estimates obtained for the selected products indicate that organic products are positively valued in Argentina, since consumers are willing to pay price premiums to acquire these products of better quality. Such results are undoubtedly conditioned by the real price premiums charged in the domestic market, which, in turn, are conditioned by the incidence of export prices, as the foreign market is the main destination of organic products production in Argentina.

It is also worth mentioning that the WTP values for each of the selected organic products are explained by the consumption of organic products, health risks perceptions linked to hormone and pesticide content, regulation concerns, perceptions of organics irregular availability in the domestic market, and labels reading. Still, the relative importance of these factors is different when WTP is explained for each product.

This study verifies that those consumers whose income is above US$ 500 are worried about products quality as well as about health risks connected to pesticide-residue and hormone-treated product. The high real price premiums condition the purchase of these healthy-perceived products, even when respondents express their desire to acquire them. These consumers know what organics stands for, they perceive products scarcity and irregular store availability, and they would be willing to increase consumptions if these products were cheaper. The price premiums in the market depend on the product type but, regarding the analysed products, they range between 6% and 298%.

The effect of regulation programs on consumers' WTP may not seem to be conditioned by the degree of product processing. On the other hand, the concern consumers express regarding current regulatory and controlling bodies is worth noticing as well as their preference for a public system.

To conclude, the scarcity of organic products in the domestic market as well as of high price premiums are identified as the most difficult obstacles to overcome when it comes to organic domestic consumption expansion in Argentina.

The involvement of general food retailers in the organic food market is of major importance and should be encouraged in order to increase organic products market share. Therefore, an increase in production levels is a must together with reductions in production, processing and/or trading costs, which, in turn, translate into sale price reductions, and into an increase of organic products consumption. Lower distribution costs constitute a contributing factor which reduces price premiums by involving general food retailers.

Most countries with lower consumer price premiums have a common national label, and such label recognition is usually high. Clear recognition is a pre-requisite if organic products are to break free from niche product status. This is another key issue Argentina still has to address if it wishes to expand in the organic domestic market.

As mentioned in other studies, pull strategies should be applied to promote organic market growth. To do so, the organic market actors must convince themselves that there is a growing consumer demand for organic food and that any efforts made to increase organic products supply will enhance their competitiveness.

Argentinean current system devotes most of its resources to those enterprises and actors already inserted in the global economic system, and do not contribute to smallholders' farms inclusion through regional development programs, thereby strengthening the asymmetric distribution of benefits. The potential growth of the domestic market should be encouraged as a step towards targeting foreign markets (Rodríguez, 2005).

Given that scenario, the government goal should be to support already operating markets, assuring an equal development of both supply and demand. As consumers claim, research, consumer food education and counselling programs should be further supported. In Argentina, efficient government actions need be directed towards a stricter control system; a better coordination between public and private organisations; and a long-term planning for the organic sector.

Acknowledgements

Support for this research was provided by the Agencia Nacional de Promoción Científica y Tecnológica (FONCyT) [National Agency for Technologic and Scientific Promotion] and the Universidad Nacional de Mar del Plata [National University], Argentina (PICTO-9810).

References

Agresti, A., 2002. An introduction to categorical data analysis. Canada: John Wiley & Sons INC.

Antle, J., 1999. Benefits and costs of food safety regulation. Food Policy, 24: 605-623.

Baker, G., 1999. Consumer preferences for food safety attributes in fresh apples: market segments, consumer characteristics, and marketing opportunities. Journal of Agricultural and Resource Economics, 24: 80-97.

Blend, J. and E. Van Ravenswaay, 1998. Consumer demand for ecolabelled apples: Survey Methods and descriptive results. Staff Paper 98-20. Dept. of Agricultural Economics, Michigan St. University.

Brewer, K., 1999. Design-based or prediction-based inference? Stratified random vs. stratified balanced sampling. International Statistical Review, 67: 35-47.

Carmona-Torres, M. and J. Calatrava-Requena, 2006. Bid design and its influence on the stated willingness to pay in a contingent valuation study. Contributed paper prepared for presentation at the International Association of Agricultural Economists Conference, Gold Coast, Australia, August 12-18, 2006. Available at: http://agecon. lib.umn.edu/cgi-bin/pdf_view.pl?paperid=22558&ftype=.pdf.

Chen, K., M. Ali, M. Veeman, J. Unterschultz and T. Le, 2002. Relative importance rankings for pork attribute by Asian-origin consumers in California: applying an ordered Probit Model to choice-bases sample. Journal of Agricultural and Applied Economics, 34: 67-69.

Chow, S., 2002. Issues in statistical inference. History and Philosophy of Psychology Bulletin, 14: 30-41.

Commission Européenne, 2005. Direction Générale de L'Agriculture et du Développement Rural. Organic Farming in the European Union. Facts and Figures. Report G2 EW – JK D (2005). http://ec.europa.eu/agriculture/qual/ organic/facts_en.pdf.

Eom, Y., 1994. Pesticide residue risk and food safety valuation: A random utility approach. American Journal of Agricultural Economics, 76: 760-771.

Farina, T. and S. de Almeida, 2003. Consumer perception on alternative poultry. International Food and Agribusiness Management Review, 5 (2). Available at: http://www.ifama.org/tamu/iama/nonmember/ OpenIFAMR/OpenIFAMR.htm

Gil, J., A. Gracia and M. Sánchez, 2000. Market segmentation and willingness to pay for organic products in Spain. International Food and Agribusiness Management Review, 2: 207-26.

Goldberg, I. and J. Roosen, 2005. Measuring consumer willingness to pay for a health risk reduction of salmonellosis and campylobacterosis. Paper prepared for presentation at the 11th Congress of the European Association of Agricultural Economist, Copenhagen, Denmark, August 24-27.

Govindasamy, R. and J. Italia, 1999. Predicting willingness-to-pay a premium for organically grown fresh produce. Journal of Food Distribution Research, 30: 44-53.

Gracia, A., J.M. Gil and M. Sánchez, 1998. Potencial del mercado de los productos ecológicos en Aragón. Gobierno De Aragón.

Halbrendt, C., L. Sterling, S. Snider and G. Santoro, 1995. Contingent valuation of consumers' willingness to purchase pork with lower saturated fat. In: J. Caswell (Ed.) Valuing food safety and nutrition, pp. 319-339.

Hanemann, W., 1989. Welfare evaluations in contingent valuation experiments with discrete responses: Reply. American Journal of Agricultural Economics, 71: 1057-1061.

Hanemann, W., 1984. Welfare evaluations in contingent valuation experiments with discrete responses. American Journal of Agricultural Economics, 66: 332-341.

Hartman and New Hope, 1997. The evolving organic marketplace. Hartman and New Hope Industry Series Report. Washington D.C.

Henson, S., 1996. Consumer willingness to pay for reductions in the risk of food poisoning in the UK. Journal of Agricultural Economics, 47: 403-420.

IFOAM (International Federation of Organic Agriculture Movements), 1998. The Mar del Plata Declaration. Available at: http://www.ifoam.org/press/positions/pdfs/Declaration_Mar_del_Plata_1998.pdf.

INDEC (Instituto Nacional de Estadísticas y Censos), 2005. Incidencia de la pobreza y la indigencia en 28 aglomerados urbanos. Resultados 1º Semestre de 2005. Información de Prensa. ISSN 0327-7968. Buenos Aires, 22 de setiembre de 2005. Available at: http://www.indec.gov.ar.

INDEC (Instituto Nacional de Estadísticas y Censos), 2001. Censo Nacional de Población y Vivienda 2001. Resultados definitivos para la Ciudad de Buenos Aires. Available at: http://www.indec.gov.ar.

Jordan, J. and A. Elnagheeb, 1991. Public perception of food safety. Journal of Food Distribution Research, 22: 13-22.

Kinsey, J., 1993. GATT and the Economics of food safety. Food Policy, 18: 163-176.

Kuchler, F., K. Ralston and J. Tomerlin, 2000. Do health benefits explain the price premiums for organic foods?. American Journal of Alternative Agriculture, 15: 9-18.

Lancaster, K., 1966. A new approach to consumer theory. Journal of Political Economy, LXXIV: 132-157.

Lee, K. and C. Hatcher, 2001. Willingness to pay for information: An analyst's guide. Journal of Consumer Affairs, 35: 120-140.

Lohr, L., 2001. Factors affecting international demand and trade in organic food products. Economic Research Service/USDA/WRS-01-1.

Menard, S., 2000. Coefficients of determination for multiple logistic regression analysis. American Statistical Association, 54: 17-24.

Michelsen, J., U. Hamm, E. Wynen and E. Roth, 1999. The European market for organic products: growth and development. Organic farming in Europe: economics and policy. Vol. 7.

Misra, S., L. Huang and S. Ott, 1991. Consumer willingness to pay for pesticide free fresh produce. West Journal of Agricultural Economics, 16: 218-227.

Pearson, D., 2001. How to increase organic food sales: results from research based on market segmentation and product attributes. Australasian Agribusiness Review, 9: paper 8.

Portney, P., 1994. The contingent valuation debate: why economists should care. Journal of Economic Perspectives, 8: 3-17.

Richman, N. and C. Dimitri, 2000. Organic foods: niche marketers venture into mainstream. Agricultural Outlook, June-July, 11-14.

Rodríguez, E., 2005. The domestic and foreign markets of organic products in Argentina. Executive summary presented to the International Workshop 'How can the poor benefit from the growing markets for high value agricultural products?', CIAT, Cali, Colombia, October 2005.

Rodríguez, E. and V. Lacaze, 2005. Consumer preferences for organic food in Argentina. Handbook of the 15th Organic World Congress of the International Federation of Organic Agricultural Movements (IFOAM), September 20-23, Adelaide, South Australia, Australia.

Rodríguez, E., B. Lupín and V. Lacaze, 2006. Consumers perceptions about food quality attributes and their incidence in Argentinean organic choices. Poster paper presented at the International Association of Agricultural Economists Conference, Gold Coast, Australia, August 12-18, 2006. http://agecon.lib.umn.edu/cgi-bin/pdf_view.pl?paperid=22222&ftype=.pdf.

Rodríguez, E., B. Lupín and V. Lacaze, 2005. Las percepciones de calidad de los consumidores de alimentos diferenciados. Trabajo presentado en la XXXVI Reunión Anual de la Asociación Argentina de Economía Agraria, Adrogué, Buenos Aires, octubre de 2005. Resúmenes de Trabajos y Comunicaciones, p. 31.

Rodríguez E., N. Gentile, B. Lupín and L. Garrido, 2003. El mercado interno de alimentos orgánicos: Perfil de los consumidores argentinos. Revista de la Asociación Argentina de Economía Agraria Nueva Serie, VI, N° 1, Otoño.

Ryan, T., 1997. Modern regression methods. Canada: John Willey & Sons INC.

Schonlau, M., R. Fricker and M. Elliot, 2002. Conducting research surveys via e-mail and the web. Available at: http://www.rand.org/publications/MR/MR1480.

SENASA (Servicio Nacional de Sanidad y Calidad Agroalimentaria), 2007. Situación de la producción orgánica en Argentina durante el año 2006. Available at: http://www.senasa.gov.ar/Archivos/File/File827-2006.pdf.

Thompson, G., 1998. Consumer demand for organic produce: what we know and what we need to know. American Journal of Agricultural Economics, 80: 113-118.

Underhill, S. and E. Figueroa, 1996. Consumer preferences for non-conventionally grown produce. Journal of Food Distribution Research, 27: 56-66.

Van Ravenswaay, E., 1995. Public perceptions of agrichemicals. Council for Agricultural Science and Technology, Iowa.

Van Ravenswaay, E. and J. Wohl, 1995. Using contingent valuation methods to value the health risks from pesticide residues when risks are ambiguous. In: J. Caswell (Ed.) Valuing food safety and nutrition, pp. 287-317.

Weaver, R., D. Evans and A. Luloff, 1992. Pesticide use in tomato production: consumer concerns and willingness-to-pay. Agribusiness, 8: 131-142.

Direct marketing to tourist hotels: a study on horticultural market in Fiji

C. Salvioni

Abstract

This paper reports a study of the horticultural chain in Fiji. The objective of the research was to understand how the domestic horticultural supply could meet the demand currently generated by the tourism sector. If policy interventions are to be directed at strengthening backward economic linkages between tourism and local food supplier, a better understanding of factors driving farmers marketing choice is required. This paper analyses the decisions of farmers to engage in direct selling to hotels. Hotels pay premium prices for quality fresh products and direct marketing can allow farmers to retain the highest possible portion of this premium without sharing it with intermediaries. I apply a logit model to data collected in the chain study. The results suggest that quality and, more specifically, the use of Integrated Pest Management (IPM) and post harvest technologies play an important role in growers' choice of direct marketing to hotels. In addition, the decision is affected by ethnicity, by whom (growers or extension staff) is in charge of the choice of the marketing channel, by the involvement of the husband in the selling activity, by distance from the market and by the ownership of means of transportation.

Keywords: horticultural markets, direct marketing, discrete choice models, Fiji

1. Introduction

Fiji, as many other Small Island Developing (SID) countries, faces several disadvantages related both to its reduced size and to being made up by many small islands. These disadvantages include, among others:
- limited resources, which lead to undue specialisation;
- excessive dependence on international trade and hence vulnerability to global developments;
- costly public administration;
- insufficient infrastructures -including transportation and communication;
- limited institutional capacities;
- domestic markets too small to provide significant scale economies; and
- reduced volumes available for exports, sometimes from remote locations, which lead to high freight costs and reduced competitiveness.

The performance of Fiji's economy since 1999 has been irregular, also due the political and constitutional instability experienced between 1999 and 2001. According to recent estimates (NZIER, 2007) the 2000 coup caused a 39.4% fall in visitor arrivals; a 33.1% fall in investment and a 3.5% increase in the real interest rate. A moderate countervailing factor was a 9.9% increase in government expenditure. In percentage terms, the biggest estimated long-run effects were a 36% decrease in informal sector wages, a 24% decrease in exports, an 8% decrease in real GDP and a 7% decrease in real national welfare. Although it is too early to be certain, the effects of the 2006 military coup are likely to be quantitatively similar, because the external and internal shocks are likely to be the same. However, it is very unclear at the moment to predict what the quantitative impacts will be.

In addition, the Fijian economy continues to face uncertainties due to unsolved political issues[69] and the impact of restructuring its agriculture and manufacturing sectors in response to changes in international trade arrangements[70].

The crisis of the sugar sector, started even before the EU price reductions[71], has already caused a sharp drop in the contribution of this sector to GDP -from 11.3% in 1995 to 6% in 2005. This, in turn, has made the overall contribution of the agricultural sector to total GDP decline[72] and the tourism and textile become the largest GDP contributors.

Among sectors performing well, tourism is the leading one: it presently is the country's largest source of economic growth, investment and foreign exchange earnings. The multiplier effects of tourism growth, though, are limited by the high dependence of tourism industry on imported supplies.

By increasing backward economic linkages between tourism and local food suppliers, Fiji can:
- increase benefits from tourism development;
- improve benefits distribution; and
- reduce the pressure on the national balance of payments.

First of all, the increase in the demand for local fresh food supplies by hotels and resorts can generate positive direct, indirect and induced impacts on domestic agricultural production, hence on farm incomes. At present, food, as well as most other supplies and services used in the tourism sector, are brought in from overseas. This means that the growth in the tourism sector demand for food creates less than the total income that would be generated if this demand were satisfied by domestic supply. The substitution of domestic food supplies to the imported ones, can make the tourism multiplier increase, in this way enhancing benefits from tourism development.

In addition, stronger backward economic linkages between tourism and the domestic agricultural sector can increase not only the level but even the distribution of benefits from tourism growth. At present, only the population living in the coastal areas where the tourism industry is located is gaining from tourism, If the hotels and restaurants demand for domestic fresh food increases, then even the livelihoods of rural, often poor, agricultural population can improve.

Finally, the substitution of the current import flow of fresh food with a domestic supply can help reducing the widening deficit in the balance of payments[73], thus contributing to increasing the country's macroeconomic stability. The Strategic Development Plan 2007-2011 has addressed this import substitution stance and has indicated in (a) the enhancement of tourism industry and agriculture sector linkages to match demand and supply, and (b) the promotion of food safety and quality, the two leading strategies to meet the objective of reducing the value of food imports from $ 370m in 2006 to $ 260m in 2011 (Ministry of Finance and National Planning, 2007).

---

[69] Such as constitutional reform and leasehold land issues.

[70] Mainly related to the loss, in 2005, of its garment quota with the USA and the progressive erosion of its preferential access to the EU market, where up to 60% of Fiji's sugar production (173,000 tons/year) has been sold in the past years.

[71] EU has planned a price reductions from 5% in 2006/2007 to up to 39% by 2009/2010.

[72] From 19% in 1989 to 13% in 2005.

[73] The annual food import bill continues to rise.

Previous research (Bennett *et al.*, 1999) has however pointed out that these linkages cannot be assumed to emerge alone – they must be actively facilitated. This points to the need of establishing how these linkages could be best put in place.

If policy interventions are to be directed at strengthening backward economic linkages between tourism and local food supplier, a better understanding of farmers production and marketing choices is required. This paper aims to contribute in this regard by using the findings of a horticultural chain study recently implemented in Fiji to identify the factors affecting farmers' choice of selling directly to hotels.

The rest of the work is organised as follows. In the next section, the methodology used to carry out the horticultural chain study is presented. We then present the major findings of the study in regard to the demand for horticultural products of the tourism sector and the supply of the Fijian horticultural. In the last paragraph we report the estimation results from applying a discrete choice model to identify the factors affecting the farmers' decision to engage in direct sales to hotels.

2. The horticultural chain study

The study implemented by FAO-UN[74]/INEA[75]/SPC[76] was meant to understand the nature and relevance of the existing constraints of the local production and distribution of high quality horticultural productions so as to meet the tourism demand for fresh food in Fiji. The research was implemented in the context of three chain studies carried out throughout the South Pacific region by the FAO regional project: 'Support to the Regional Programme for Food Security in the Pacific Islands Countries' (GTFS/RAS/198/ITA).

The intent of the study was to solicit information from agricultural, trading and tourism firms in Fiji on the existing constraints to link the tourism industry demand to the domestic horticultural production.

More specifically, the objectives of the study were:
- to investigate the present domestic demand for the targeted products and the existing constraints to the development of their national supply so as to replace (at least partly) their current import flows;
- to prepare a strategy to overcome identified bottlenecks and to assure the full exploitation of the detected potentials.

The study followed a 'participatory' approach by which INEA proposed first drafts of both the overall methodology and the specific tools (questionnaires) to be used to collect information, which were then validated by regional (SPC and FAO-SAPA) and national counterparts (country coordinator for Fiji).

The four horticultural products (mango, papaya, tomato and carrot) targeted with the study were selected by the local Ministry of Agriculture in the light of their relevance within the fresh-agricultural products basket currently demanded by the Fiji tourism industry.

---

[74] Food and Agriculture Organization of the United Nations.

[75] Italian Institute of Agricultural Economics.

[76] South Pacific Commission.

Four surveys were then implemented to study the state of facts of the four targeted crops. To this end, four types of questionnaires were elaborated, one for each of the relevant operator within the chain (horticultural producers, domestic traders, importers and tourism operators). The questionnaires were worked out based on secondary information collected through an 'ex-ante' assessment and were validated, prior to their use, through field testing.

Although changing according to the operator investigated, the questionnaires targeted primary information related to: the enterprise; production or procurement of the investigated crops; harvest and post-harvest issues; domestic marketing or trading (importing) issues and, only in the case of the producer's questionnaires, matters dealing with certification schemes, financing and extension services.

The surveys were carried out in May-July 2006 by a private consultant (working as national coordinator) assisted by 24 interviewers/data collectors. The investigated samples (Table 1) consisted of 238 farmers, 100 tourism operators and 55 traders – out of which 5 were importers.

Table 1. Sample dimension by survey.

Survey	Interviews
Producers	238
Tourism operators	100
Hotels	46
Restaurants	33
Supermarkets	21
Traders	50
Importers	5
Total	393

3. Demand for horticultural products of tourism sector

The tourism sector is the country's largest source of economic growth, investment and foreign exchange earnings. Visitor arrivals (Figure 1) have been growing over the last decades, although with some fluctuation due to political turmoil. Medium term prospects were very encouraging, although they might need to be reconsidered in the light of the 2006 coup.

Tourism earnings are by far the largest foreign exchange earner and reflect the number of visitor arrivals as well as their length of stay. Food and beverage consumption are a significant part of tourist expenditures. According to recent investigations (Berno, 2006), around 15% of visitor expenditure in Fiji is currently spent on food.

In addition, it has been estimated that, although nearly half of hotel purchases are from local providers, two thirds of overall food import expenditures destined to meet the tourism sector demand is for products that could be grown in Fiji (Berno, 2006).

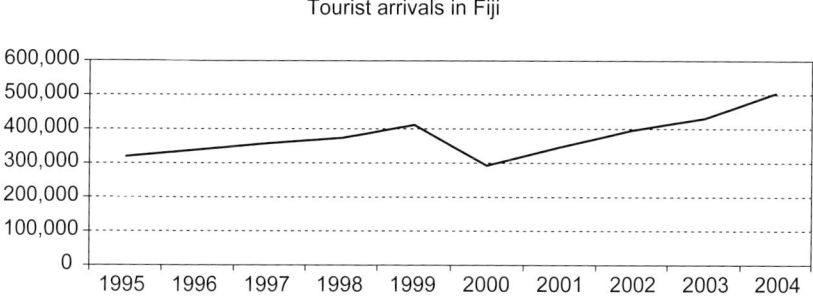

Figure 1. Tourists annual arrivals. Source: Fiji Island Bureau of Statistics.

On the basis of the projected growth in tourist arrivals, the value of the tourism sector demand for food could increase significantly. If domestic production fails to respond to this increasing demand, the tourist industry growth will be accompanied by even higher levels of food imports.

The survey carried out among tourism sector operators has shown that almost all the hotels and restaurant requirements of papaya and mango are met by domestic supplies. However, in the case of tomato, hotels and restaurants have a clear preference for imported products while nearly all hotels surveyed sourced practically all their requirements of carrots from importers (Figure 2).

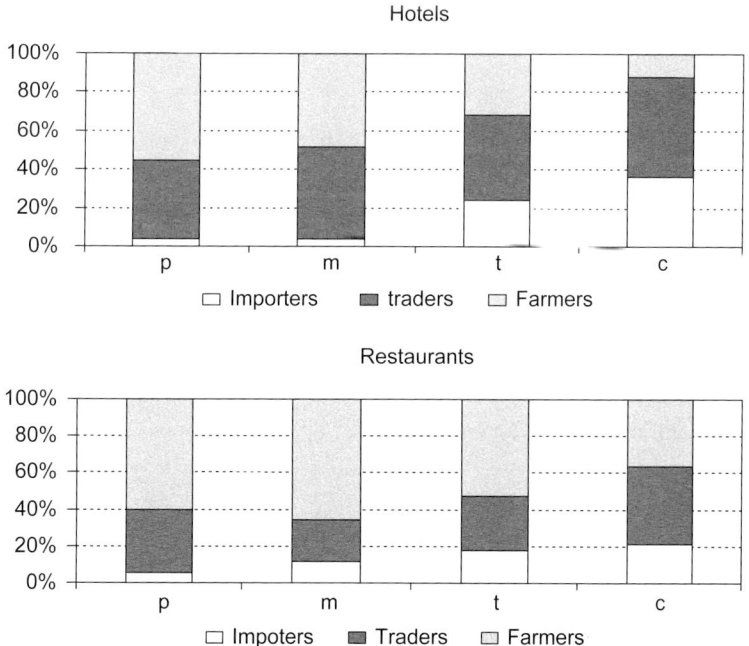

Figure 2. Hotels and restaurant: origin of purchases (percentage).

The survey also confirmed that the two major reasons for preferring imported fresh produces are unavailability and inconsistency of local high quality fresh horticultural supplies. More specifically in the case of carrots the problem is mainly the lack of local supply, while in the case of the other products the major problem is reported to be the poor quality of local supply.

4. The horticultural sector in Fiji

The agriculture sector, excluding sugar, contributes around 6% to GDP; accounts for around 14% of agriculture exports and for 15% of total food imports and sustains 54% of the total country's population.

The total value of horticulture production (around $ 50 million) is growing quite rapidly either in terms of contribution to the total agricultural value added and to exports.

Most of the commercial horticultural supply is originated in the Ba region and the Sigatoka Valley. Even though this latter area is smaller than the Ba region, it is however a larger supplier of vegetables to the nearby tourism sector.

A random sample of 252 farmers were interviewed and data on 238 of them were used for the descriptive analysis (Table 2).

Farms in which papaya and tomato are grown have on average bigger size (respectively 17.3 and 13.2 ha) than those in which mango and carrots are grown (9.6 and 2.8 ha).

Most of farmers investigated make use of traditional production technologies (Table 3) with the exception of all tomato producers in the Nadroga / Navosa province who make use of Integrated Pest Management. In addition, a tomato grower is presently converting its production into organic.

With regards to certification schemes, none of the farmers interviewed indicated that they make use of organic, fair trade, EUREP-GAP or other certification schemes.

The most commonly cited production constraints (Table 4) for the four crops were issues related to availability of improved varieties, seeds and credit, and to pests and diseases. At the same time, it is worth noting that the land issue was cited amongst the least relevant production constraints. For Indian farmers, in fact, land tenure may not have been a problem because the lease agreements have been worked out to their satisfaction. For Fijian farmers, however, the question related to

Table 2. Farmers interviewed by province and crop.

Province	Fruit and vegetables					
	Papaya	Mango	Tomato	Carrot	Papaya + tomato	Total
Ba	22	40	21	10	0	93
Nadroga / Navosa	38	0	55	0	23	116
Ra	1	10	18	0	0	29
Total producers	61	50	94	10	23	238

Table 3. Technology used by crop.

Production technology	Papaya	Mango	Tomato	Carrot	Total
Traditional	83	50	91	10	234
IPM	1	0	26	0	27
Organic	0	0	0	0	0
Under transition	0	0	1	0	1
Total	84	50	118	10	262

Note: since a producer may be growing 2 or more of the targeted commodities, the numbers in the table exceeds the number of producers (238) interviewed.

Table 4. Ranking of production problems by crop.

Production problem	Papaya	Mango	Tomato	Carrot
Lack of improved varieties	1	4	6	4
Fertilisers / chemicals not available or of bad quality	2	11	3	4
Certified seed too expensive	3	5	7	2
Certified seed not available from local dealers	4	7	4	3
Too many pests and diseases	5	1	2	6
Lack of specific credit lines	6	3	8	1
Lack of technical advice	7	9	5	8
Lack of suitable land	8	2	9	9
Land tenure	9	10	10	10
Lack of water for irrigation	10	6	1	7
Inadequate harvesting technology	.	12	.	.
Trees too scattered and grow wild	.	8	.	.

Notes: relevance of the matter decreases from 1 to 12.

this issue in the questionnaire did not delve into the issue of communal land and thus did not allow to explore this matter satisfactorily despite the fact that land tenure, especially in terms of security, may be an issue among these growers.

Papaya and tomato growers prefer to market directly (Table 5), whilst carrot and mango producers make greater use of intermediaries. The choice of the marketing channels by carrots growers is mainly explained by the localisation of farms producing this crop in the upper part of the Sigatoka Valley, where the temperature is cooler. The distance from the final market and the poor infrastructures (roads) makes it difficult for these farmers to have direct contacts with the final demand, and it explains why they rely on wholesalers in village/town to sell their products. Distance is often at the basis of the decision of many mango growers to rely on intermediaries more than on direct selling. Commercial farms producing this crop are in fact limited to two main sites, both on the drier western side of the main island

The interviewed producers identify price fluctuations (Table 6) as the most important problem in marketing for almost all the targeted commodities, with the exception of carrots for which

Table 5. Marketing channels by crop.

Selling practice	Papaya (%)	Mango (%)	Tomato (%)	Carrot (%)
Through intermediaries				
Through intermediaries when the production is still in the field/on the tree	17	28	7	0
Through wholesalers in village/town markets	9	36	20	90
Through retailers in village/town markets	9	1	11	10
Direct sales				
Sell directly in my farm/in front of my house/ on the road side	18	8	13	0
Sell directly in village/town markets (personally or family members)	24	21	25	0
Sell directly to supermarkets	1	0	12	0
Sell directly to hotels and/or restaurants	10	1	11	0
Other	12	5	1	0
Total	100	100	100	100

Table 6. Ranking of marketing problems by crop.

Marketing problems	Papaya	Mango	Tomato	Carrot
Inadequate or too expensive post-harvest technology	6	7	6	7
Inconsistency of supply flows	5	6	8	6
Lack of transport	8	8	7	2
Final buyers too far away from the production areas	7	5	5	3
Unreliable demand from the tourism sector	2	3	3	7
Low quality of supplies	4	1	2	5
Lack of market information	3	3	4	1
Market price fluctuations	1	1	1	4
Other	9	9	9	9

the distance from the market and lack of means of transportation are more important problems than prices. As a matter of fact, the municipal market wholesale prices show a seasonal pattern, but virtually no sign of instability. The market price fluctuation problem seems to be in some way related to the unreliable demand from the tourist sector. This is ranked as one of the top three marketing problems identified by the survey. The horticultural growers reported that hotels and restaurants appear to have difficulties in estimating their demand of food, despite the fact that most of their guest numbers are known in advance, and that hotels often suddenly increase their requirements or cancel their orders. This results in tensions on the local markets that have to absorb the surplus of supply and of demand resulting from the changes in hotel requirement. The suggestions given by growers to overcome this problem were to extend the use of contract farming with the final purchasers (hotels) to secure sales and to put in place a system to secure markets for vegetables as is done with sugar.

Lack of market information is considered by the producers one of the most important problems in the marketing of mango, papaya and tomato, while the low quality of supply is reported to be a problem mainly in the marketing of mango and tomato.

The lack of market information and the price fluctuations are reported to be among the major marketing problem also by carrots growers. In addition to this, the marketing of carrots is negatively affected by the lack of transports and distance of farms from the market. Information collected during the survey have also shown that the wholesale prices of local production are constantly higher of those of the imported produce. For this reason, substitution of imported supplies does not appear to be an achievable objective in the short time.

5. Determinants of direct sales to hotels

The data collected in the survey were used to better understand what are the determinants in the establishment of direct linkages between horticultural growers and hotels and restaurants, that is what are the variables that are affecting the probability of farmers to directly sell their products to the tourism sector agents. Hotels and resorts pay premium prices for quality fresh produce and direct selling can allow farmers to retain the highest possible portion of this premium without sharing it with intermediaries.

5.1 Modelling the choice of direct selling to hotel

Discrete choice models can be used to analyse farmers' decision of direct selling to hotels within a utility maximisation framework. In these models the observed choice is considered an expression of a continuous latent variable reflecting the propensity to choose a specific option amongst diverse alternatives.

The basic assumption here is that farmer's choices are driven by a random utility model (RUM). Random utility models are founded on the assumption that agents undertake an action based on a marginal cost/marginal benefits calculation derived from the utilities achieved with their choice. The utilities are not observable, but the observed choice reveals which one provides the greater utility. In this context, binary choice logit/probit models are usually employed.

In the binary specification model I propose to estimate the direct selling to hotels choice probability 'y_i' denotes the categories –not selling to hotels and restaurants ($y_i = 0$), directly selling to hotels and restaurants ($y_i = 1$). The binary model estimates the probability that a farmer choices to selling directly to hotels and restaurants, using household and farm characteristics as regressors. The estimated coefficients are used to calculate the marginal effects - that is, the change in predicted probability associated with changes in the explanatory variables.

5.2 Definition and description of data

After excluding observations missing data for the variables to be used in the model, I have a sample of 197 farms. Table 7 provides definitions and descriptive statistics of the variables used in the estimation. The exogenous variables used to explain the farm behaviour refer first of all to the characteristics of the farm household (ethnicity, dimension, participation of individual members in the marketing activities of farm products); of the operator (age, education and level of involvement in farming activity); of the farm (dimension, level of production of target crop, distance from the market, use of own/buyer's means of transportation, use of own criteria or those suggested by the extension staff in the choice of the marketing channel).

Table 7. Description of variables used in the model.

Variable	Description	Farms selling to hotels		Farms not selling to hotels	
		Mean	Std.Dev.	Mean	Std.Dev.
Ethnicity	Ethnicity of the operator (0 Fijian; 1 Indian)	0.35	0.49	0.61	0.49
Family_size	Number of components	2.92	1.55	3.28	1.61
Age1	Operator age (years)	47.31	11.44	47.07	10.52
Educ1	Operator education (primary=1; secondary=2; university=3)	1.92	1.32	1.68	0.94
Employ1	Operator employed exclusively or part time in the farm (exclusively=0; partially=1)	0.08	0.27	0.18	0.38
Lfarm	Total land (ha) logarithms	2.58	1.81	3.77	4.19
Lprod	Used land (ha) logarithms	8.04	18.88	12.38	35.08
P_pln	Quantity of papaya produced (ln)	3.01	4.26	2.16	3.66
M_pln	Quantity of mango produced (ln)	0.30	1.54	1.91	3.20
T_pln	Quantity of tomatoes produced (ln)	4.74	3.88	2.87	3.61
P_harv_tecn	Pre-cool, selecting and grading (0=no; 1=yes)	0.73	0.45	0.66	0.47
IPM	Use of IPM (0=no; 1=yes)	0.27	0.45	0.08	0.27
Transp_own	Own transport used (0=no; 1=yes)	38.08	40.50	30.50	44.97
Transp_buyer	Buyer's transport used (0=no; 1=yes)	25.00	31.78	29.62	43.68
Distance	Distance from selling place (Km)	7.85	9.11	24.30	37.28
Sell_extens	Purchasers choice recommended by extension staff (0=no; 1=yes)	0.31	0.47	0.13	0.34
Sell_personal	Purchasers choice: personal (0=no; 1=yes)	2.08	1.41	2.58	1.05
Sell_son	Son participates in the production, marketing activities (0=no; 1=yes)	0.15	0.37	0.13	0.34
Sell_husband	Husband participates in marketing activities (0=no; 1=yes)	0.58	0.50	0.67	0.47
Sell_wife	Wife participates in marketing activities (0=no; 1=yes)	0.15	0.37	0.23	0.42
Sell_nonfam	Non family person participates in marketing activities (0=no; 1=yes)	0.12	0.33	0.21	0.41
N		26		171	

Farmers of the Fijian ethnic group tend to engage more in direct deliveries to hotels as compared to those with an Indian origin. Family size has a normal distribution, with an average of 4 persons. Operators engaged in direct selling to hotels do farm more on a full-time basis than those not engaged, while there are no statistically significant differences in age and education of the two categories of farmers.

Farms who sell their products to hotels are on average smaller than the rest of the sample in terms of land, both total and used land, and, in the case of mango, of level of productions. In addition, they tend to use more their own means of transportation and less buyer's ones. Another difference is that famers engaged in direct selling rely more on extension recommendations than on their own preferences in the selection of purchasers and marketing channels. Additionally, they tend to present a higher participation of their sons into the marketing activities. Finally, they more frequently make use of IPM techniques.

5.3 Results

Table 8 shows the estimated coefficients, the marginal effects[77], the likelihood ratio test and two selection criteria (AIC and BIC) of the full model (model 1) and of a restricted one (model 2). The full model contains all the variables that I supposed to have an influence on the decision process. The restricted one excludes variables Age1, Employ, Tomat_prod, Sell_nonfamily, Sell_wife, Sell_son, total land, Family_size, utilised Land, education and Mango_prod. I formally tested the null hypothesis that the coefficients of these variables are jointly equal to zero by using a LR test. A LR test shows a test statistic of LRT = 16.07 which is much larger than conventional

---

[77] The marginal effect of a continuous independent variable x is the partial derivative, with respect to x, calculated at the mean of the independent variable. For a dummy variable the marginal effect is calculated as the difference of the prediction function at one and its value at zero.

Table 8. Results of logit model (dependent: probability of direct selling to hotels).

	Model 1		Model 2			
	Estimated	Std. er.	Estimated	Std. er.	Marg. effect	Std. er.
IPM	3.337	1.282	3.563	1.000	0.398	0.220
Sell_extension	1.314	0.930	-0.616	1.084	0.080	0.070
P_harv_tecn	2.074	1.068	1.525	0.827	0.036	0.020
Papaya_prod	0.399	0.267	0.140	0.074	0.004	0.003
Transport_own	0.031	0.013	0.017	0.009	0.000	0.000
Transport_buyer	-0.018	0.011	-0.018	0.009	-0.001	0.000
Distance	-0.055	0.029	-0.070	0.031	-0.002	0.001
Sell_personal	-0.636	0.336	-0.508	0.268	-0.014	0.009
Sell_husband	-1.004	0.792	-1.210	0.607	-0.043	0.028
Ethnicity	-3.173	1.078	-2.129	0.745	-0.081	0.043
Age1	0.048	0.030				
Employ	-1.519	1.119				
Tomat_prod	0.320	0.288				
Sell_nonfamily	-1.082	0.999				
Sell_wife	-0.918	0.892				
Sell_son	0.874	0.959				
Total land	-0.011	0.013				
Family_size	-0.192	0.231				
Utilised Land	0.056	0.119				
Education	0.075	0.308				
Mango_prod	-0.080	0.344				
Constant	-3.889	2.617	1.585	0.765		
Log likelihood	-41.332		-49.368			
AIC	126.665		120.736			
BIC	198.783		156.796			

critical values of a χ^2_{11} distribution, and hence, I can reject the null that the restrictions do not apply. This means that the excluded variable do not jointly explain the choice of direct marketing to hotels. For this reason I will comment the results only of the restricted model. The smaller values of both the Akaike and the Schwarz Information Criteria confirm the constrained model is the best fitting one.

Model 2 correctly predicts the choice to directly selling to hotels in 92.35 percent of cases (Table 9). The results show that variables referred to quality, that is the use of IPM and of post harvest technologies -such as cooling facilities during or immediately after the harvesting operations, selecting and grading, present the highest impact on the probability of selling to hotels. It seems important to note that quality is at some extent linked to ethnicity given that 7 out of the 13 tomato growers selling directly to the tourist sector make use of IPM technologies and all of them are Fijans.

Ethnicity is the only statistically significant variable among those relating to the family. This confirms the hypothesis made in the past (Macedru, 2003) about the importance of ethnicity in determining farmers' attitude for commercial agriculture. The negative sign means that Indians farmers have a lower probability than Fijian farmers to participate in direct marketing to hotels. This may be explained with the fact that Indians, usually more trade oriented than Fijians, have already good consolidated relationships with intermediaries and wholesalers

The participation of the husband, who is the farm operator, in the selling activity has a negative though small decreasing effect on the probability to be engaged in direct marketing. I also find a positive effect when farms rely on the extension staff suggestion about the purchaser choice. On the contrary, when the selection is made on the basis of personal criteria in the probability decreases.

In addition. the choice of selling directly to hotels is (slightly) preferred by big producers of pawpow. On the contrary, the level of production does not seem to influence the choice in the case of the other three crops.

I also find that when the farm is very far from the market, farmers are less likely to sell directly to hotels. In addition, when producers rely on the use of buyers' means of transportation the probability decreases, while when they use their own means to transport their products to the market the probability to sell directly to hotels increases. This suggests that hotels do not usually arrange for transportation. Given the lack of public transportation, those farm households who live far away from the market and who do not own a mean of transportation are in fact excluded by the chance to sell directly to hotels..

Table 9. Actual and predicted observations (Model 2).

		Predicted		
		D=0	D=1	Total
Actual	D=0	14	3	17
	D=1	12	167	179
	Total	26	170	196

6. Final remarks

The information collected in the horticultural chain study confirmed that the demand for fresh products of the Fijian tourism sector is constantly increasing and that, due to the inconsistency of a local high quality horticultural supply, this demand is largely met by imports. The high propensity to import limits the multiplier effect of tourism growth and, in addition, prevents the spreading of economic benefits stemming from tourism activities to the rural population. In order to amplify the multiplier effect of tourism growth, emphasis must be placed on increasing the size and quality of the domestic agricultural supplies meant to meet the demand of Fiji's hotels and resorts.

The chain study conducted by FAO-UN[78]/INEA[79]/SPC has so far confirmed that the main issues to be tackled for the targeted products are still relating to an increase in the consistency of their supplies and to traditional aspects of quality _ mostly referring to grading, etc. – while no interest was detected as far as more sophisticated aspects of quality are concerned – such as geographic indications, organic or fair trade kind of certifications. There is still a need to develop production and to improve distribution networks for papaya and mango, so as to gather supplies from different local producers into the volumes required by the tourism sector. However, in the case of tomatoes, efforts should also be made in improving the overall quality of this product. As for carrots, it is quite unlike that domestic growers could compete with the latter in any close future.

In addition, the application of a logit model has helped us to identify the factors that affect the farmers decision to be engaged in direct marketing to hotels and restaurants and to quantify their influence on the choice. Direct selling is a particularly interesting strategy for farmers either because hotels and resorts pay premium prices for quality fresh products, and because it allows farmers to retain the highest possible portion of this premium without sharing it with intermediaries. Variables referred to quality, namely the use of IPM and of quality increasing post harvest technologies, have been found to have an high influence on the probability of selling to hotels. Results also confirmed that ethnicity is important, with Fijian more oriented to this choice as compared to Indians. In addition, the probability increases when farmers follow the extension criteria in the purchasers selection and when farmers make use of their own means of transportation to deliver their products. The decision to sell to hotel is less probable when farmers use personal criteria for the purchaser selection, when the husband participates in marketing activities, when the farm is distant from the market and when they make use of purchasers' means of transportation to deliver their products.

References

Berno, T., 2006. Sustainability on a plate: linking agriculture, food and the tourism industry. Christchurch Polytechnic Institute of Technology. Mimeo. Fiji.

Bennett, O., D. Roe and C. Ashley (Eds.), 1999. Sustainable tourism and poverty elimination study. UK Department for International Development (DFID)/ Overseas Development Institute (ODI). UK: London.

Macedru, A., 2003. Use of participatory approaches in the South Pacific. In Farmer participatory methods for coconut genetic resources in Asia-Pacific region. Available at: http://www.ipgri.cgiar.org/publications/HTMLPublications/545/ch3.htm.

---

[78] Food and Agriculture Organization of the United Nations.

[79] Italian Institute of Agricultural Economics.

Ministry of Finance and National Planning, 2007. Strategic Development Plan 2007 – 2011. Fiji: Suva.
NZIER (the New Zealand Institute of Economic Research Inc), 2007. What does a coup cost? NZIER Update: February, 2007.

Acknowledgements

This book is a selection of the contributions to the 105th EAAE seminar on "International marketing and international trade of quality food products" that was held in Bologna 8th to 10th March, 2007. There were presentations from many researchers from all around the world. All contributions in this book have been double-blind reviewed and updated before publishing.

The seminar was organized by the Department of Agricultural Economics and Engineering in cooperation with the third meeting of the BEAN-QUORUM project (Building a Euro-Asian Network for Quality, Organic, and Unique food Marketing – TH/Asia-Link/006) funded by the European Union's Asia-Link Programme. The Asia-Link Programme is dedicated to the promotion of regional and multilateral networking among higher education institutions in European Union Member States and South Asia, South-East Asia and China. The BEAN-QUORUM project is a co-operation initiative aimed at creating a network of Asian and European higher education institutions that are interested in the marketing issues regarding quality food. A description of the BEAN-QUORUM network is available on the web site http://www.bean-quorum.net.

The editors wish to express their gratitude to the following people who generously provided advice on manuscripts as reviewers of the papers selected for publication in this book, in a special issue of the Journal of Food Products Marketing, and in a special double issue of the Journal of International Food and Agribusiness Marketing:

Sedef Akgüngör, Dokuz Eylül University, Faculty of Business, Department of Economics, Kaynaklar Buca, Izmir, Turkey.
Sven Anders, University of Alberta, Department of Rural Economy, Edmonton, Alberta, Canada.
Matthieu H. Arnoult, Reading University, Department of Agricultural & Food Economics, Reading, Berkshire, United Kingdom.
Cemal Atici, Adnan Menderes University, Department of Agricultural Economics, Aydin, Turkey.
Giuseppe Attanasi, Toulouse School of Economics, Laboratoire d'économie des ressources naturelles (LERNA), Toulouse, France.
Fernando Balsevich, Universidad Americana, Department of Business and Economics, Asuncion, Central, Paraguay.
Alessandro Banterle, University of Milan, Department of Agricultural, Food and Environmental Economics, Milan, Italy.
Manuel Belo Moreira, Higher Institute of Agronomy - Technical University of Lisbon, Department of Agrarian Economics and Rural Sociology, Lisbon, Portugal.
Stefano Boccaletti, Università Cattolica del Sacro Cuore, Istituto di Economia Agro-Alimentare, Piacenza, Italy.
Julián Briz, Universidad Politecnica Madrid, Agricultural Economics and Social Sciences, Madrid, Spain.
Luca Camanzi, Alma Mater Studiorum-University of Bologna, Department of Agricultural Economics and Engineering, Bologna, Italy.
Marija Cerjak, University of Zagreb, Faculty of Agriculture, Department of Agricultural Marketing, Zagreb, Croatia.
Teresa Del Giudice, University of Naples, Department of Agricultural Economics and Policy, Portici, Naples, Italy.
Rachael Dettmann, United States Department of Agriculture, Economic Research Service (ERS-USDA), Washington, D.C., United States of America.

Carolyn Dimitri, United States Department of Agriculture, Economic Research Service (ERS-USDA), Washington, D.C., United States of America.

Christian Fischer, Massey University, Agribusiness, Supply Chain Management and Logistics Division, IFNHH, Auckland, New Zealand.

Mogens Fosgerau, Technical University of Denmark, Department of Transport, Kgs. Lyngby, Denmark.

Luigi Galletto, University of Padova, Dipartimento Territorio e Sistemi Agroforestali (TESAF), Legnaro, Padova, Italy.

Jean-Philippe Gervais, North Carolina State University, Agricultural and Resource Economics, Raleigh, North Carolina, United States of America.

Jose M. Gil, Centre for Research on Agro-food and Development Economics-UPC-IRTA (CREDA), Castelldefels, Barcelona, Spain.

Julie Guthman, University of California, Santa Cruz, Community Studies, Santa Cruz, California, United States of America.

Rainer Haas, BOKU, University of Natural Resources and Applied Life Sciences, Vienna, Institute of Marketing & Innovation, Vienna, Vienna, Austria.

Getu Hailu, University of Guelph, Department of Food, Agricultural and Resource Economics, Guelph, Ontario, Canada.

Sheryl Hendriks, University of KwaZulu-Natal, African Centre for Food Security, Pietermaritzburg, KwaZulu-Natal, South Africa.

Jill E. Hobbs, University of Saskatchewan, Department of Bioresource Policy, Business & Economics, Saskatoon, Saskatchewan, Canada.

Wuyang Hu, University of Kentucky, Department of Agricultural Economics, Lexington, Kentucky, United States of America.

Lionel Hubbard, Newcastle University, School of Agriculture, Food and Rural Development, Newcastle upon Tyne, Tyne and Wear, United Kingdom.

Helen H. Jensen, Iowa State University, Department of Economics, Ames, Iowa, United States of America.

Terhi Latvala, Pellervo Economic Research Institute, Helsinki, Uudenmaan Lääni, Finland.

Margaret Loseby, Università della Tuscia, Dipartimento di ecologia e di sviluppo economico sostenibile (DECOS), Viterbo, Italy.

Maria L. Loureiro, Universidade de Santiago de Compostela, Departamemto de Fundamentos da Análise Económica, Santiago de Compostela, A Coruña, Spain.

Jordan Louviere, University of Technology, Sydney, Centre for the Study of Choice (CenSoC), Sydney, New South Wales, Australia.

Giulio Malorgio, Alma Mater Studiorum-University of Bologna, Department of Agricultural Economics and Engineering, Bologna, Italy.

Andrea Marchini, University of Perugia, Department of Economics and Food Sciences, Perugia, Italy.

Massimiliano Mazzanti, University of Ferrara, Economics Institutions & territory, Ferrara, Italy.

Mirko Moro, University College Dublin, Environmental Policy, Dublin, Ireland.

Rodolfo M. Nayga, University of Arkansas, Department of Agricultural Economics and Agribusiness, Fayetteville, Arkansas, United States of America.

Tomas Nilsson, University of Alberta, Department of Rural Economy, Edmonton, Alberta, Canada.

Alessandro Olper, University of Milan, Department of Agricultural, Food and Environmental Economics, Milan, Italy.

Willis Oluoch-Kosura, University of Nairobi, Department of Agricultural Economics, Nairobi, Kenya.

Vania Paccagnan, University IUAV of Venice, Planning Department, Venice, Italy.

Daniel Pick, United States Department of Agriculture, Economic Research Service (ERS-USDA), Washington, D.C., United States of America.

Marco Platania, University of Catania, Department of Education Sciences, Catania, Italy.

Moises A. Resende Filho, Universidade Federal de Juiz de Fora, Departamento de Análise Econômica, Juiz de Fora, Minas Gerais, Brazil.

Brian J. Revell, Harper Adams University College, Directorate, Newport, Shropshire, United Kingdom.

Philip E. Rodgers, Erinshore Economics Limited, Saxilby, Lincoln, United Kingdom.

Elsa M. Rodriguez, Universidad Nacional de Mar del Plata, Centro de Investigaciones Económicas, Mar del Plata, Buenos Aires, Argentina.

Rocco Roma, University of Bari, Department of Agricultural economics and policy, rural evaluation and planning, Bari, Italy.

Fabio Maria Santucci, University of Perugia, Department of Economics and Food Sciences, Perugia, Italy.

Günter Schamel, Free University of Bozen-Bolzano, School of Economics and Management, Bozen-Bolzano, Italy.

Eugenia Serova, Food and Agriculture Organization of the United Nations (FAO), Investment Centre, Rome, Italy.

Diogo Souza Monteiro, University of Kent, Kent Business School, Wye Campus, Ashford, Kent, United Kingdom.

W. Bruce Traill, University of Reading, Department of Agricultural and Food Economics, Reading, Berks, United Kingdom.

Nina Urala, Kuulas Millward Brown, Helsinki, Uudenmaan Lääni, Finland.

Mark Vancauteren, Tilburg University, Econometrics & operation research, Tilburg, The Netherlands.

Wim Verbeke, Ghent University, Department of Agricultural Economics, Gent, Flanders, Belgium.

Davide Viaggi, Alma Mater Studiorum-University of Bologna, Department of Agricultural Economics and Engineering, Bologna, Italy.

Elena Viganò, Università degli Studi di Urbino "Carlo Bo", Dipartimento di Economia e Metodi Quantitativi, Urbino, Pesaro-Urbino, Italy.

Helga Willer, Research Institute of Organic Agriculture, Communication Department, Frick, Aargau, Switzerland.

Cesare Zanasi, Alma Mater Studiorum-University of Bologna, Department of Protection and Value-enhancement of agricultural products and food, Bologna, Italy.

Raffaele Zanoli, Università Politecnica delle Marche, Dipartimento di Ingegneria Informatica, Gestionale e dell'Automazione (DIIGA), Ancona, Italy.

About the authors

Sedef Akgüngör
Dokuz Eylül University, Faculty of Business, Department of Economics, Buca, Izmir, Turkey; sedef.akgungor@deu.edu.tr.

Osman Aydogus
Ege University, Faculty of Economics and Administrative Sciences, Department of Economics, Bornova, Izmir, Turkey; osman.aydogus@ege.edu.tr.

R. Funda Barbaros
Ege University, Faculty of Economics and Administrative Sciences, Department of Economics, Bornova, Izmir, Turkey; funda.barbaros@ege.edu.tr.

Giovanni Belletti
University of Florence, Faculty of Economics, Department of Economics, Via delle Pandette 9, 50127 Firenze, Italy; giovanni.belletti@unifi.it.

Odd Jarl Borch
Bodø Graduate School of Business, N-8049 Bodø, Norway; Odd.Jarl.Borch@hibo.no.

Julián Briz
Universidad Politecnica Madrid, ETS Ingenieros Agrónomos, Avda. Complutense s/n, 28040 Madrid, Spain; julian.briz@upm.es.

Tunia Burgassi
University of Florence, Faculty of Economics, Department of Economics, Via delle Pandette 9, 50127 Firenze, Italy; tunia.burgassi@unifi.it.

Luca Camanzi
Alma Mater Studiorum-University of Bologna, Department of Agricultural Economics and Engineering, Viale Giuseppe Fanin 50, 40127 Bologna, Italy; luca.camanzi@unibo.it.

Domenico Carlucci
University of Bari, Department of Agricultural Economics and Policy, Evaluation and Rural Planning, via Amendola 165/a, 70126 Bari, Italy; carlucci@agr.uniba.it.

Isabel De Felipe
Universidad Politecnica Madrid, ETS Ingenieros Agrónomos, Avda. Complutense s/n, 28040 Madrid, Spain; isabel.defelipe@upm.es.

Marie-Noëlle Duquenne
University of Thessaly, Department of Planning and Regional Development, Pedion Areos, 38334 Volos, Greece; mdyken@prd.uth.gr.

Marian García
The University of Kent, Kent Business School, KBS Room 116, Canterbury, Kent CT2 7PE, United Kingdom; M.Garcia@kent.ac.uk.

About the authors

Cristina Grazia
Alma Mater Studiorum-University of Bologna, Department of Agricultural Economics and Engineering, Viale Giuseppe Fanin 50, 40127 Bologna, Italy; c.grazia@unibo.it.

Catherine Greene
U.S. Department of Agriculture, Economic Research Service, 1800 M Street NW, Rm S4051, Washington, DC 20036, USA; cgreene@ers.usda.gov.

Shelly Grow
U.S. Department of Agriculture, Economic Research Service, 1800 M Street NW, Rm S4051, Washington, DC 20036, USA.

Jon Hanf
Leibniz-Institute of Agricultural Development in Central and Eastern Europe, Theodor-Lieser-Str. 2, 06120 Halle (Saale), Germany; hanf@iamo.de.

Meike Henseleit
Justus Liebig University of Giessen, Institute of Agricultural Policy and Market Research, Senckenbergstrasse 3, 35390 Giessen, Germany; meike.henseleit@agrar.uni-giessen.de.

Mohamed Ismaili
University Moulay Ismail of Meknes, Department of Biology, Faculty of Sciences, BeniMhamed BP 11201, Meknes, Morocco; ismailih2000@yahoo.fr.

Masaru Kagatsume
Kyoto University, Division of Natural Resource Economics, Graduate School of Agriculture, Kitashirakawa Oiwake-cho, Sakyo-ku, 606-8502 Kyoto, Japan; kagatume@kais.kyoto-u.ac.jp.

James Kirwan
Macaulay Institute and Countryside and Community Research Institute, Craigiebuckler, Aberdeen AB15 8QH, United Kingdom.

Sabine Kubitzki
Justus Liebig University of Giessen, Institute of Agricultural Policy and Market Research, Senckenbergstrasse 3, 35390 Giessen, Germany; sabine.kubitzki@agrar.uni-giessen.de.

Rainer Kühl
Justus Liebig University of Giessen, Institute of Agricultural Policy and Market Research, Senckenbergstrasse 3, 35390 Giessen, Germany; Rainer.Kuehl@agrar.uni-giessen.de.

Victoria Lacaze
Universidad Nacional de Mar del Plata, Facultad de Ciencias Económicas y Sociales, Funes 3250, Mar del Plata (Buenos Aires) B7602AYJ, Argentina.

Aykut Lenger
Ege University, Faculty of Economics and Administrative Sciences, Department of Economics, Bornova, Izmir, Turkey; aykut.lenger@ege.edu.tr.

Renaud Lunardo
REPONSE Laboratory, Troyes Champagne School of Management, 217 Avenue Pierre Brossolette BP 710, 10002 Troyes Cedex, France; renaud.lunardo@groupe-esc-troyes.com.

Beatriz Lupín
Universidad Nacional de Mar del Plata, Facultad de Ciencias Económicas y Sociales, Funes 3250, Mar del Plata (Buenos Aires) B7602AYJ, Argentina.

Giulio Malorgio
Alma Mater Studiorum-University of Bologna, Department of Agricultural Economics and Engineering, Viale Giuseppe Fanin 50, 40127 Bologna, Italy; giulio.malorgio@unibo.it.

Elisabetta Manco
†, University of Florence, Firenze, Italy.

Andrea Marescotti
University of Florence, Faculty of Economics, Department of Economics, Via delle Pandette 9, 50127 Firenze, Italy; andrea.marescotti@unifi.it.

Davide Menozzi
University of Parma, Faculty of Agriculture, Area delle Scienze - Campus, via Langhirano, 43100 Parma, Italy; davide.menozzi@unipr.it.

Cristina Mora
University of Parma, Faculty of Agriculture, Area delle Scienze - Campus, via Langhirano, 43100 Parma, Italy; crismora@unipr.it.

Alessandro Pacciani
University of Florence, Faculty of Economics, Department of Economics, Via delle Pandette 9, 50127 Firenze, Italy; alessandro.pacciani@unifi.it.

Lorenzo Paluan
ICEA - Institute for Ethic and Environmental Certification, Strada Maggiore 29, 40125 Bologna, Italy; l.paluan@icea.info.

Agata Pieniadz
Leibniz-Institute of Agricultural Development in Central and Eastern Europe, Theodor-Lieser-Str. 2, 06120 Halle (Saale), Germany; pieniadz@iamo.de.

Maurizio Prosperi
University of Foggia, Dipartimento Pr.I.M.E., Via Napoli 25, 71100 Foggia, Italy; prosper169@supereva.it.

Jofi Puspa
Justus Liebig University of Giessen, Institute of Agricultural Policy and Market Research, Senckenbergstrasse 3, 35390 Giessen, Germany; Jofi.Puspa@ernaehrung.uni-giessen.de.

Meri Raggi
Alma Mater Studiorum-University of Bologna, Department of Statistics, Via delle Belle Arti 41, 40126 Bologna, Italy; meri.raggi@unibo.it.

About the authors

Ingrid H.E. Roaldsen
Nordland Research Institute, 8049 Bodø, Norway; ingrid.roaldsen@nforsk.no.

Elsa Rodríguez
Universidad Nacional de Mar del Plata, Facultad de Ciencias Económicas y Sociales, Funes 3250, Mar del Plata (Buenos Aires) B7602AYJ, Argentina; emrodri@mdp.edu.ar.

Paula Rossi
Kyoto University, Division of Natural Resource Economics, Graduate School of Agriculture, Kitashirakawa Oiwake-cho, Sakyo-ku, 606-8502 Kyoto, Japan; paularossi@argentina.mbox.media.kyoto-u.ac.jp.

Cristina Salvioni
University G. d'Annunzio of Chieti-Pescara, DASTA, Via della Pineta 4, 65129 Pescara, Italy; salvioni@unich.it.

Fabio G. Santeramo
University of Napoli "Federico II", Department of Agricultural Economics and Policy, via Università 96, 80055 Portici (NA), Italy; fabiogaetano.santeramo@unina.it.

Silvia Scaramuzzi
University of Florence, Faculty of Economics, Department of Economics, Via delle Pandette 9, 50127 Firenze, Italy; silvia.scaramuzzi@unifi.it.

Antonio Seccia
University of Bari, Department of Agricultural Economics and Policy, Evaluation and Rural Planning, via Amendola 165/a, 70126 Bari, Italy; seccia@agr.uniba.it.

Bill Slee
Macaulay Institute and Countryside and Community Research Institute, Craigiebuckler, Aberdeen AB15 8QH, United Kingdom; b.slee@macaulay.ac.uk.

Ramona Teuber
Justus Liebig University of Giessen, Institute of Agricultural Policy and Market Research, Senckenbergstrasse 3, 35390 Giessen, Germany; ramona.teuber@agrar.uni-giessen.de.

Davide Viaggi
Alma Mater Studiorum-University of Bologna, Department of Agricultural Economics and Engineering, Viale Giuseppe Fanin 50, 40127 Bologna, Italy; davide.viaggi@unibo.it.

George Vlontzos
University of Thessaly, Department of Planning and Regional Development, Pedion Areos, 38334 Volos, Greece; georgevlontzos@yahoo.gr.

Cesare Zanasi
Alma Mater Studiorum-University of Bologna, Department of Protection and Value-Enhancement of agricultural products and food, via F.lli Rosselli 107, 42100 Reggio Emilia, Italy; cesare.zanasi@unibo.it.

Keyword index

A

acceptance	182
adaptations	155-157, 162, 164
advantage	63
– comparative	94, 151
– competitive	16, 89, 192, 245
– revealed comparative	101, 104, 110
advertising	64
affective factors	272
agro-food system	133
AIDA model	177
appropriationism	135
a priori segmentation	172
asymmetric information	47
authenticity	281, 284, 285, 290, 292, 293
awareness	236

B

bar coding system	89
benchmarking	92, 93
biodiversity	64
bottlenecks	317
brand	46, 50, 53, 55, 57, 69, 192
– collective	63
– private	50
– protection	43
– reputation	215
(non-)branded products	77, 185, 189
branding strategies	63
British Retail Consortium	186, 194

C

case study	155
certification	90, 91, 136, 224, 240, 245, 247, 320, 327
club good	204
cluster analysis	175, 180
clustering	171
code of practice	203, 207, 214
collective	
– marketing	203
– marks	55
– organisations	205, 221
– quality strategy	195, 197
– reputation	47
co-marketing	72
communication	168, 170, 175, 182
competitive	
– disadvantage	191

G

H

I

Q

R

S